AF372129

THE SOLAR SYSTEM

To Joan Margaret

Too swift arrives too tardy as too slow.
SHAKESPEARE

La vérité consiste dans les nuances.
RENAN

Frontispiece: Comet Arend Roland 1957 III. Perihelion passage, April 8. 03117 U.T.

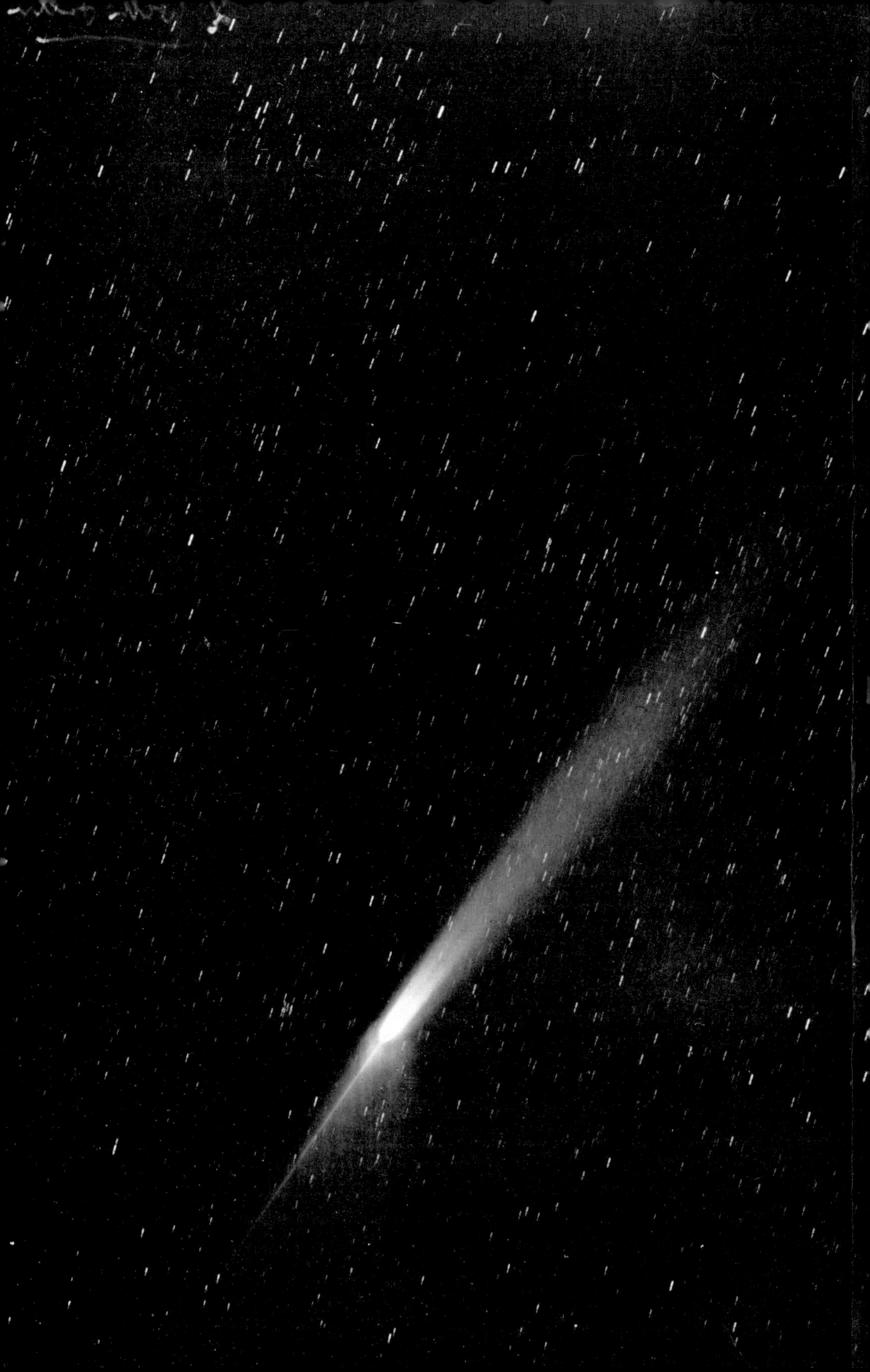

THE SOLAR SYSTEM

Frank W. Cousins

Drawings by Malcolm Chandler

JOHN BAKER LONDON

Published in 1972 by
JOHN BAKER (PUBLISHERS) LTD
4, 5 & 6 Soho Square
London W1V 6AD

ISBN 0 212 98392 X

Printed in Great Britain by
W & J MACKAY LIMITED
Chatham

Contents

Illustrations		7
Acknowledgements		13
Introduction		15
1	The solar system in classical times	19
2	The solar system in the Renaissance	33
3	Models of the solar system	62
4	The telescopic appearances of the components as seen from the Earth and Earth-launched space craft	70
5	The planets and their satellites	73
6	The atmospheres of the planets	112
7	The interiors of the planets	120
8	Interplanetary dust and gas	124
9	On the nature of the Sun	129
10	On the Earth/Moon system	150
11	On comets, minor planets, meteors, meteorites and tektites	183
12	The scale of the solar system and the law of planetary and satellite distances	213
13	The stability of the solar system	229
14	The origin of the solar system	235
15	Life in the solar system and beyond	254
	Appendices	
	A. *Abbreviations used in this book*	285
	The elements of planetary orbits	286
	B. *Biographical Notes*	287
	C. *Catalogue of comets*	291
	Index	293

Illustrations

PLATES

Between pages 64–5

1. The geocentric system of the world as conceived by some of the Greek philosophers.
2. The circles of the Sun and the Moon and the ovoid or elliptiform orbit of Mercury's epicycle according to Azarquiel (eleventh century).
3. The title page of the *De revolutionibus* of Copernicus.
4. Cracow at the time of Copernicus.
5. The Preface and Foreword to the *De revolutionibus* of Copernicus.
6. Kepler's explanation of the structure of the planetary system.
7. Celestial globe of Atlante Farnesiano, *c.* 300 BC.
8. Celestial globe by Martin Bylica of Olkusz, *c.* 1480.
9. The Golden Jagellonian Globe, *c.* 1510.
10. Armillary sphere.
11. English orrery, *c.* 1781.
12. A typical orrery.
13. A tellurion, a model of the Earth/Moon system, from a drawing, *c.* 1793.
14. The famous astronomical clock of Strassburg Münster.
15. A reconstruction of the clock of Giovanni de Dondi, *c.* 1364.
16. Rowley's tellurion, *c.* 1712.
17. 'The Orrery', by Joseph Wright, A.R.A., 1766.
18. Astronomical globe on the back of Pegasus.
19. An English tellurion signed, *W. & S. Jones, Fecerunt,* 153 *Holborn, London.*
20. Newton's Orrery, 1860.
21. Desagulier's Cometarium.
22. Rev. William Pearson's mean motion orrery (1813).
23. Schematic diagram of Pearson's orrery.
24. Armillary sphere of Sanctucci della Pomarance, made between 1588–93.
25. The Leyden 'Sphaera' by Stevin Tracey of Rotterdam.
26. Orrery by T. Heath, *c.* 1757.

Between pages 112–13

27. Orrery by Francis Du Commun, *c.* 1813.
28. The house of Eisinga called 'De Ooyevaar' (The Stork) after the stork chiselled in the gable.
29. The orrery of Eise Eisinga in the ceiling of his house at Franeker, Friesland, Netherlands.

30. English orrery by Tho. Tompion & Geo. Graham, London, *c.* 1710.
31. Front view of David Rittenhouse Orrery.
32. Rittenhouse Orrery (rear view).
33. Joseph Pope's Orrery, *c.* 1787.
34. Orrery of Le Verrier, *c.* 1847.
35. Mechanized solar system model made by the Sondes Place Research Institute and displayed at the Festival of Britain in 1951.
36. Urameton Orrey, 1955, designed and made by the author.
37. The 100 in. (254 cm.) reflecting telescope of the Mount Wilson Observatory, U.S.A.
38. The Jodrell Bank 250 ft. (7620 cm.) radio telescope, Cheshire, England.
39. The 48 in. $\times$ 72 in. f/2.5 Schmidt reflecting telescope at the Palomar Observatory.
40. The 200 in. (508 cm.) Hale reflecting telescope.
41. The dish of the Arecibo Ionospheric Observatory's radio telescope at the Palomar observatory in Puerto Rico.
42. The moon at the gibbous phase.
43. Mercury as a 'morning star', 1968, November 4, 13h. 25m. U.T., exhibiting the gibbous phase (70 per cent).
44. Mercury: 1969, January 18, 17h. 30m.; 1969, April 29, 19h. 50m.; 1969, May 8, 20h. 10m. April 29, diameter 6·6 seconds of arc, phase 0.582, distance 1.066 a.u. (18 in. OG. $\times$ 300). Drawings by J. B. Murray.
45. Drawings of Venus.
46. Mars in blue light, Mars in red light.
47. Venus at the gibbous and crescentic phase.
48. Sequence of photographs from Mariner 6 showing the spacecraft's approach to Mars on 1969, July 29–31.
49. Named Martian markings.
50. The close approach to Mars by the Mariner 6 spacecraft.
51. Mariner 7 photograph showing the edge of the south polar cap on Mars.
52. Mariner 7 spacecraft, 1969, August 5. A television relay picture of the south polar cap of Mars.

Between pages 160–1

53. Close approach to Mars by the Mariner 6 spacecraft on 1969, July 30.
54. Close approach to Mars by the Mariner 6 spacecraft on 1969, July 30.
55. Jupiter, 1964, November 1,1h. 29m. U.T.
56. Saturn, 1969, September 6. (30 cm. Cassegrain.)
57. Saturn.
58. Comet Ikeya-Seki, 1965. Perihelion passage October, 21.
59. Meteor trail in Pleiades, 1932, January 11.
60. Transit of Mercury across the Sun's disc, 1970, May 9, 8h. 22m. U.T.
61. Mars taken with the 200 in. Hale reflecting telescope at Palomar.
62. (a) Mercury at the third contact, 1970, May 9, 12h. 9·50m. U.T. (b) Mercury superimposed on the Sun's disc, 1970, May 9, 11.11h. U.T. (c) Mercury superimposed on the Sun's disc, 1970, May 9, 11.04h. U.T. (d) Transit of Mercury, 1970, May 9, 10.57h. U.T. Mercury's disc is about 10·9 seconds of arc on the photograph.
63. Jupiter. A photograph taken with the 200 in. Hale reflecting telescope.

64. Saturn. A photograph taken with the 200 in. Hale reflecting telescope.
65. The sun-god at Sippar, Babylonia, on a stone tablet by Nabu-Apal-Iddina (ninth century BC).
66. A solar prominence, 1969, April 30, 10h. 40m. U.T., 30,000 miles (48,270 km.) high.
67. One of the twenty-four great wheels to the Sun-god at Konarak, Temple of the Sun.
68. The Sun's disc seen from Greenwich at Sun spot maximum 1947, May 24.
69. The face of the Sun, 1967, May 29.
70. 'Butterfly' diagram showing changes in spot latitude during the solar cycle.
71. A sun spot group, 1959, September 10.
72. Solar granulations and sunspot, 1959, August 17.
73. Burning device with a great liquid lens built in 1774 to the order of the Academie Royale des Sciences, Paris.
74. Laboratoire de'energie solaire, Citadelle de Mont Louis.
75. The largest radio telescope at Effelsberg, West Germany, for the Max-Planck Institute for Radio Astronomy.
76. Laboratoire de l'Energie Solaire Citadelle de Montlouis, France.
77. The first photograph of the far side of the moon.

Between pages 208–9

78. The far side of the Moon—a view taken from the Apollo XI command ship in lunar orbit.
79. Landing site of the Apollo XII mission in the Ocean of Storms.
80. The Apollo XII landing vehicle 'Intrepid' seen descending towards the Moon from the mother command ship 'Yankee Clipper'.
81. The Imbrium collision area after Urey.
82. The Earth seen from above the Moon on the Apollo XI mission.
83. Model of contiguous craters made by fluidization.
84. Lunar craters Azophi and Abenezra.
85. Pseudo-Thebit, made by fluidization.
86. Lunar crater Thebit (diameter 48 km.) in the south polar area of the Moon.
87. Single crater showing terracing.
88. Cauchy crater fault and scarp.
89. Air photograph of the fault scarps of the south-east slope of the Silali Volcano, Baringo district, Kenya.
90. Apollo XI lunar sample fines.
91. Apollo XI lunar sample fines.
92. Apollo XI lunar sample fines.
93. (a) Scanning electron micrographs of part of a dumbbell shaped glass bead. (b) Same particle at higher magnification. (c) Spherical bead showing cratering at the lower edge of the photograph. (d) Spherical beads with partially embedded particles.
94. Comet Ikeya 1963 I. Perihelion passage, March 21.
95. Comet Mrkos 1957 V. Perihelion passage, August 1.
96. Comet Arend Roland 1957 III. Perihelion passage, April 8.
97. Comet Bennett, 1970 April 7.
98. Comet Bennett 1969 I. The photograph was taken at Archallagan Observatory, Foxdale, Isle of Man, 1970, April.

Illustrations

99. Exploding Andromedid Meteor, Butler 1895, November 23. (2 in. lens.)
100. Forty-three meteors from the phenomenal Leonid shower at Kitt Peak, 1966, November 17, 12.00 hrs. U.T.
101. The Great Arizona or Barringer Meteorite Crater near Winslow, Arizona.
102. Tektites.
103. Shatter cones in mississagi quarzite along the southern margin of the Sudbury Basin, Canada.

Between pages 256–7

104. Meteorite from Canyon Diablo, Coon Butte, Arizona.
105. Ahnighito, the largest of the Cape York irons.
106. Allende meteorite, Chihuahua, Mexico.
107. Lost City meteorite, Oklahoma.
108. Standing on the slope of the Hadley Delta, David Scott photographs surface features of the Moon with a 70 mm. camera.
109. The Apollo XV mission—surface features of the Moon.
110. A close study of the floor of a crater termed by the Apollo XV crew as 'relatively fresh'.
111. The Apollo XV mission. Command and service modules in lunar orbit, viewed from the lunar module.
112. The Mare Orientale on the far side of the Moon.
113. The Moon photographed by Apollo VIII, 1968 November 24.
114. Aratus of Soli, Greek didactic poet.
115. Anaximander of Miletus, 601–547 BC.
116. A representation of Plato.
117. Aristotle.
118. Socrates, *c.* 471 ?469–399 BC.
119. Thales of Miletus, 640–546 BC.
120. Metrodorus, second century BC.
121. Anaxagoras, *c.* 500–428 BC.
122. Hipparchus of Nikaia, *c.* 180–126 BC.
123. Pythagoras, born *c.* 576 BC.
124. Eudoxus, *fl.* fourth century BC.
125. Nicolaus Copernicus (or Koppernigk), 1473–1543.
126. Tycho Brahe, 1546–1601.
127. Johann Kepler (Keppler), 1571–1630.
128. Galileo Galilei, 1564–1642.
129. Sir Isaac Newton, 1642–1727.
130. Eise Eisinga, 1744–1828.
131. John Couch Adams, 1819–92.
132. Urbain Jean Joseph Leverrier, 1811–77.
133. Johann Gottfried Galle, 1812–1910.
134. Professor J. Challis, 1874.
135. Edmund Halley.
136. Joseph Louis Lagrange, 1736–1813.
137. Pierre Simon Laplace, 1749–1813.
138. George Biddell Airy, 1801–92.

FIGURES

1. Lemniscate a quartic curve discovered by Jacques Bernoulli (Acta Eruditorum, 1694) and later investigated by G. C. Tagnano.   24
2. The equivalence of the epicyclic and eccentric systems.   26
3. The world system according to Polydore Vergill (c. 1470–1555).   28
4. A simplified drawing of the Ptolemaic system of the world.   29
5. Part of a Ptolemaic model showing the 'objectionable' non-uniform circular motions.   36
6. Number of circles for each planet.   40
7. The Copernican system of the world in its simplest form.   41
8. The Copernican system of the world with the known satellites of Jupiter and Saturn portrayed; the heliocentric system.   41
9. The world system of Tycho Brahe.   44
10. The five regular Platonic polyhedra known to the ancient world and the two stellated dodecahedra discovered by Kepler.   46
11. Kepler's drawing of the planetary orbits showing the enormous epicycle of Mars.   47
12. Properties of an elliptical orbit.   49
13. The epicycle as a generator of curves.   54
14. The ovoid orbit of Mercury from Peurbach's *Theoricae noue planetarum, 1482*.   56
15. Mercury's orbit according to Ptolemy.   57
16. Motion of Venus according to Ptolemy.   58
17. Motion of Venus according to Copernicus.   58
18. Newton's law of gravitation.   60
19. Oppositions of Mars from the Earth, 1939–90.   78
20. Distances of Jupiter's satellites.   93
21. Section through Saturn's satellite system.   98
22. Diagram of the varying elevations of Earth and Sun in relation to the ring-plane and the ring-phases of Saturn in 1920–1.   103
23. Saturn's rings and their shadow as seen from the Earth in 1966.   103
24. Diurnal path of the Sun about Mercury.   114
25. Variations of pressure with height in the atmospheres of Mercury, Earth and Venus.   117
26. Explanation of variations in the Zodiacal light.   126
27. Schematic view of the layers of the Sun.   132
28. Schematic diagram of the radiation belts about the Earth.   144
29. Concave orbit of the Moon.   151
30. Variation of area of land mass heated by the Sun with seasonal change.   152
31. Vector diagram of angular momentum of the Moon's orbital motion and the Earth's rotation.   167
32. Density of Earth, Moon and Mars.   167
33. Orbital velocity relationship for the planets of the solar system.   170
34. Sequence of forms in lunar origin by fission from the Earth during formation of the Earth's core.   173
35. Orbit of a short-period comed at an schematic illustration of the capture process.   192

Illustrations

36. Groupings of the minor planets (asteroids).   195
37. The Trojan asteroids, 1960 June 20, with their mean motion per day.   197
38. The orbital period relation $T_n = T_0 A^n$ applied to the planetary system.   227
39. The orbital period relation $T_n = T_0 A^n$ applied to the satellite system of Jupiter.   227
40. The orbital period relation $T_n = T_0 A^n$ applied to the satellite system of Saturn.   227
41. The orbital period relation $T_n = T_0 A^n$ applied to the satellite system of Uranus.   227
42. Main theories of the origin of the solar system according to Kant, 1755.   240
43. Laplace, 1796.   241
44. Chamberlin-Moulton, 1900.   242
45. Jeans-Jeffreys, 1917.   244
46. Berlage, 1930.   245
47. Lyttleton, 1936.   247
48. Alfvén, 1942.   248
49. Weizsäcker, 1944.   249
50. Hoyle, 1944.   251
51. The elements used to define the orientation in space of an elliptic orbit.   286

Permission to reproduce the plates in this book was kindly granted by the following: Trustees of the British Museum, pls. 2, 65, 102; The Ronan Picture Library, pls. 6, 136, 138; National Museum, Naples, pl. 7; Professor Dr Karol Estreicher, Jagellonian University, Cracow, pls. 8 & 9; Frank Partridge & Sons, London, pl. 10; M. Alain Brieux, pls. 11, 34; Museum and Art Gallery, Derby, pl. 17; Metropolitan Museum of Art, New York, pl. 18; Christie's, London, pl. 19; Science Museum, London, pls. 20–2; Museo di Storia della Scienza, Florence, pl. 24; Rijksmuseum voor de Gescheidenis der Natuurwetenschappen, Leiden, pl. 25; M. Andre Curtit, Musée d'horlogerie, Ville de La Chaux-de-Fonds, pl. 27; Dr F. Maddison, Museum of the History of Science, Oxford University, pl. 30; Mr Howard C. Rice jr, pl. 31; Harvard College Library and Carolyn E. Jabeman, pl. 33; Central Office of Information, London, pl. 38; Cornell University, New York, pl. 41; The Observer, pls. 43, 44, 47, 55, 57; Mount Wilson and Palomar Observatories, pls. 46, 61, 64; NASA, Washington DC, pls. 48, 50, 53, 54, 78, 79, 80, 82, 88, 108–13; Mr Alan McClure, Mount Pinos, California, pl. 58; Dr H. R. Soper, Onchan, Isle of Man, pls. 60, 62, 63, 97, 98; Mr W. M. Baxter, 66, 69, 71; Allen & Unwin Ltd, pl. 67; Project Stratoscope Directors of Princeton University, pl. 72; Dover Books, pl. 81; Dr R. L. Waterfield, pls. 95, 96; Dr Trevor Gebbie, pl. 101; American Museum of Natural History, pl. 104; Fitzwilliam Museum, Cambridge, pl. 114; Dr I. Grafe, pls. 115, 120; The Mansell Collection, pls. 116, 120–4; Mr Anthony West, pl. 117 from the collection of Walter Leaf; Royal Astronomical Society, pl. 131; Bodleian Library, Oxford, pl. 135.

Acknowledgements

In making acknowledgements, I would first recall that I wrote much of this book and thought long about its parts in the peculiar tranquillity afforded by the London to Brighton train. During its compilation I have travelled some 30,000 miles and a word of praise for that unique establishment, British Rail, should be gratefully recorded.

The book has not been written without the help of many friends. Where I have been acutely conscious of major difficulties I have sought assistance and, as many in similar circumstances will know, men of learning and vast abilities did not refuse to help me.

I desire to record my deep appreciation of the kindness and special skills put at my service by my friends Professor Herbert Dingle, Dr J. G. Porter and Mr R. H. Stevens. These gentlemen read the manuscript and commented on its clarification, a time-consuming and valued service. I am similarly indebted to Dr Carl A. Rouse, who read my chapter on the nature of the Sun and made suggestions concerning its accuracy. My good friend Mr Malcolm Chandler prepared, from my sketches, all the drawings but one, and readers who have a keen eye for elegant line and typography will know how much the book prospers from his talent. Mr P. Robinson with no less skill prepared the very complex drawing of my table orrery. A special word must be said in praise of my secretary, Miss Margrit Schirmer, who unwearyingly typed the manuscript; of my wife Joan and Mr R. H. Stevens, who read the galley proofs and gave me many valued comments upon the final form of the work.

I have had much generosity shown to me from many sources in the gathering of the photographs and to the following I am keen to acknowledge my debt: The Trustees of the British Museum; R. S. Clarke and R. E. McCrosky of the Smithsonian Institution; Bartholomew Nagy, Professor of Geochronology of the University of Arizona; V. R. Boscarino, Assistant to the Director of Princeton University Observatory; Carolyn E. Jakeman of the Houghton Library of Harvard University; Howard C. Rice, Jr, of Princeton University Library; Anthony West of Blackheath, London; the National Aeronautics and Space Administration of Washington D.C.; Ralph Kazarion of Cornell University; Audouin Dollfus, Director of Observatoire de Paris, Meudon; John Murray of the University of London Observatory; Dr

Acknowledgements

I. Grafe of the Phaidon Press London; MM. Curtit and Reinwald of the Musée d' Horlogerie, Ville de la Chaux-de-Fonds; Francis Maddison, Curator of the Museum of the History of Science, Oxford; Harold L. Davis, Editor of *Physics Today*; H. E. Dall of Luton; W. M. Baxter of Acton; the Directors of the Mansell Collection, London; W. E. Fox of Newark; J. Hedley Robinson of Teignmouth; A. W. Heath of Nottingham; P. B. Doherty of Stoke-on-Trent; John Hiscott of Canterbury; Cdr H. R. Hatfield of Sevenoaks, Kent; Mrs Colin Ronan of The Ronan Picture Library, Cowlinge, Newmarket, Suffolk; Dr H. R. Soper of Onchan, Isle of Man; and Mr Howard Nelson, Assistant Keeper of the Department of Oriental Printed Books and Manuscripts of the British Museum, Bloomsbury.

Finally I have to record my thanks to John Baker* and Donald Kidd who set me the formidable task of writing the book, for their continual encouragement and the publication of it in such a handsome form.

FRANK W. COUSINS
Westminster, London
1971

*It is with profound regret that I record the death of my friend on 1971 November 25, one day before the book proofs appeared on his desk. I pay my tribute to him in these pages. No longer shall we meet in 'Arcadia'.

Oh Ioné—into many a green valley drifts the appalling snow.

F.W.C.

Introduction

In writing a small book on the solar system I am very conscious of the vast amount of labour and thought that have been expended to give us our present conception of it.

In this book I seek to gather into a small space a body of learning that is very widely scattered and some of it singularly inaccessible, thereby to exhibit the main features of the remarkable system to which our Earth inexorably belongs.

It obviously is not possible for any one person to have a comprehensive first-hand knowledge of the entire range of disciplines now used to treat of the solar system, and a work of this kind is essentially a compilation from the literature.[1]

I have given many references and footnotes[2] in an attempt to cater for the serious student who wishes to turn to the source material and yet not spoil a straightforward reading of the text for anyone who chooses not to be encumbered with too much detail.

It is to be hoped that there will always be a demand on librarians for the great books of the past from which our present knowledge stems, and at the same time a similar demand for the learned papers and journals of our own day in which is published the work that is being done by our contemporaries to widen our horizons. It is in this manner that one may seek to obtain a knowledge of recent advances, tempered with a good sense gathered from the deeper and older perspectives of the history of science thereby to find revealed the *sub-stratum* of the system of the world that has always contained sufficient intrinsic subtleties, secrets and surprises to challenge the best brains of any age.

We are told that every Attic burgher was familiar with the Attic culture, yet in our age of universal education many western ideas aim at something that only the few can comprehend and science especially rises from an elementary ground plan to higher regions that are inaccessible to the layman. From this we have our modern popularizations of science, which are without intrinsic value, often *détraquée* and much falsified;

---

[1] The literature is vast. *Nature* first published in 1869 takes some fifty feet of shelf space and *Physics Abstracts* that ran at 22,000 in 1910 have risen asymptotically to some 100,000 in 1970.
[2] I have attempted to define some of the less well known terms. An inexpensive specialized dictionary is the paperback *The New Dictionary and Handbook of Aerospace*, edited by Marks, R. W., Bantam Books Inc., 1969.

indeed this is to be seen in works on the solar system, for those works that are of value in understanding it are of immense erudition and not cast in a form to appeal to the general reader. There is, however, much that can be enjoyed provided it is reduced to manageable terms.

An appreciation of the solar system has not been the prerogative solely of astronomers and philosophers, in European thought at least, ever since Dante wrote his *Divina Commedia*.[3] In the days that have elapsed since Dante, much has been achieved in unravelling the system of the world and in presenting an intellectual picture of it that is thought to reveal its 'intrinsic' nature; yet undoubtedly it also is heavily weighted by human subjectivity.

It is now generally accepted that the solar system comprises the Sun and the lesser bodies that move about him as a centre of gravitational attraction. Our planet Earth is but one of these. The members of the system may be classified under four heads. Firstly there is the Sun, in which resides the greater part[4] of all the matter of the solar system and which is the only intrinsically incandescent body of the system. It is now known that the Sun is a star and shines by emitting its own light. Secondly there are the planets which consist of nine known major bodies and a large number of minor planets moving about the Sun at various fixed distances that change but little except from the vicissitudes occasioned by the perturbations of their elliptical orbits. Thirdly, there are the satellites of the planets, which are often termed secondary bodies since they are found to revolve about a primary body, i.e. one of the major planets, and thus accompany the major planet in it circumbendibus[5] with and about the Sun. A fourth class of bodies, which vary widely from ephemeral phenomena to those with a recorded periodicity of many years, comprises the meteors and the comets.

The second class, that is to say the planets, is conveniently subdivided into two broad divisions; the inferior planets, Mercury and Venus, and the superior planets Mars, Jupiter, Saturn, Uranus, Neptune and Pluto. The terms *inferior* and *superior* define the planets in terms of their distance from the Sun in relation to the distance of the Earth from its great luminary. The inferior planets are closer to the Sun than the Earth, and the superior further away. The minor planets lie within the realm of the superior planets, but owing to their very small size and their great

---

[3] We are unable to fix with precision the date (probably *c.* 1302–3) when Dante first entered on the work of writing the *Commedia*. He was born in A.D. 1265 and had reached the 'halfway' point of the three score years and ten that he, with the Psalmist (PS. XC.10) recognized as the normal span of man. (For Dante's idea of the Ptolemaic cosmos see Sayers, Dorothy L., *The Divine Comedy, Hell*, The Penguin Classics, pp. 292–95).

[4] The Sun is more than 700 times as massive as all the other bodies of the solar system combined (see page 237).

[5] *Circumbendibus.* A creation of Rabelais to mean going round wheeling about. One must remember that the Sun is also in motion within the galaxy and the planets therefore move in very complex paths. The Earth in fact takes a helicoidal movement within the galaxy.

16

number they are usually considered independently and I have adopted this convention.

A most remarkable feature of the solar system, indeed one that distinguishes it from all other known systems in the universe, is the symmetry in the disposition of its major bodies and the symmetry of their motion. All the bodies move in orbits which are wholly or partly conic sections and the majority of the major planets move in Keplerian ellipses, which are so nearly circular in form that the unaided eye would not notice the deviation from that form. The Sun is situate at one of the foci of these ellipses and so the displacement of the planet from the seemingly circular path is often detectable, owing to the planet's change in diameter, especially in the case of Mercury and Mars. The majority of all these bodies revolve about the Sun and about their primaries in the same direction and their rotation on their axes is also in harmony with the direction of their revolutions; there are some exceptions which pose a pretty problem and these will be discussed in the chapters that follow.

Chapter 1

The solar system in classical times

O God, I could be bounded in a nut-shell,
and count myself a king of infinite space.
SHAKESPEARE

The Sun and the five naked eye planets (Mercury, Venus, Mars, Jupiter and Saturn) known from antiquity appear to the eye of a watcher on Earth as wandering stars. They oscillate back and forth in the ecliptic with a preponderantly direct motion, that is to say in a direction from west to east. It is difficult to describe the motions correctly even when we have a knowledge of the actual arrangement of the system of the world, but without it the task is one of extreme difficulty—it is small wonder that the process of elucidating the true form of the system was to extend over many centuries and severely tax the intellectual skill of those who concerned themselves with the problem.

Classical antiquity runs over a period of some fourteen centuries from Homer to Damascios and embraces an incredible variety of thought.

The Greek conception of the solar system is predominantly anthropocentric except for the truly remarkable aberrations to be found in the heliocentric system of Heraclides and Aristarchus, which antedates the ideas of Copernicus and Kepler by seventeen centuries. In all, some twelve philosophers raised the intellectual edifice of the solar system in classical times to one of surprising sophistication. It is with lasting and deep regret that we have to record that very little of the writings of these remarkable men has come down to us.

We are forced to rely on reports, often only fragmentary, and written by those who wished to refute the case that they propounded. In our chosen subject this serious want of information is felt immediately when we attempt to deal with the thought of the first of the Ionian philosophers, Thales of Miletus, who was born in *c.* 640 B.C. and died in *c.* 546 B.C. He appears to have left no writing behind him and to have gained his main insight into scientific things from his Egyptian mentors. We are indebted to Aristotle[1] for a reference to his view. Thales is un-

[1] Aristotle, *Metaphysics* (Everyman Edition 1000), p. 58.

aware of the sphericity of the Earth and considers it to be 'like a piece of wood' on the ocean (*terrarum orbem aqua sustineri*), the water of which is the origin of everything. The pupil of Thales, Anaximander,[2] who according to the computations of Apollodorus, was born in 611 B.C. and died in 547 B.C., is a philosopher of much subtlety. He derives from a primal chaos[3] of vague and limitless extent, a cylindrical earth[4] poised equidistant from surrounding orbs of fire that had originally clung to it, as the bark of a tree, until their constituents were severed and they parted into several wheeling, fire-filled bubbles of air. But as the numberless and endless had been the prime cause of the motion into separate existences and individual forms, so also, according to the just award of destiny, these forms would, at an appointed season, suffer the vengeance due to their earlier act of separation and return into the vague immensity whence they had issued. Thus the world, the central figure, would disappear into the 'indeterminate'. This was a poetic and metaphysical construction of some beauty, but one which was not hampered by any appeal to direct observation of the cosmos; indeed, the free intellectual rein was such that Anaximander put the Sun as the most distant of the celestial bodies with the fixed stars nearer to us.

The ideas of Anaximander were carried forward by his pupil Anaximenes of Miletus,[5] who espoused the view that everything is air at different degrees of density. In this way he 'formed' the Earth as a broad disc floating on the circumambient air. Similar condensations produced the Sun which generated its heat from the flaming state derived from the velocity of its motion. He is said to have remarked that the firmament turns round the Earth 'like a hat round the head', which suggests that he may have seen the Earth as an hemisphere.

The elucidation of the form of the solar system was next advanced by the Pythagorean school. Its founder, Pythagoras,[6] was a pupil of Pherecydes and Hermodamus. In their view the solar system comprised a globular[7] Earth, self-supported in empty space, revolving about a central luminary, as also did the Sun and the other planets.[8] The central luminary was termed Hestia the watchtower of Zeus. Around the central

---

[2] See *Dissertation sur la philosophie d'Anaximandre* in the Mémoires de l'académie des sciences de Berlin 1815.

[3] Anaximander's ἄπειρον (a word untranslatable into any other western tongue) is that which possesses no 'number', no measurable dimensions or definable limits and therefore no being; the measureless and negation of form.

[4] Its height was one-third of its breadth.

[5] See Schmidt, *Dissertatio de Anaximensis psychologia* (Jena 1869). (Anaximenes lived in the latter half of the sixth century B.C.)

[6] Pythagoras, 586–*c.* 500 B.C. He was not a philosopher according to the pre-socratics; he was a saint, prophet and founder of a fanatical religious society.

[7] The idea of a spherical Earth is generally attributed to Parmenides of the renowned southern Italian school, the Eleatic.

[8] This point is often misunderstood, the Pythagoreans did not teach a heliocentric system. The great Dr Jowett is so misled, see *The Dialogues of Plato*, vol. III, 3rd ed., p. 388. Boeckh, *De Platonico systemate coelestuim globorum et de vera indole astronomial Philolaicae* (Heidelberg 1810). See also Boeckh's, *Kleine Schriften*, vol. III (Berlin 1866).

fire revolved ten bodies, Antichthon,[9] the counter-earth, the Earth, the Moon, the Sun, the five planets and the heaven of the fixed stars. Pythagoras then equated his discoveries of number in the musical harmonies with the distances of the revolving orbs from the central fire— the famous doctrine of 'the harmony of the spheres'. The velocities of the several orbs depended upon their distance from the centre, the slower and nearer giving out a deep musical note, the swifter a high note, and the concert of the whole emitting the cosmic octave undetectable by mere mortals.

Pythagoras is thought to have been the first to realize that the stars of the morning sky are those also of the evening sky, a point which few may readily come to without the benefit of education, and of which some in our present age are not yet aware.[10] He had what is best described as a mighty and truly religious intuition, which enabled him to see with complete inward certitude that in number, as the sign of complete demarcation, lies the *essence* of everything actual which is cognized, and this was to form the base of a formidable cult[11] embracing the entire world and the interspaces which instead of being filled with the impressive symbol of space is for him the nonent (τὸ μὴ ὄν).

Heraclitus[12] (540–475 B.C.) takes nothing from Pythagoras, indeed his thought is a regression. He is closer to Anaximander. To him everything is and is not, all things are and nothing remains. In his conception of the system of the Sun he believes it to be re-created daily and of a very small diameter. In this his thought resembles that of Metrodorus,[13] who accepted the plurality of worlds and held that the stars are formed from day to day from moisture. He says 'we know nothing, no, not even whether we know or not'. Anaxagoras was born in 500 B.C. at Clazomenae in Asia Minor. He is said to have died in Lampsacus in *c.* 428 B.C. Some authorities assert that Socrates was amongst his disciples. His influence stems in the main from his astronomical and mathematical eminence. His observations of the celestial bodies led him to form new theories of the universal order. As with all innovators he upset the popular faith and he had to leave Athens for Lampsacus. He boldly attempted to give a scientific account of eclipses, meteors, rainbows and the Sun, which latter he described as a mass of blazing metal, larger than the Peloponnesus. The heavenly bodies were thought to be masses of stone torn out of the Earth and ignited by rapid rotation. His cosmological ideas are complex

[9] The antichthon was invented by Philolaus, it is a hypothetical second Earth on the opposite side of the Sun.

[10] Emerson records that the man in the street does not know a star in the sky.

[11] The Pythagoreans were of a sober spirit, cold and filled with a pedantic ecstasy. They were responsible for the total destruction of Sybaris (*c.* 510 B.C.), an event that shocked the classical world.

[12] See Bywater, I., *Heracliti Ephesii reliquiae* (Oxford 1877), English Edition by G. T. W. Patrick, Baltimore 1889.

[13] Some of Metrodorus' writings are in *The Treasury of Mathematics* (vol. II), Midonick, H. (Penguin Books 1968).

and profound, indeed he is thought by some to have advanced his thought to the point where he directly paved the way for the realization of the atomic theory.[14] In harmony with the Pythagorean notation he put the 'planets' in the order: Moon, Sun, Venus, Mercury, Mars, Jupiter and Saturn.

The next philosopher upon the stage, to advance a novel system of the world, is Empedocles of Agrigentum (490–430 B.C.). He is the democrat of antiquity, statesman, prophet, physicist, physician and reformer. His system of the world is remarkable for the idea of twin celestial hemispheres, one of fire and one of air. Within these hemispheres he had twin Suns, one in one hemisphere and the other in the second hemisphere, a reflection from the rounded Earth. The planets are fiery masses moving freely in space beyond the orbit of the Moon. The axis of rotation of the twin hemispheres is not easy to determine, but it cannot have coincided with the axis of the sphere of the stars. He is said to have placed the Moon twice as far from the Sun as from the Earth.[15]

For all the undoubted importance of Plato's thought we need not remain long in discussing his views on the solar system. In the *Phaedrus* he describes the universe as a sphere, and in the *Phaedo* the Earth is said to stand in the midst of heaven. The daily rotation of the heavens is dealt with in the 10th book of the Republic in which a complex machinery for moving the heavenly spheres is described. In the *Timaeus* he gives the distances of the planets derived from the two geometrical progressions 1, 2, 4, 8; 1, 3, 9, 27. He places the planets in the Pythagorian order: Moon (1), Sun (2), Venus (3), Mercury (4), Mars (8), Jupiter (9), Saturn (27). By the interpolation of the numbers of fractional form he derives an arithmetical musical scale of four octaves and a major sixth.[16] In both the *Timaeus* and the *Republic* we find Saturn described as the slowest-moving planet, which is difficult to understand since the heaven of the fixed stars makes one rotation in twenty-four hours, and owing to the proper motion of the planets the outermost will return to the meridian the soonest, and exhibit the greatest angular velocity. Plato had no knowledge of the actual dimensions of the planetary orbits and hence the actual orbital velocity was inaccessible to him. We must conclude, for all his insight (and he has left us the most noble edifice of the human intellect in his theory of ideas) that he retains the daily rotation of the heavens about a static globular Earth.

Eudoxus of Cnidus (*c.* 408–355 B.C.) offers us a very different example.

[14] See Fairbanks, A., *The First Philosophers of Greece* (1898). Schaubach, E., *Fragments of Anaxagoras* (Leipzig 1827). Parmentier, L., *Euripide et Anaxagore* (1892).
[15] Aëtius II, 31. *Note:* Aëtius summarized the opinions of the Greek philosophers on natural philosophy. These *Placita* are reproduced in the p.s. Plutarchean *Epitome* and in Strobaeu's *Eclogae* edited by H. Diels in *Doxographia Graeci* (1879). They form one of our most important sources of the opinions of the philosophers whose works have perished.
[16] See Boeckh, *Ueber die Bildung der Weltseele im Timaeos des Platon* (1807), Kleine Schriften III, pp. 109–80.

He stands in the foremost rank of the Greek mathematicians, indeed Plato is said to have proposed him as one of only two men who could solve the Delian problem of the duplication of a cube.[17] Eudoxus is the first philosopher to advance a system of the world in which each planet has a separate piece of machinery, albeit invisible, peculiar to itself. He does not rely, as did his precursors, on the planetary spheres being dragged round by the diurnal motion of the sphere of the fixed stars.

The system of Eudoxus is remarkably elegant, as one would expect from one who is without a peer in the mathematical arts as applied to the macrocosmos. Eudoxus explains his revolutionary ideas in a book entitled *On velocities*. This work is lost to us, but we have both Aristotle's *Metaphysics*[18] and Simplicius' *Commentaries* to Aristotle's book on the heavens as guides to the system of Eudoxus. In more recent times the ideas of Eudoxus have been examined and commented upon with sublime skill by the great Italian astronomer Schiaparelli.[19] Eudoxus accepts, as do all his precursors, that the planets move in circular orbits. His system is novel for its placing of each of the planets on a sphere co-operating with other spheres, the centres of which all coincide with the centre of the Earth. These are the famous homocentric spheres of Eudoxus. Every celestial body is placed on the equator of a sphere that revolves with uniform motion around its axis. To explain the complex arcs of retrogression and other anomalous movements of the planets as seen in the natural heavens, Eudoxus assumed that the said poles were carried by a second concentric sphere, the poles and the rotational motion of which did not coincide with those of the first sphere. The second sphere was carried by yet a third and in some instances a fourth concentric sphere *mutatis mutandis*. Any sphere which did not itself carry a celestial body was termed a starless sphere and was a metaphysical part of the cosmic assemblage. Eudoxus found that he needed three spheres to explain to his satisfaction the motions of the Sun, three for the Moon, four for each of the five planets, and one for the stars: in all twenty-seven homocentric spheres. The system is, needless to say, complex of analysis and the table below page 25 may be of interest to those who wish to follow this system a little more thoroughly. Dreyer[20] shows, for example, that a planet fixed to the fourth sphere is endowed with four motions, the daily one, the orbital one along the zodiac, and the other two in the planets' synodic period.[21] These motions conspire to give the planet an

---

[17] The point is made by Plutarch in his *De genio Socratis*, cap. VIII. The other mathematician of equal ability is Helikon.
[18] Aristotle, *Metaphysics*, Everyman Edition No. 1000, 1956, p. 355.
[19] Schiaparelli, *Le sfere omocentriche di Eudosso*, publication of the Observatory of Brera in Milan, IX (1876).
[20] Dreyer, J. L. E., *History of the Planetary Systems from Thales to Kepler* (Cambridge 1906).
[21] *Synodic period*. An interval of time between two similar positions of the Moon or a planet, relative to the line joining the Earth and Sun; hence the time from one conjunction, opposition, or quadrature to another, and the period of the phases of the Moon or a planet.

apparent motion against the stars in the form of a spherical lemniscate (Fig. 1) having the polar equation $r^2 = a^2 \cos 2\theta$. The longitudinal axis of the lemniscate lies along the zodiac and its length is equal to the diameter of the circle described by the pole of the fourth sphere. The longitudinal axis of this spherical lemniscate revolves round the ecliptic in the sidereal period[22] of the planet and this is able to produce the difficult arc of retrogression seen in nature, a feat of no mean geometrical skill.

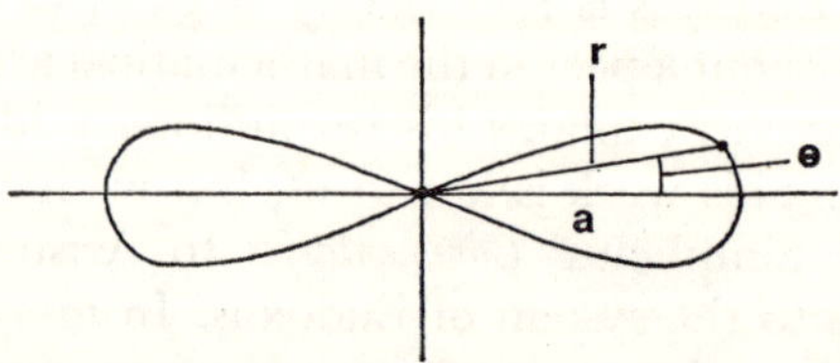

Fig. 1. Lemniscate aquartic curve discovered by Jacques Bernoulli (*Acta Eruditorum*, 1694) and later investigated by G. C. Tagnano. The lemniscate is the locus of a point that moves so that the product of its distances from two fixed points is constant and equal to the square of half the distance between these points; it is a particular form of Cassini's oval. Its polar equation is $r^2 = a^2 \cos 2\theta$.

This elegant system was improved by Calippus, primarily by adding a fifth sphere to the system of spheres for the planets, Mercury, Venus and Mars, to bring about a retrograde motion[23] to the planets' paths without distorting the synodic period; and two further spheres each to the systems of the Sun and the Moon to explain their observed unequal motions in longitude. This refinement increased the total number of spheres to thirty-four. For the first time in history observational astronomy appears on the scene and the solar system is no longer solely subjective. The intellectual system is now required to match the world model and the attempt to find the 'true' physical system of the world makes its début.

As we have already remarked, the diurnal motion of the Earth and its revolution about the Sun are alien to all the early systems, indeed even Eudoxus has a static earth. Yet this modern system of the world came to an early but ephemeral flowering in the minds of two men in classical times—Heracleides and Aristarchus. Some scholars argue that the diurnal rotation of the Earth was anticipated by Anaximander and Plato,[24] but this is not easily supported when one considers the complete systems that they espoused. In contradistinction both Heracleides and Aristarchus teach unambiguously the diurnal rotation and the revo-

---

[22] *Sidereal period.* The interval between two successive positions of a celestial body in the same point with reference to the fixed stars; applied to the Moon and planets to indicate their complete revolution of the heavens as against their synodic revolution relative to the line joining the Earth and the Sun.

[23] *Retrograde motion.* Apparently or actually contrary to the order of the signs; directed from east to west.

[24] See Grant McColley, 'The theory of the diurnal rotation of the Earth', *Isis*, 26, 392 (1936–7).

THE WORLD SYSTEM OF EUDOXUS

The system comprises a multiplicity of concentric spheres that defy accurate representation by a drawing.

The following chart shows something of the system:

Fixed star sphere makes one revolution in 24 hours.

Moon. 3 spheres		
Sphere I (outermost)	Sphere II	Sphere III
Rotation E →W in 24 hours.	Rotation W →E. Axis of Zodiac. Period 223 lunations.	Rotation W →E. Axis inclined to Zodiac. Period 27 days to account for $18\frac{1}{2}$ year motion of the nodes of the lunar orbit.

Sun. 3 spheres		
Sphere I (outermost)	Sphere II	Sphere III
Rotation 24 hours.	Rotates along the Zodiac with slow direct motion.	Rotates once in a year inclined to the Zodiac.

Planets. 4 spheres			
Sphere I (outermost)	Sphere II	Sphere III	Sphere IV
Rotates once in 24 hours.	Uniform motion along the Zodiac. 3 outer planets have sphere with a period equal to their sidereal period. Mercury and Venus have a sphere rotating once in a year.	Poles are on sphere II at opposite signs of the Zodiac. Period is equal to the planet's synodic period.	Moves in the opposite direction to sphere III, but with same period. The planet is fixed to the equator of this sphere.

lution about the Sun. The heliocentric system for which mankind had to wait another seventeen centuries was born and cast aside for reasons that are not now easy to adduce. We know that Aristarchus was accused of impiety and this mean act is attributed to Cleanthes the Stoic. It would seem, then, that the mould of the classical mind was unable to see that the many anomalies of the planetary system, especially the changes in the brightness of Mars and Venus that could not be explained by concentric spheres, as put forward by Eudoxus, would be resolved by the *simple* expedient of making the Earth move about the Sun. Whatever the truth of the matter the modern system did not capture the mind of man at this stage and we must enter further ingenious labyrinths before we once again discover the light.

The solar system

In an attempt to explain *inter alia* the changing brightness of Mars and Venus the classical mind set itself to develop the use of eccentric[25] orbital motions, and in the hands of three intellectual giants, Ptolemy, Hipparcus and Apollonius, the epicyclic system was invented and developed. This system enabled the positions of the planets to be computed, but the search for the 'true' system of the world faded into insignificance for many long years.

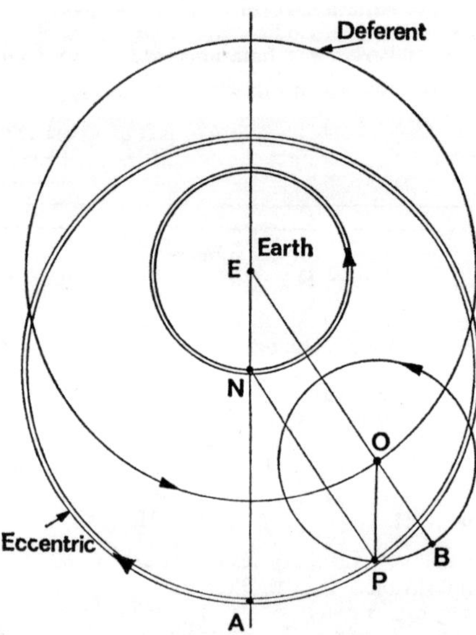

Fig. 2. The equivalence of the epicyclic and eccentric systems. The planet P describes an epicycle about O on the deferent. The planet P also describes an eccentric; the radii of the eccentric and the deferent are equal (EO = NP). Note that the motion of the eccentric is retrograde and the epicycle direct. EO is parallel to NP, let them swing round to remain so; then since $A\hat{N}P = A\hat{E}O = P\hat{O}B$, OP swings round at a uniform rate relative to OB, and P describes an epicycle about O, while O uniformly describes a deferent about E.

For some time two systems attracted attention, that of movable eccentrics and that of the epicycle.[26] It can be shown by simple geometry that the two systems offered identical paths for the planet (Fig. 2). The epicycle was finally adopted since it allows a more ready explanation of the stationary points and the arcs of retrogression; further the movable eccentrics were only applicable to the outer planets and the epicycle had

[25] *Eccentric.* In Ptolemaic astronomy: a circle or orbit not having the earth precisely in its centre.
[26] *Epicycle.* A circle, having its centre on the circumference of a greater circle. In the Ptolemaic system of astronomy each of the 'seven planets' was supposed to revolve in an epicycle, the centre of which moved along a greater circle called a deferent. This conception, though superseded as a physical explanation, describes with approximate correctness the relative motion of a planet when the Earth is assumed as fixed; and it is therefore still occasionally used for this purpose by modern astronomers.

26

to be used to explain the phenomena of the inner planets. The dual system offended the simplicity concept, so often thought for sheer elegance to lie nearest to the heart of truth, that there is little wonder that the unity of the epicyclic system succeeded and was generally accepted. Using the epicyclic system Hipparchus[27] of Nicaea (*fl.* 146–126 B.C.) was able to represent the apparent motions of the Sun to within an error of less than one minute of arc. He was also able to explain the first inequality of the Moon's motion, the so-called equation of the centre,[28] which in modern terms is caused by the elliptical orbit of the Moon about the Earth and is the difference between the true and mean anomalies.[29] We should reflect on the fact that Hipparchus worked and taught three centuries before Ptolemy, yet Ptolemy worked much as if he was the immediate disciple of his illustrious precursor. Nothing speaks more highly of the lucidity which Hipparchus brought to these notoriously difficult matters.

Claudius Ptolemy's incunabula is empty and we have very little knowledge of his true nationality. We can fix his period from the observations of the heavens that he made in the reigns of Hadrian[30] and Antoninus Pius. He is said by Olympiodorus to have lived for forty years in the elevated terraces of the temple of Serapis, at Canopus near Alexandria. Ptolemy's great work is the Mathematical Synthesis, universally known as the *Almagest*, of which Books IX to XIII on the planetary motions have the most appeal to us here in our enquiry into the system of the world. These books take the brilliant work of Hipparchus and develop it.

Ptolemy rejects the system of Aristarchus (to our eternal loss) and erects in its place the epicyclic system[31] (Fig. 3) which has ever afterwards carried his name—the Ptolemaic system of the world—which was to hold undisputed the intellectual stage for the following sixteen centuries.

We should not fail to remark that Ptolemy wrote one other work equal in power to the *Almagest*, his *Geōgraphice Hyphēgesis*, in which he discusses the size of the Earth and the then known world. Unfortunately

[27] One only of his many works has survived, a Commentary on the *Phaenomena* of Aratus and Eudoxus, published by P. Victorius at Florence 1567 and included in D. Petavius' *Uranologium*, Paris 1630, a new edition of which was published by Carolus Manitius (Leipzig 1894).

[28] *Equation of the centre.* In astronomy 'equation' is the action of adding to or subtracting from any result of observation or calculation such a quantity as will compensate for a known cause of irregularity or error.

[29] See Brown, E. W., *An Introductory Treatise on the Lunar Theory*, 1896 (Dover, reprint N.Y. 1960, p. 35).

[30] His first observation is made in the eleventh year of Hadrian *c.* A.D. 127. Some authorities give it as the ninth year in view of a lunar eclipse referred to in the Almagest IV. 9. It is unclear, however, whether Ptolemy observed it personally. The best estimate of his working life's duration is *c.* AD 139–61.

[31] See Hoyle, F., *Astronomy*, London 1962, appendix on the Epicyclic Constructions of Hipparchus and Ptolemy, pp. 306–9.

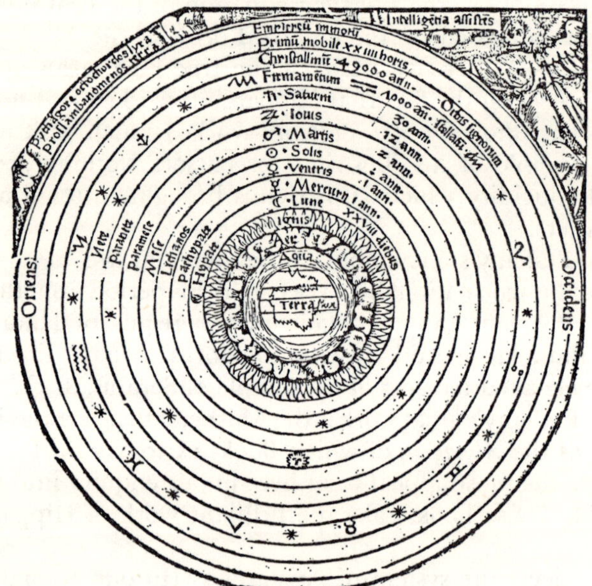

Fig. 3. The world system according to Polydore Vergil (*c.* 1470–1555), from *De inventoribus rerum*, 1499; the geocentric system.

he preferred the estimate of Poseidonios to that of Eratosthenes,[32] the more accurate of the two. He also wrote on optics and astrology. His breadth of knowledge and his diversity are truly astonishing. It was the Arabic translation of the *Mathematikē Syntaxis* that was given the name of *Almagest* and this work was eventually criticized by the most advanced astronomers of the Muslim world, who noticed from their own researches into positional astronomy the increasing difficulty of reconciling the phenomena of the heavens with the positions computed from the epicycles of the Ptolemaic system (Fig. 4).

[32] See Sarton, G., *Ancient Science and Modern Civilisation*, Univ. of Nebraska Press 1954. Diller, A., 'Ancient Measurements of the Earth', *Isis, 40*, 6 (1949).

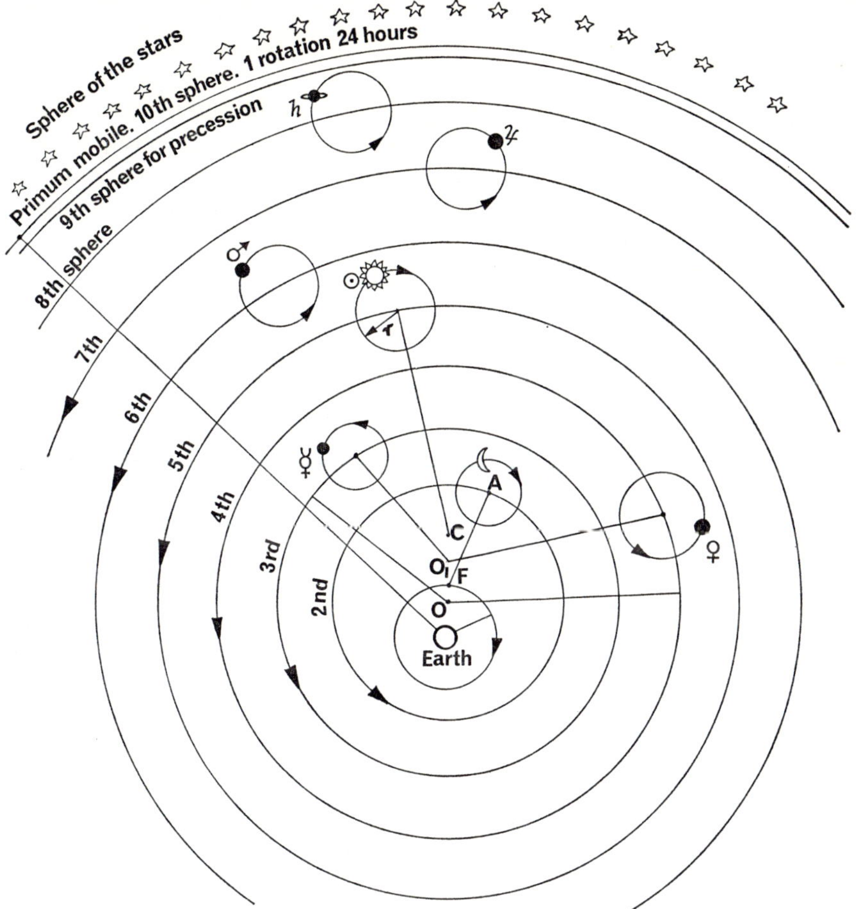

4. A simplified drawing of the Ptolemaic system of the world.

Periods of revolution are:

Planet	Centre of epicycle on the eccentric	Centre of planet on the epicycle
Mercury Venus	A sideral year	Synodic period
Mars Jupiter Saturn	Zodiacal period of planet	Synodic period

Venus, Mars, Jupiter, Saturn
O Stationary centre of deferent
O_1 Epicycle moves about this point, $EO_1 = 2EO$.

Moon
Epicycle moves about A on eccentric deferent centre F, which revolves from east to west about E.

Sun
As with Hipparcus the Sun is on an eccentric centre C, based on eccentricity CE/CA, $r = OE$, period one tropical year.

The solar system

This brief recapitulation of the ideas of the world that come to us out of classical times must now be closed, not with a little regret, for the subject has a fascinating quality not readily found in any other field. We should not, however, pass from this domain without reflecting on its variety and pathos.

Spengler[33] helps us to see its lasting meaning and deep significance in a passage of great insight.

We have already observed that, like a child, a primitive mankind acquires (as part of the inward experience that is the birth of the ego) an understanding of number and *ipso facto* possession of an external world referred to the ego. As soon as the primitive's astonished eye perceives the dawning world of *ordered* extension, and the *significant* emerges in great outlines from the welter of mere impressions, and the irrevocable parting of the outer world from his proper, his inner, world gives form and direction to his waking life, there arises in the soul—instantly conscious of its loneliness—the root-feeling of *longing* (Sehnsucht). It is this that urges 'becoming' towards its goal, that motives the fulfilment and actualizing of every inward possibility, that unfolds the idea of individual being. It is the child's longing, which will presently come into the consciousness more and more clearly as a feeling of constant *direction* and finally stand before the mature spirit as the *enigma of Time*—queer, tempting, insoluble. Suddenly, the words 'past' and 'future' have acquired a fateful meaning.

It is a deep world-fear which never leaves the higher man, the believer, the poet, the artist—that makes him so infinitely lonely in the presence of the alien powers that loom, threatening in the dawn, behind the screen of sense-phenomena. The element of direction, too, which is inherent in all 'becoming', is felt owing to its inexorable *irreversibility* to be something alien and hostile, and the human will-to-understanding ever seeks to bind the inscrutable by the spell of a name. It is something beyond comprehension, this transformation of future into past, and thus time, in its contrast with space, has always a queer, baffling, oppressive ambiguity from which no serious man can wholly protect himself.

With the close of the truly remarkable study of the solar system by the Greeks a long period of time, some fourteen centuries, pass before we come to the Renaissance, the subject of our next chapter. We should not, however, pass over the Middle Ages without some comment, for it is a period sometimes referred to as the Dark Ages, and Sarton[34] reminds us that it is nowhere more dark than in our appalling ignorance of it.

In the Middle Ages I can discover no significant contribution to the solar system, but that is not to say that keen minds did not exist and

[33] Spengler, O., *Untergang des Abendlandes*, (*The Decline of the West*,) Knopf, New York 1939, vol. I, pp. 78, 79.
[34] Sarton, G., *Introduction to the History of Science* (5 vols), Carnegie Institution of Washington 1948. See also Delambre, *Histoire de L'Astronomie du Moyen Age*, Paris 1819.

attempt to grapple with the problem of the Earth, Sun, Moon and planets. It was a period singularly difficult for the intellectual spirit to flourish in. Theology was the core of science and science and religion were inseparable, so that there was a very different centre of gravity in intellectual pursuits from what we see today. Publication was in manuscript form, travel was slow, translations of many important works made the tortuous passage from Greek into Syriac into Arabic into Latin and then into our own tongue. To see that the age was not wholly sterile in scientific astronomical and astrological works one has but to recall that Arabic[35] was for a long period the *lingua franca* of science, and such names as Jābir-ibn Haiyān, al-Kindi, al Khuārizmi, al Farghāni, al Rāzi and Omar Khayyám, all of whom flourished between 950 and 1100 A.D., are of themselves sufficient to destroy the validity of such an approach.

Strange to say nothing contributed more to the decline of scientific thinking in the Middle Ages than the Roman indifference to its claims, and it is said that Roman utilitarianism combined with theological expediency and then theological domination, refused to permit a scientific revival in those centuries of which we speak.

With the advent of the twelth century Ptolemy's *Almagest* was made available in Latin, and Sarton observes that 'during the medieval period every astronomer, whether Jewish, Christian or Muslim, might be assumed to be familiar with Ptolemaic astronomy, directly or indirectly —we might even say that every one was a Ptolemaist'.

The fourteenth century witnessed the ravages of the Black Death which began in 1348 and the conquests of Tīmur Lang. These two tribulations alone claimed millions of men; not a country was free from plague or war for a whole year. Then came the Hundred Years War with its inroads upon the human spirit in the West. In 1300–50 there was, for example, no astronomer in the bishopric of Liége.[36] Henry Bate, canon of Malines, practised the art, as did John of Saxony (1327–55), and in England we had Richard Swineshead and Thomas Bradwardine both Merton alumni.

No one was wholly free of astrological studies in those days, and John of Ashendon produced a famous treatise on the subject in 1348. The calendar was reformed a little by Jean de Meurs. Dante, deriving his ideas on astronomy from al Farghāni, was the first western writer to refer to the Southern Cross. The main theorist was Giles of Rome (*d.* 1316) and Hugh of Città di Castello wrote (1337) a commentary on one of the most popular textbooks of the Middle Ages, *The Sphere* of John of

[35] See Hell, J., *The Arab Civilization*, Heffer, Cambridge 1926, pp. 87–91, translated by S. Khuda Bukhsh from the professor's distinguished monograph *Die Kultur der Araber*, a work of singular scholarship yet short, accurate and felicitous in diction. See also Suter, H., *Die Mathematiker and Astronomen der Araber*, Vienna 1917; Spengler, O., *Untergang des Abendlandes*, vol. II, p. 227 *et seq.* (*The Decline of the West*, vol. II, pp. 189–323); Pingree, D., *Gregory Chioniades and Palaeologan Astronomy*, Dumbarton Oak Papers, *18*, 135, 1964.
[36] Liége was a great centre of learning in the eleventh century.

Sacrobosco. Pietro d'Abano in 1310 produced an elaborate astronomical treatise defending Ptolemaic theories and in 1391 Chaucer gave us his celebrated treatise on the astrolabe.

With the advent of the printing press the ideas of classical times concerning the system of the world found a wide audience. Indeed by 1543, the date of the publication of Copernicus' *De revolutionibus*, most important ancient works on astronomy were available to scholars in printed editions[37]

[37] Johnson, F. R., *Astronomical thought in Renaissance England* 1937.

The dates of the first printed editions of the most important ancient works dealing with astronomy are as follows:

Ptolemy's *Almagest*: First edition of an epitome (Peurbach and Regiomontanus), Venice, 1496; first complete edition, a Latin translation from the Arabic, Venice, 1515; first Latin translation from the Greek, by George of Trebizond, Venice, 1528; first edition of Greek text, by Simon Grynaeus, Basel, 1538.

Aristotle's works: First complete edition, with Greek text, by Aldus Manutius, Venice, 1495–98; a vast number of editions of Latin translations had been published much earlier.

Plato: Ficino's Latin translation of the works, Venice, 1482; first Greek edition, Venice, 1513.

Chapter 2

The solar system in the Renaissance

Happy the man who swamped in this sea of Error
Still hopes to struggle up through the watery wall;
What we don't know is exactly what we need
And what we know fulfils no need at all.
GOETHE

The ways of Heaven are dark and intricate,
Puzzled in Mazes and perplex'd with errors:
Our understanding traces 'em in vain,
Lost and bewilder'd in the fruitless search;
Nor sees with how much art the windings run,
Nor where the regular confusion ends.
ADDISON

The authority of Aristotle's teachings was so immense that from the fourth century B.C. up to the time of Ptolemy and beyond, it was difficult to break the firm hold that his ideas had over the human intellect.

Aristotle had adopted the system of Eudoxus and had relied on the heavenly bodies being carried by the spheres, which always remained invisible to the human eye and which themselves always operated with uniform circular motions. We have already seen that Ptolemy adopted the metaphysical approach of Aristotle, but took a number of liberties with the motions of the heavenly bodies, introducing the epicycle, the deferent, the eccentric and, what is alien to uniform motion in a circle, an angular motion about any point inside the circle of motion not necessarily the centre—known as the equant;[1] something which would have horrified the 'pure' mind of Aristotle. Ptolemy did not deny the spheres but he appears conveniently to have ignored them; yet the spheres were much too tenacious of human affections to wither and die, and in the

[1] *Equant.* A circle imagined by the ancient astronomers for the purpose of reducing the planetary movements to consistency with the hypothesis that celestial motion must be uniform in velocity.

33

fair city of Venice in 1538 there was published the principal work of Girolamo Fracastoro,[2] his *Homocentrica* in which an ill-defined system of seventy-nine[3] spheres was made to account for the behaviour of the planets.

The basic theme of Eudoxus was also advanced and embellished (but not quite so extravagantly as with Fracastoro) by Giovanni Battista Amici[4] in a small obscure work published in Venice in *c.* 1536 with the flamboyant title: *De motibus corporum caelestiū iuxta principia peripatetica sine excentricis et epicyclis*. He avers that nature does not know the epicycle or the eccentric and in his system the perfection of the world is restored to the Aristotelian mould with some sixty-seven spheres. Yet, at the same time, we have Calcagnini[5] (1479–1541) announcing that the whole heavens do *not* revolve in one day, but that it is the movement of the Earth which produces the effect of the heavens revolving from east to west. In an essay written before 1525 '*Quod caelum stet, terra moveatur, uel de perenni motu terrae*', he places the Earth at the centre and is, as far as our knowledge is reliable, the only man alive[6] at that time to teach the diurnal rotation of the Earth before the great work of Copernicus was published.

The difficulties of equating the phenomena of the planetary motions with the Ptolemaic system became increasingly difficult and the systems of Fracastoro and Amici did not go to the heart of the matter. The Ptolemaic system became more and more a mere calculating device. We have some evidence that Ptolemy himself so regarded it, since his lunar theory required, as one example of the inherent anomalies, that the distance and thus the apparent diameter of the Moon should vary by 100 per cent during the span of a single month and yet this was never observed to occur. Notwithstanding the complexities, the elegant geometry of the epicyclic system, augmented by numerous geometric refinements, enabled Apollonius, Hipparcus and Ptolemy to construct a system adequate to deal with the crudely measured phenomena of their period.

It is against this confused intellectual climate walled-in by a rigid code of respect for the ancient culture that we must view the work of the philosophers and astronomers of the renaissance,[7] especially that of Copernicus. We have already remarked that it was with not little

[2] Fracastoro was professor of philosophy at Padua; born *c.* 1483, died *c.* 1553.
[3] The system of Fracastoro gave 8 spheres for the stars and planets. 6 for the precession, 10 for Saturn, 11 for Jupiter, 9 for Mars, 6 for the Sun, 11 for Venus, 11 for Mercury, 6 for the Moon and one for the sublunary realm.
[4] Not to be confused with his namesake the Italian astronomer and microscopist, professor of mathematics at Modena 1786–1863.
[5] Caelio Calcagnini was professor in the University of Ferrara.
[6] The diurnal rotation of the Earth had been taught to the early Greeks, hence Aristotle's active and extended attack upon it in *De Caelo*, II, 13, 14.
[7] The term renaissance is difficult to define; see the article by J. A. Symonds, *Encyclopaedia Britannica*, 11th edition.

audacity that Muslim astronomers[8] criticized the Ptolemaic system of the world from their own researches into positional astronomy. Indeed some of them were such advanced geometers as to find a substitute for the epicycle and retain the convention of uniform circular motion. Now, at the centre of European civilization, attempts were being made to resuscitate the homocentric spheres of Eudoxus, and yet the major advance, on which future ages of astronomers were to build, did not come from the centre but from 'a quiet student at the shore of the Baltic[9] on the very outskirts of civilization'.

Nicolaus Copernicus[10] was born on 1473 February 19 at the Polish town of Torun on the Vistula. He studied at Cracow University from 1491 until 1494 when he went to the Universities of Bologna, Padua and Ferrara. We know that his sojourn at Bologna brought him into friendship with the famous professor of astronomy Domenico Maria Novara. Much of what subsequently took place and the reasons for it are now pure conjecture, since Copernicus leaves us very little on which to form a firm opinion. Without doubt his deep thinking was silently done and in his short yet penetrating *Commentariolus*[11] written at Frombork (*c.* 1575) he sets out, with economy and lucidity, the system of the world that was to be embodied in the work on which he laboured for some thirty-six years and which he saw in proof only on his death bed—*De revolutionibus*.

Copernicus, by an inward genius now difficult to appraise, changed the geostatic system of Ptolemy for a heliostatic[12] one; yet despite this bold break with tradition he is still sufficiently an Aristotelian, and enslaved by his age, to be unable to free himself from the epicycle. We should not today come quickly to the view that his reluctance to dispense with the epicycle is a fault, since it can be shown that the epicycle is one of the most versatile devices[13] known to the geometer and one which enables him to make some attempt to encompass Nature's numerous subtleties as seen in the planetary motions. But the cast of Copernicus' mind is such that it will not allow him the licence to use the Ptolemaic equant

[8] See Kennedy, E. S., and Roberts, V., 'The Planetary Theory of Ibn-al-Shatir', (*c.* 1304–75), *Isis*, *50*, 227–35, 57, 208–19, based on Arabic Treatise NIHAYAT AL SUL, Bodleian MSS. Marsh 139, 290, 501, Hunt 547.
Roberts, V., 'Solar and Lunar Theory of Ibn-al-Shatir', *Isis*, *48*, 428–32.
Faud Abbud, 'The Planetary Theory of Ibn-al-Shatir', *Isis*, *53*, 492–9.
Kennedy, E. S., 'Late Medieval Planetary Theory', *Isis*, *57*, 365.
[9] The actual town is Frombork on the shores of the Zalew Wislany which opens on to the Gulf of Danzig 19°30′ E. 54°18′ N.
[10] His name was Koppernigk. Some thirty signatures are known. The Jagellonian University of Cracow (see Rybka, E., *Four hundred years of the Copernican heritage*) give it as Nikalaj Kopernik.
[11] A beautiful inexpensive edition of the celebrated *Commentariolus* is available through Dover Books Inc.; see Rosen, E., *Three Copernican Treatises*, 2nd ed., 1959. This work is based on the text of Leopold Prowe taken from a collation of the Vienna and Stockholm MSS.
[12] The Ptolemaic system was not geocentric, since the Earth was not at the centres of the planetary motions but at the centre of the sphere of the fixed stars. In the Copernican system the Sun was eccentric to the paths of the planets.
[13] See Norwood Russell Hanson, 'The Mathematical Power of Epicyclical Astronomy', *Isis*, *51*, 150–8.

The solar system

and he retains the old metaphysical requirements of uniform circular motion and the invisible spheres.[14] What precisely he owes to Arabian astronomers will never be made clear, but today it is known that both Copernicus and Ibn-al-Shatir used a Tusi couple[15] in their plan for the orbit of Mercury. (See Fig. 5.)

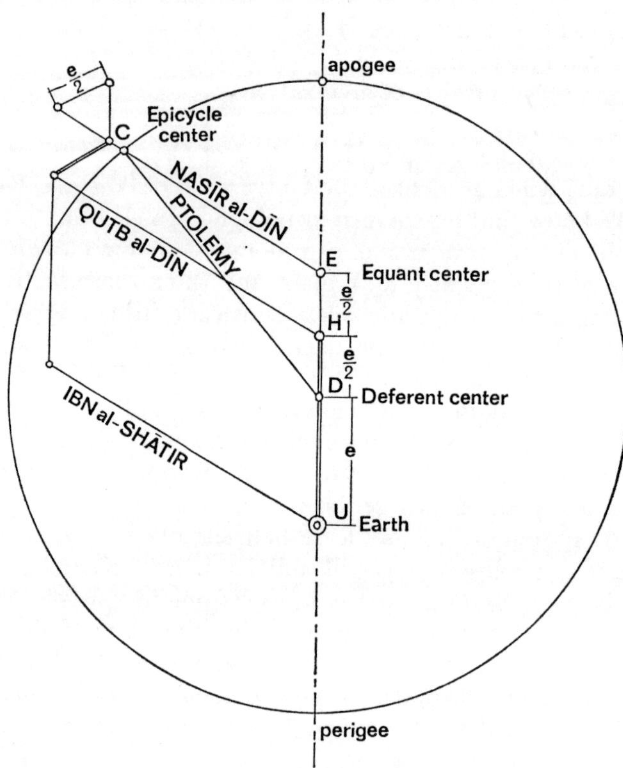

Fig. 5. Part of a Ptolemaic model showing the 'objectionable' *non*-uniform circular motions.

Copernicus saw that many anomalies would be removed once the Earth has a diurnal rotation and an annual revolution about the Sun, yet his system is equally as anthropocentric as that of Ptolemy; indeed, if we follow Copernicus we abandon a crude anthropocentricism of our senses in favour of a more ambitious anthropocentricism of our reason. If now we enquire for the reasons that may have excited Copernicus to exchange his terrestrial station for a solar one, it seems possibly to reside in a

[14] The full title of his great work is *De Revolutionibus Orbium Coelestium* (On the revolutions of the Heavenly *Spheres*).
[15] A device devised by Nasir al Din al Tusi (*c.* 1201–74). See Suter, H., *Die Mathematiker und Astronomen der Araber*, 1900; also Kennedy, E. S., 'Late Medieval Planetary Theory', *Isis*, 57, p. 365.

36

human preference for an abstract theory, and Polanyi[16] has pointed out, in a work of singular erudition, that

it becomes legitimate to regard the Copernican system as more objective than the Ptolemaic, only if we accept a shift in the nature of intellectual satisfaction as the criterion of greater objectivity. This would imply that of two forms of knowledge we should consider as more objective that which relies to a greater measure on theory rather than on more immediate sensory experience.

So that the theory being placed like a screen between our senses and the things of which our senses otherwise would have gained a more immediate impression, we would rely increasingly on theoretical guidance for the interpretation of our experience, and would correspondingly reduce the status of our raw impressions to that of dubious and possibly misleading appearances.

It has certainly been shown that the intellectual satisfaction the *heliostatic* system originally provided and which gained universal acceptance for the *heliocentric* system of today, proved it to be the token of a deeper significance unknown to its illustrious originator. It was the insight of Copernicus that anticipated in part the later discoveries of Kepler, Newton, Herschel, Hubble and Baade; because the *rationale* of *his* system while still seen 'as through a glass darkly' was yet an intimation of an intrinsic reality incompetely revealed to his intellect. In the *Commentariolus* (*c.* 1530)[17] Copernicus gives us a sketch of his hypotheses, yet surprisingly he makes no reference to Aristarchus of Samos.

NICHOLAS COPERNICUS[18]
SKETCH OF HIS HYPOTHESES FOR THE
HEAVENLY MOTIONS

Our ancestors assumed, I observe, a large number of celestial spheres for this reason especially, to explain the apparent motion of the planets by the principle of regularity. For they thought it altogether absurd that a heavenly body, which is a perfect sphere, should not always move uniformly. They saw that by connecting and combining regular motions in various ways they could make any body appear to move to any position.

Callippus and Eudoxus, who endeavored to solve the problem by the use of concentric spheres, were unable to account for all the planetary movements; they had to explain not merely the apparent revolutions of the planets but also the fact that these bodies appear to us sometimes to mount higher in the heavens, sometimes to descend; and this fact is incompatible with the principle of concentricity. Therefore it seemed better to employ eccentrics and epicycles, a system which most scholars finally accepted.

[16] Polanyi, M., *Personal Knowledge*, p. 4, Routledge and Kegan Paul, London 1958. Professor Polanyi, F.R.S., held *inter alia* the chairs of physical chemistry and social studies at Manchester. This work is the fruit of nine years philosophic enquiry into the whole range of his previous wide experiences.
[17] A. Birkenmajer in *Le premier système héliocentrique imaginé par Nicolas Copernic*, La pologne au vii Congrès International des Sci Hist, I, 1933, pp. 91–7, would date the *Commentariolus* before 1515.
[18] From Edward Rosen's translation.

The solar system

Yet the planetary theories of Ptolemy and most other astronomers, although consistent with the numerical data, seemed likewise to present no small difficulty. For these theories were not adequate unless certain equants were also conceived; it then appeared that a planet moved with uniform velocity neither on its deferent nor about the centre of its epicycle. Hence a system of this sort seemed neither sufficiently absolute nor sufficiently pleasing to the mind.

Having become aware of these defects, I often considered whether there could perhaps be found a more reasonable arrangement of circles, from which every apparent inequality would be derived and in which everything would move uniformly about its proper centre, as the rule of absolute motion requires. After I had addressed myself to this very difficult and almost insoluble problem, the suggestion at length came to me how it could be solved with fewer and much simpler constructions than were formerly used, if some assumptions (which are called axioms) were granted me. They follow in this order.

Assumptions

1. There is no one centre of all the celestial circles or spheres.

2. The centre of the earth is not the centre of the universe, but only of gravity and of the lunar sphere.

3. All the spheres revolve about the sun as their mid-point, and therefore the sun is the centre of the universe.

4. The ratio of the earth's distance from the sun to the height of the firmament is so much smaller than the ratio of the earth's radius to its distance from the sun that the distance from the earth to the sun is imperceptible in comparison with the height of the firmament.

5. Whatever motion appears in the firmament arises not from any motion of the firmament, but from the earth's motion. The earth together with its circumjacent elements performs a complete rotation on its fixed poles in a daily motion, while the firmament and highest heaven abide unchanged.

6. What appear to us as motions of the sun arise not from its motion but from the motion of the earth and our sphere, with which we revolve about the sun like any other planet. The earth has, then, more than one motion.

7. The apparent retrograde and direction motion of the planets arises not from their motion but from the earth's. The motion of the earth alone, therefore, suffices to explain so many apparent inequalities in the heavens.

Having set forth these assumptions, I shall endeavor briefly to show how uniformity of the motions can be saved in a systematic way. However, I have thought it well, for the sake of brevity, to omit from this sketch mathematical demonstrations, reserving these for my larger work. But in the explanation of the circles I shall set down here the lengths of the radii; and from these the reader who is not unacquainted with mathematics will readily perceive how closely this arrangement of circles agrees with the numerical data and observations.

Accordingly, let no one suppose that I have gratuitously asserted, with the Pythagoreans, the motion of the earth; strong proof will be found in my exposition of the circles. For the principal arguments by which the natural

38

philosophers attempt to establish the immobility of the earth rest for the most part on the appearances; it is particularly such arguments that collapse here since I treat the earth's immobility as due to an appearance.

We should not fail to observe that the Copernican system is almost always inadequately portrayed today, since most writers are content to reproduce simply the diagram that Copernicus himself displayed in *De revolutionibus* which, while adequate for his contemporaries well versed in the classical geometry, is today a singularly unsatisfactory presentation.[19] I have accordingly made an attempt in the accompanying drawing, Fig. 6 (see also Fig. 7 and Fig. 8), to show the 'complete' system of Copernicus in which some thirty-four[20] circles (against some twenty[21] for the Ptolemaic system) combine to explain the gyrations of the Earth, Moon and planets. It is still surprisingly complex, but a great advance in simplicity on the proposed eighty spheres of his Paduan contemporary Girolamo Fracastoro. Copernicus' system does possess a great advantage over its rivals in that the first inequalities—the retrograde motions of all the planets—are now due to a single movement of the Earth. At the same time the parallactic motion gave a criterion for assigning relative distances to the planets since the amplitude of the retrograde motion could be equated with the distance of the planet exhibiting it. Further, for the first time it was made clear why it was in Nature that the superior planets should be in perigee at opposition, and why the periods of their epicycles in the Ptolemaic system had been so curiously related to the period of the revolution of the Earth about the Sun.

Despite all these advances he placed the Sun in the centre, it seems, to aid calculation, and so as not to confuse the reader too much in relation to Ptolemy, he referred everything not to the Sun *but to the centre of the Earth's orbit*, for this reason the correct designation of his system is geocentric-heliostatic. By passing the line of nodes of each planet and also the line of apsides through the Earth at the centre of her orbit the eccentricities are reckoned from a point, the distance of which from the Sun measures the Earth's eccentricity and the sphere, which contains it, has no thickness.[22] In this we see something of the immense difficulties

---

[19] See Stowikowski, J., *La Signification des figures de Copernic et de Keppler dans la nature*, Warszawa, 1903 (B.N. 8563 g 20), for an unusually detailed analysis. (The British Museum lists all o Kepler's works under the spelling KEPPLER.)
[20] The system of the *Commentariolus*. In *de Revolutionibus* the number was increased to forty-eight.
[21] Several learned commentators refer to eighty circles; see Dingle, H., *Copernicus*, The Observatory, 65, p. 51, June 1943. Johnson, F. R., *Astronomical Thought in Renaissance England*, Baltimore 1935, p. 102, l. 19. In my view they confuse Ptolemy's system as adjusted in the sixteenth century with that of Fracastoro. Since 'discovering' this point I have studied Koestler's *The Sleepwakers* and noted that he has difficulty with it. He thinks the Ptolemaic system requires forty circles whereas Peurbach gives only twenty-seven. Apparently the late Professor Koyré has commented on this difficulty but not in a published document. We should keep in mind that John Tolhopf in the fifteenth century and Novara in the thirteenth had reduced the circles of Ptolemy to twenty-eight and thirty-four respectively.
[22] See Dreyer, J. L. E., *A History of Astronomy from Thales to Kepler*, p. 377, Dover 1953.

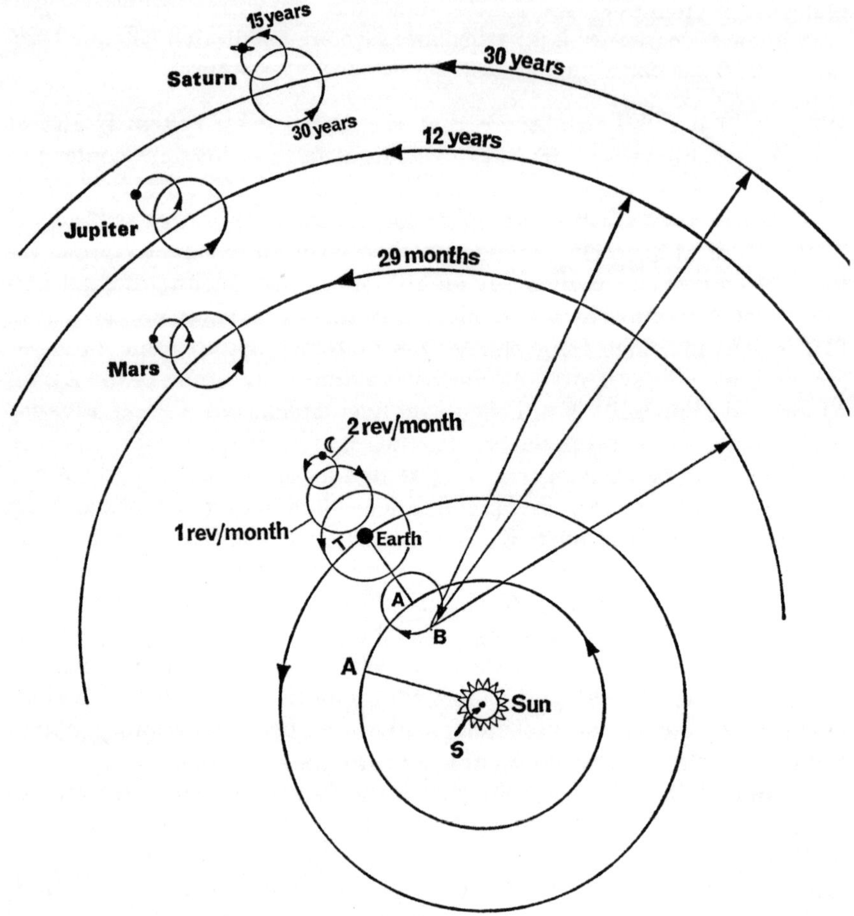

Fig. 6. Number of circles for each planet: Mercury 7, Venus 5, Earth 3, Moon 4, Mars, Jupiter and Saturn 5 each, *total* 34. Note the earth has three motions in order to remove the ninth and tenth spheres of Ptolemy. B revolves about A in 3434 years. A revolves about the sun in 53,000 years. B is the centre of the earth's orbit. If BT is unity SA = 0·0368, AB = 0·0047. B is fixed relative to the sun.

Note: The deferents of the planets are centred on the earth's orbit at point B, the system is heliostatic and geocentric. The diagram is not to scale. The orbits of Mercury and Venus are not shown.

Fig. 7. The Copernican system of the world in its simplest form as shown in Copernicus' great work *De Revolutionibus*.

Fig. 8. The Copernican system of the world with the known satellites of Jupiter and Saturn portrayed; the heliocentric system.

Note: For a detailed analysis see Neugebauer. O., *Vistas in Astronomy, 10,* 89 (1968).

net, in quo terram cum orbe lunari tanquam epicyclo contineri diximus. Quinto loco Venus nono mense reducitur. Sextum denicp locum Mercurius tenet, octuaginta dierum spacio circū currens, in medio uero omnium residet Sol. Quis enim in hoc

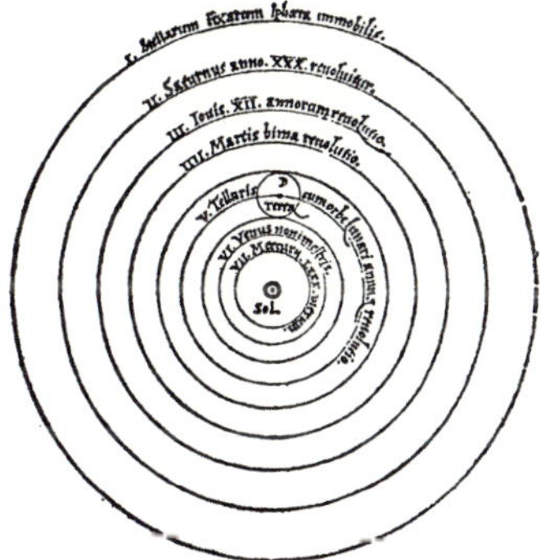

pulcherimo templo lampadem hanc in alio uel meliori loco po neret, quàm unde totum simul possit illuminare? Siquidem non inepte quidam lucernam mundi, alij mentem, alij rectorem uocant. Trimegistus uisibilem Deum, Sophoclis Electra intuentē omnia. Ita profecto tanquam in solio re gali Sol residens circum agentem gubernat Astrorum familiam. Tellus quoq; minime fraudatur lunari ministerio, sed ut Aristoteles de animalibus ait, maximā Luna cū terra cognationē habet. Concipit interea à Sole terra, & impregnatur annuo partu. Inuenimus igitur sub hac

7

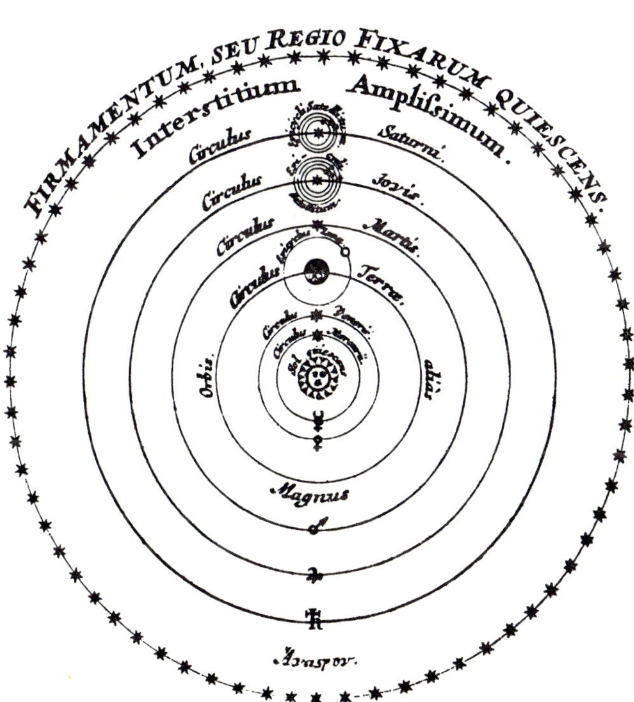

8

The solar system

faced by Copernicus in making such revolutionary changes in a system of ancient learning to which he had a particular affinity.

The major work of Copernicus, *De revolutionibus*, was published in 1543, and Kepler's book on Mars,[23] showing that planets move in elliptical orbits within rigorous control of a heliocentric system, was not published until 1609. The period between these two events is of outstanding interest to our subject. The Copernican system, contrary to what may have been expected, did not capture the intellects of those most able to form a judgment; indeed the fight for its acceptance was to be a hard and difficult one over the following centuries. The main difficulty did not lie, however, in the dynamical problems of the new system, despite the fact that the diurnal rotation of the Earth was exceedingly reticent in declaring itself to the senses and was held to be contrary to Scripture.[24] It was long held that such a rotation would produce a continuous wind of great velocity and that an arrow fired vertically upward should fall at a distance of many miles from its starting point. To these very real objections Copernicus had little to offer before the time of Galileo and the disputations were much confused and the ideas on these subjects ill-formed. The church took no part in this and for a long time Copernican ideas were not hindered by its long spiritual arm. The more significant heresy which called forth all its later ire was not based on grounds of dynamical astronomy at all, but lay in the following five innovations which are explained in great depth by Lovejoy:[25]

1. The assumption that other planets of the solar system are inhabited by sentient creatures.
2. The shattering of the outer walls of the medieval world.
3. The conception of the fixed stars as Suns similar to our own Sun and surrounded by planetary systems.
4. The supposition that the planets of these outer worlds have conscious inhabitants.
5. The assertion of the infinity of solar systems (for which Giordano Bruno was burnt by the Church).

These views challenged the entire drama of the Incarnation and Redemption, as Thomas Paine[26] clearly saw in his famous phrase 'to suppose that every world in the boundless creation had an Eve, an apple, a serpent, and a Redeemer'. We shall examine these profound

[23] *Astronomia Nova αιτιολογητος Seu. Physica Coelestis, tradita commentariis de Motibus stella Martis.*
[24] One has but to recall Joshua command, 'Sun stand thou still upon Gideon and thou Moon in the valley of Ajalon', Joshua 10^{12}. For a remarkable analysis see Maunder, E. W., *The Astronomy of the Bible*, 1909 and 'The Quarterly Bulletin of the London Grand Rank Association', New Series No. 32/71, May 1971, p. 4.
[25] Lovejoy, A. O., *The Great Chain of Being*, Harper Torchbooks, 6th ed., 1965 (The William James Lectures of 1933).
[26] Paine, T., *Age of Reason*, ch. 13.

problems later—but for the moment we must return to the dynamical problems and the resolution of the geometry of the planetary system of the Sun. This resolution was the work of two men of widely different backgrounds, nationalities, natures and abilities; that they were to join together in this great enterprise at a time of major political disruptions, after years of personal tragedies and wanderings, is one of those 'happy' quirks of nature that some would seek to align with a well-ordered world. I refer to Tycho Brahe,[27] an assiduous observer of the heavens, coming from wealthy Danish stock, and Johannes Kepler,[28] a poor German, a mathematical genius of the highest order and cast in the mystical Pythagorean mould.

Tycho Brahe (1546–1601) was born of a noble line and studied at the University of Copenhagen (where he observed the partial eclipse of the Sun of 1560 August 21). He subsequently left for the University of Leipzig and later Wittenberg, Rostock and Basle. In the ancient free city of Augsburg he met the celebrated Professor of Philosophy at Paris, Pierre de Ramée,[29] an advanced thinker who suffered for his distaste of the all-powerful Aristotelian philosophy of that time and his desire to see it overthrown for a *de novo* approach to the system of the world, based on firm mathematical foundations.

In 1576 Tycho commenced the construction of his famous observatory at Uraniborg on the island of Hveen at a site 160 feet above the sea. Here he remained for twenty-one years in what was the most splendidly equipped[30] observatory in the western world, and, by assiduous observations of the heavens, gathered the data that were to provide mankind with the labyrinth of figures from which future deductions could be made. Here he observed Mars throughout its apparitions, compiled his catalogue of 777[31] stars for the end of the year 1600 and printed it in his *Progymnasmata*. Thus his main work was achieved in the sanctuary of Hveen before he was forced to leave it following a dispute with the successor of his benefactor, Frederick II.[32]

Before we make the next historical step, however, we should pause to

[27] See Dreyer, J. L. E., *Tycho Brahe*, Dover Books 1963, an unabridged and corrected reprint of the original of 1890.
[28] See Caspar, M., *Kepler*, Abelard Schuman, London 1959, translated by C. Doris Hellman.
[29] de Ramée was made to apologize to the Parliament of Paris for his disrespect to the Aristotelian philosophy. He perished with many of his Huguenot contemporaries at the massacre of St Bartholomew, 1572.
[30] The observatory had many fine and unusual instruments, see *Tycho Brahe's Descriptions of his instruments and scientific works*, Astronomiae Instauratae Mechanica, Wandesburgi 1598, translated Strömgren, H. R., and København, B., 1946; also *Tycho Brahe's System of the World*, by Marie Boas and A. Rupert Hall, R. A. S., Occasional Notes no. 21, 1959. Nothing of these beautiful instruments remains today, neither at Prague nor Copenhagen. The instruments at Prague claimed to be his are according to Dreyer not of his refined workmanship and the two sextants do not have his peculiar pinnules.
[31] The number is not without theological interest. Perfection in heavenly things is thought to be based on the number 7. Thus Lamech, who is an archetype of the Almighty, was 777 years of age.
[32] He died, 1588 April 4, in his fifty-fourth year. Tycho left Hveen on 1597 March 29.

The solar system

refer to the Tychonic system of the world which was developed by Tycho in about 1583 (Fig. 9) and advanced in Chapter VIII of his beautiful book entitled *On the Most Recent Phenomena of the Aetherial*

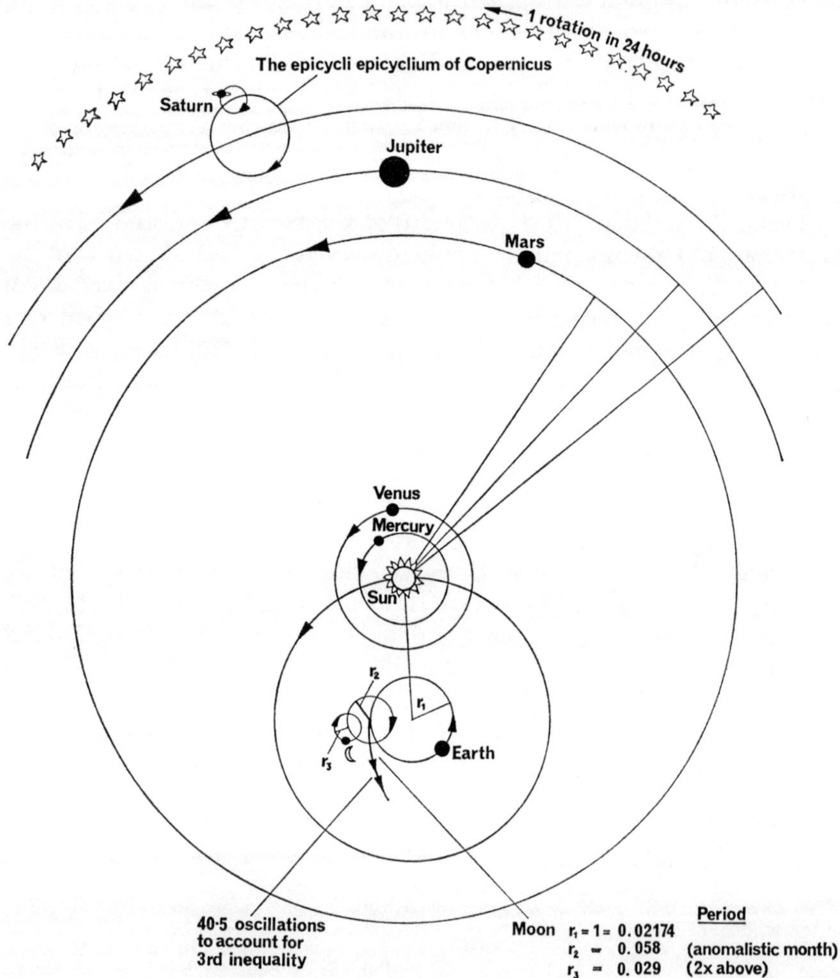

Fig. 9. The world system of Tycho Brahe.

Saturn's orbit, 12,220 × semi diameter of Earth; 235 × semi diameter of elementary world bordered by the orbit of the Moon.

Moon's distance, 52 × semi diameter of Earth (860 German miles, see Fernel's *Cosmotheoria*, Paris 1528.

Sun's distance, 20 × that of the Moon.

In Tycho's system Saturn *alone* was given the system of Copernicus; see Tycho's *Theatrum astronomicum*, vol. I.

World,[33] which was printed on his own press at Uraniborg in *c.* 1588.

It is clear that, though he rejected the motion of the Earth, he had adopted from a mathematical standpoint the system of Copernicus and, by his observation of the vast orbit of the comet of 1577, destroyed for ever the reality of the spheres of Eudoxus, Aristotle, Fracastoro and Copernicus.

Eventually Tycho died in Bohemia in 1601, but not before he had made contact with the man able to enter the labyrinth and bring to an astonished world the hidden beauties of its innermost ways, a man of unparalleled genius—Johannes Kepler.

Kepler was born on 1571 December 27, at Weil in the Duchy of Würtemberg and was of premature birth. He was much neglected in childhood and in his fourth year became a victim of smallpox, which foul disease left him with a permanent bodily weakness, poor eyesight and crippled hands. He studied at Adelberg and Maulbronn, where a brilliant examination for the degree of bachelor procured him in 1588 admittance to the University of Tübingen. Here he came under the influence of Mästlin and imbibed the Copernican philosophy. In 1594 at the age of 23 he accepted the post of provincial mathematician of Styria and took the vacant chair of astronomy at Gratz. In 1596 he saw published at Tübingen his first great work *Prodromus Dissertationium Cosmographicarum* in which he gave the intellectual world a foretaste of his Pythagorean genius by presenting his elegant but now discredited discovery of the relation of the five regular solids to the planetary distances (see Fig. 10). This early work is especially pleasing for its lucid exposition of what Kepler saw as the inherent difficulties of the Ptolemaic world system. He shows that (Fig. 11) the Ptolemaic epicycles of the outer planets are seen exactly under the same angle from the Earth as is the orbit of the Earth from a point in each of the outer planetary orbits, and he goes on to explain how this produces the enormous epicycle of Mars and why that of Jupiter is so much smaller, a point that no one previously had been able to explain away. He also remarks on the inability of the Ptolemaic system to deal with the fact that the Sun and Moon never enter a retrograde motion and that the periods of the inner planets should equal that of the Sun.

In 1600 Kepler fled from Catholic persecution in Gratz. He was given only 'six weeks three days' to quit the land, having been brought with more than a thousand other citizens before the Reformation commissioners to answer questions on his creed. His name is the fifteenth entry in a list of sixty-one men who refused to compromise themselves and accepted banishment. It is said that it is an ill wind that blows no good and posterity can draw comfort from these savage events that forced him to journey to Prague with his family and there by one of the

---

[33] The full title is *Tychonis Brahe Dani De Mundi aetherei recentioribus phaenomenis Liber secundus qui est de illustri Novembris anno MDLXXVII usque in finem Januarii sequentis conspectae Uraniburgi cum Privilegio.*

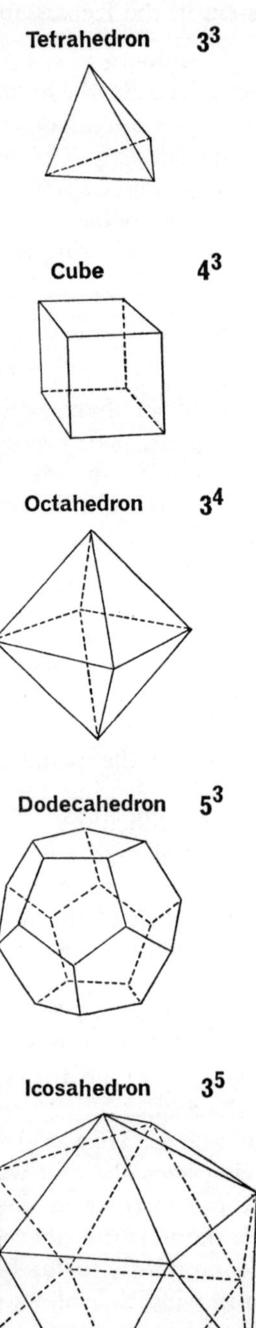

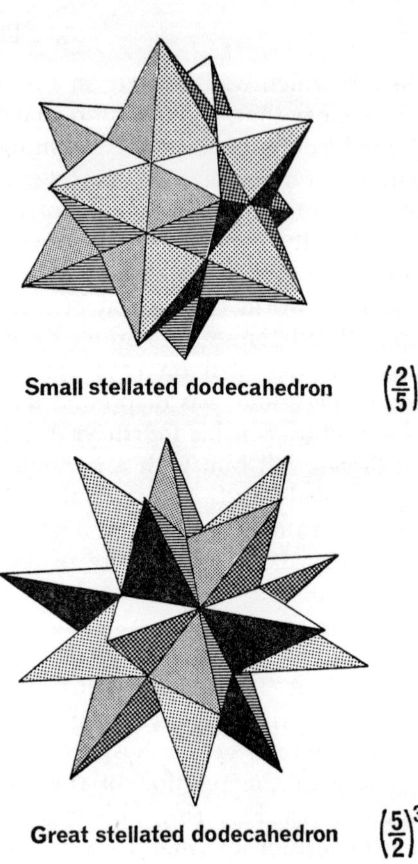

Fig. 10. The five regular Platonic Polyhedra known to the ancient world and the two stellated dodecahedra discovered by Kepler. Two further polyhedra were later discovered by Poinsot (1777–1859), they are the great icosahedron and the great dodecahedron. The figures after the titles of the solids are the Schläfli symbol which gives number of sides in the regular face and as an index the number of faces at each vertex. For details see: Cundy, H. M. & Rollett, A. P. *Mathematical Models*, Oxford 1961.

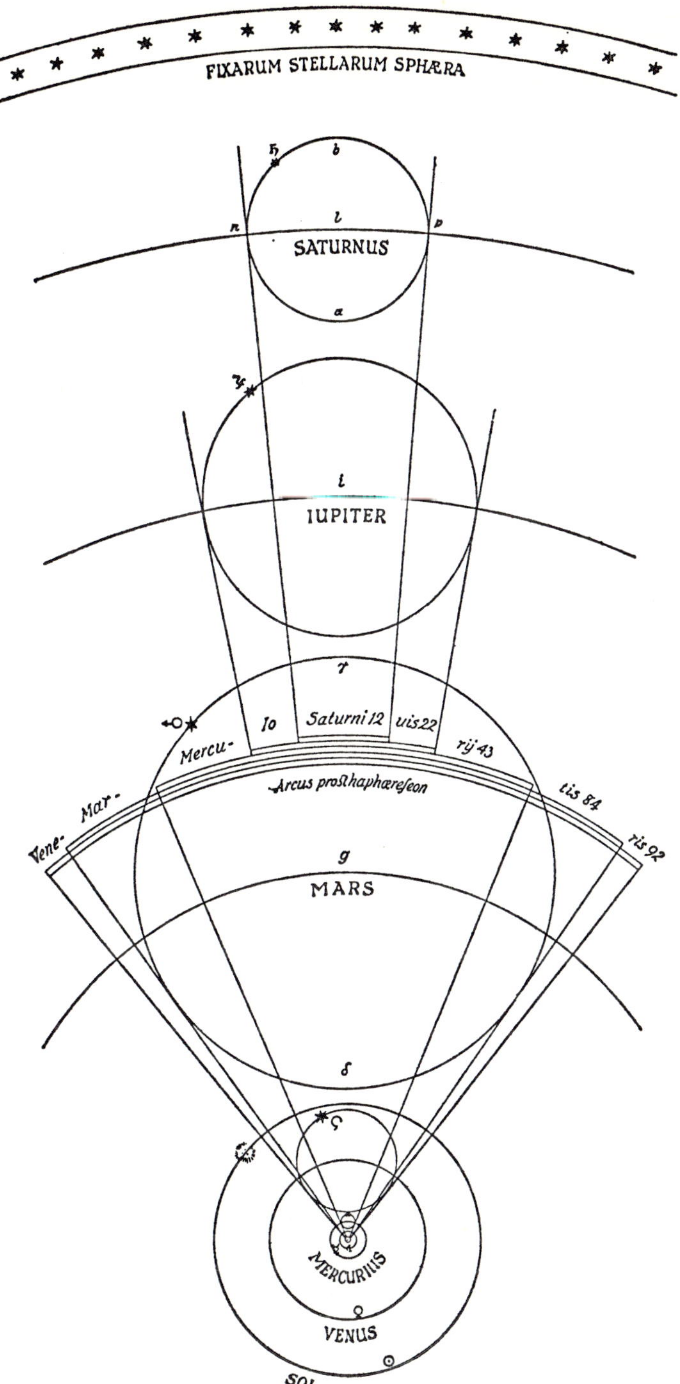

Fig. 11. Kepler's drawing of the planetary orbits showing the enormous epicycle of Mars.

most remarkable events of scientific history come to work with another outcast and wanderer, Tycho Brahe. Tycho died unexpectedly in 1601 and Kepler succeeded him as Imperial Mathematician in Prague. His illustrious biographer the late Professor Max Caspar, says

In order that the paths in life of these two great astronomers, who uniquely supplemented one another, could unite, it was necessary that both be displaced from their widely separated residences, in order to meet at the court of an emperor whom history reproaches for having neglected the affairs of government for the sake of his astrological and alchemical bents. Kepler himself expresses his conviction of the rule of a divine decree in these events when he writes: 'If God is concerned with astronomy, which piety desires to believe, then I hope that I shall achieve something in this domain, for I see how God let me be bound with Tycho through an unalterable fate and did not let me be separated from him by the most oppressive hardships.'

With Tycho's death Kepler became custodian of the invaluable but labyrinthine astronomical treasure amassed by that assiduous observer at Hveen and by sure insight set himself to deal with the problem of the Martian orbit, which had baffled Longomontanus.[34] In working with it he surmounted what today by modern analysis we term an elliptical integral and a transcendental equation.[35] He came by immense labour to the view that the orbit could not be circular and this led him to the ovoid, egg-shaped, orbit with the blunt end at aphelion. In 1609 he was able to announce his solution of the great Martian enigma and to 'lead the captive planet to the foot of the imperial throne'. He shows how by a 'wearied and tedious method', in which he asks the reader to 'take pity on me who carried out at least seventy trials of it with the loss of much time and don't be surprised that this already is the fifth year since I have attacked Mars', that he had eventually seen by a flash of inspiration that the figure 1·00429 is equal to the secant of the greatest optical equation of Mars and that this is the secant of 5°18′ of which the tangent is equal to the eccentricity.

From these specialized mathematical juxtapositions he extended his grasp of the inner mathematical conundrum to substitute what he called the *distantia diametralis*, and the radius vector of the recalcitrant planet was captured in the elegant and simple expression:

$$r = a(1 - e \cos E)$$

where a is the semi-major axis of the ellipse
e is the eccentricity
E is the eccentric anomaly.

This equation gives the heliocentric distance of a planet moving in an

[34] Longomontanus (1562–1647), assistant to Tycho Brahe, elected 1605 to professorship in the university of Copenhagen.
[35] See Gingerich, O., 'The Computer versus Kepler', *American Scientist*, 52, 218 (1964).
Jens P. Møller, *On the Solution of Kepler's equation*, Festschrift für Elis Strömgren, Copenhagen, Munksgaard 1940, pp. 163–74.

elliptical orbit with the Sun at one focus. The position of the planet is given by:

$$r \sin v = b \sin E$$
$$r \cos v = a(\cos E - e)$$

where b is the semi-minor axis of the ellipse and v is the true anomaly. From these equations an alternative expression for the radius vector may be obtained in terms of the true anomaly:

$$r = \frac{a(1 - e^2)}{1 + e \cos v}$$

(Fig. 12 makes the matter more clear.)

In working on this problem, using multiple areas in a manner taught by Archimedes, he came also upon his second planetary law, most usually expressed as—the radius vector joining the Sun to each planet sweeps out equal areas of its ellipse in equal times, which implies that $r^2 \dfrac{dv}{dt}$ is a constant.

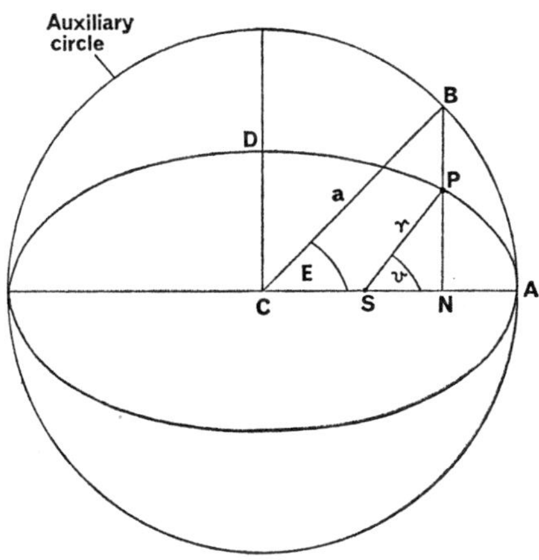

12. Properties of an elliptical orbit.

$$SP = r$$
$$CB = CA = a, CD = b$$
$$CS = ae$$
$$\frac{PN}{BN} = \frac{b}{a}$$
$$b^2 = a^2 (1 - e^2)$$

E is the eccentric anomaly†
v is the true anomaly‡
S is one focus
e is the eccentricity

Clearly $CN = CB \cos E = a \cos E = CS + SN$

$$= ae + r \cos v$$

$$r \cos v = a \cos E - ae = a(\cos E - e) \qquad (1)$$

also $\quad PN = r \sin v$ and $BN = a \sin E$

$$\frac{PN}{BN} = \frac{b}{a} = \frac{r \sin v}{a \sin E} \therefore r \sin v = b \sin E \qquad (2)$$

From (1) and (2) by squaring and adding

$$r = a(1 - e \cos E)$$

† ‡ For a full mathematical discussion of the true anomaly, the eccentric anomaly and the mean anomaly associated with undisturbed elliptic motion see Brown, E. W., *An Introductory Treatise on the Lunar Theory*, Cambridge Univ. Press 1896 (The Dover Reprint 1960, p. 29). See the Appendix for a more full discussion of the ellipse and definitions of the eccentric, true and mean anomaly. *Anomaly* per se is the angular distance of a planet or satellite from its last perihelion or perigree: so called because the first irregularities of planetary motion were discovered in the discrepancy between the actual and the computed distance.

The solar system

These remarkable revelations were put forward in his celebrated *Astronomia de Stellae Martis* (Prague 1609). This is the first modern astronomical textbook, for it contains two of the foundations of astronomical science, making known the nature of the planetary orbits about the Sun and the manner in which the planets move over them. It was this knowledge in the hand of Newton that was to lead to the theory of universal gravitation.

With all these triumphs not yet fully understood, Kepler's personal fortunes came to their nadir in 1611 with the death from smallpox of his favourite child and then the death of his wife; Rudolf, his patron, also passed away on 1612 January 20 and Kepler left Prague for Linz on the Danube, where he was to sojourn for fourteen years. In 1613 he married again and set up a more congenial home from which he was able once more to turn to the elucidation of many problems.

It is from Linz that he recalculated the year of Christ's birth,[36] made the reform of the calendar, produced his remarkable *Stereometrica Doliorum*,[37] and crowned his work with the great book *Harmonice Mundi*, which realized his Pythagorean ambitions by announcing his celebrated third planetary law[38]—that of the sesquiplicate ratio between the planetary periods and their distances, from which the complete

[36] The crucifixion is generally agreed by scholars to have occurred in A.D. 29, A.D. 30 or A.D. 33, at the culmination of Christ's ministry when he was $33\frac{1}{2}$ years of age. Those who choose to accept A.D. 33 measure backwards and place his birth in the autumn of 2 B.C. Herod is recorded as having died in 4 B.C. and this anomaly is removed by casting doubt on the veracity of this date. The scholars who favour 2 B.C. point out that Herod died at the age of 70 having been made governor in Galilee in 47 B.C. at the age of 25.* This very neatly puts the death of Herod squarely in 2 B.C. and shows how malleable history can be. Some scholars appeal to an eclipse of the Moon which Josephus records to have taken place toward the end of Herod's life†. This eclipse is often said to be that of the spring of 4 B.C.—a very minor partial eclipse indeed. I am disposed, myself, from a study of Professor Oppolzer's Canon of Eclipses, to place this event in the year 5 B.C. which produced two total eclipses of the Moon, both visible from Palestine.

* Josephus. Antiq. XIV. IX. 2.
† Ibid. XVII. VI. 4.

With prodigious labour Oppolzer put together figures for the calculation of 8000 solar and 5200 lunar eclipses from 1200 B.C. to A.D. 2161. The relevant portion of the Canon, with the results of my calculations for visibility from Palestine, is shown in the table below.

Canon of Lunar Eclipses, 2 B.C. to 8 B.C.

Julian Calendar Date		2 B.C.	3 B.C.	4 B.C.	5 B.C.	6 B.C.	7 B.C.	8 B.C.
Julian Day	1.	1720890	—	1720210	1719855	1719501	—	1718823
	2.	1720712	—	1720034	1719679	1719325	—	1718645
Magnitude	1.	Partial	—	Partial	Full	Partial	—	Partial
of eclipse	2.	Partial	—	Partial	Full	Partial	—	Full
Visible from	1.	No	—	No	Yes	No	—	Yes
Palestine	2.	No	—	Yes	Yes	No	—	No

[37] See the special note on Kepler, (page 288).
[38] The squares of the orbital periods P of the planets are proportional to the cubes of the semi major axes of the orbital ellipses. n^2a^3 is a constant, where a is the semi major axis of the ellipse, n the mean motion is $2\pi/P$. See Newton's *Principia*, Book III. Phenomenon I.II and III in Cajori's edition 1947, pp. 401–4.

50

system of the world is unable to escape, from the smallest to the largest planetary mass. In the *Harmonice Mundi* we see Kepler's innermost Pythagorean and Platonic convictions draw sustenance, and he indulges, sometimes excessively, the imagination of that inner harmony and beauty, which was his peculiar orientation and mainspring; something which is closer to the poet, yet which in him served to throw a beam of illumination upon the most intractable and arcane mathematical problems ever to assail the mind of man. But this was not all. In his *Astronomia Nova* Kepler tells us

I am much occupied with the investigations of the physical causes. My aim in this is to show that the celestial machine is to be likened not to a divine organism but rather to a clockwork . . . insofar as nearly all the manifold movements are carried out by means of a single, quite simple magnetic force, as in the case of a clockwork all motions come from a simple weight . . .

Here Kepler advances to a position not previously taken by any philosopher; his vision is equal to or greater than Newton's, since he had no precursor.

In the *Mysterium Cosmographicum*, Chapter XX, he assumes that the force for which his mind seeks is such that it reaches out to the planets to keep them in tangential motion and that this force falls inversely with the increase of distance; in his own words:

We must make one or two assumptions:
either the forces of motion *animae motrices* are inherent in the planets and are feebler the more remote they are from the Sun, or there is only one *anima motrix* at the centre of the orbits, that is, in the Sun. It drives the more vehemently the closer the moved body lies . . . to this end we will assume, as is very probable, that the moving effect is weakened through spreading from the Sun in the same manner as light.

From these considerations we can say with Max Caspar:[39]

It is a new land which is glimpsed from the position next to Kepler on his summit. He left far behind him not only Ptolemy but also Copernicus and Tycho Brahe. Perhaps it seems that it makes little difference whether the planet orbit is a circle, or an ellipse deviating little from the circular shape. Yet Kepler's prodigious step forward consists precisely in the fact that with his ellipse proposition he had overthrown for all time the two-thousand-year-old axiom, according to which every motion retrograde in itself must of necessity be a uniform circular motion. By that step he had made the orbit free for a new development of astronomy. And nothing is more difficult in science than to set aside such deep-rooted opinions. People who have not read Kepler often tell the story as though Kepler had found his laws in a purely geometrical way, so to speak by trials. It is naïve to believe that a fortress as strong as that axiom indicates can be taken by such means. No, everywhere in the solution of the problem confronting him, physical concepts were in the

39 Caspar, M., *Kepler*, London 1959.

background and drove him forward. They became more and more intimately intertwined with his astronomical thinking.

Nowadays we are so accustomed to seeing mechanical forces operating in the planetary motions that it is difficult for us to think that it was once different. And yet Kepler ran up against rejection and lack of understanding on all sides. Maestlin, Fabricius, Longomontanus and others shook their heads. Even many years later Maestlin advised his former pupil to leave physical causes and hypotheses entirely out of the question and to explain astronomical matters only according to astronomical method; geometry and arithmetic alone are the vibrations of the knowledge of the heavens. It is Kepler's greatest service that he substituted a dynamic system for the formal schemes of the earlier astronomers, the law of nature for mathematical rule, and causal explanation for the mathematical description of motion. Thereby he truly became the founder of celestial mechanics.

Yet such is the intellectual myopia of scientific men that it was 'as though Kepler had spoken into the wind'. Galileo failed completely to comprehend the celestial mechanics incorporated in the Keplerian planetary laws. Not once in his celebrated *Dialogue* concerning the systems of the world did he speak of them, and this a quarter of a century after they had been promulgated—indeed the great Galileo[40] retained the discredited Aristotelian fiction of uniform motion in a circle and it is to Kepler that we *must* award the palm for having freed astronomy from the intractable bonds of Aristotelian physics.

Newton half a century later is not unaware of Kepler's contribution, but throughout Books I and II of the *Principia* the great third law of the planetary orbits is introduced anonymously, as Holton[41] has shown, as the phenomenon of the 3/2 power while the first and second Keplerian planet laws are introduced erroneously as 'the Copernican hypothesis'.[42] Before we leave Kepler for modern times we may with profit examine the problem of the ovoid path of the planets. This has been done with great skill by Dreyer,[43] Boyer[44] and Hartner.[45]

It is now clear that Copernicus[46] was aware of both parts of what is commonly and incorrectly known as Lahire's theorem,[47] viz. that if a

[40] Much has been written on Galileo's mental myopia concerning Kepler. See the detailed notes in Max Caspar's biography. *Kepler*, London 1959, p. 137. See also the attack on Koestler: 'Arthur Koestler and his Sleepwalkers', by G. Santillana and S. Drake, *Isis, 50*, 225, and Koestler's reply, *Isis, 51, 73*.
[41] Holton, C., 'Johannes Kepler's Universe: Its Physics and Metaphysics', *American Journal of Physics*, xxiv, May 1956, pp. 340–51.
[42] *Newton's Principia*, edited by F. Cajori, University of California Press, Berkeley 1946, pp. 394–5.
[43] Dreyer, I. L. E., *History of Planetary Systems from Thales to Kepler*, Cambridge 1906.
[44] Boyer, C. B., 'Note on Epicycles and the Ellipse from Copernicus to Lahire', *Isis,, 38* p. 54.
[45] Hartner, W., *The Mercury Horoscope of Marcantonio Michiel of Venice, A study in the history of Renaissance Astrology and Astronomy*, Vistas in Astronomy, vol. I, pp. 84–138.
[46] See Copernicus, the quadricentennial edition of *De revolutionibus orbium caelestium libri*, vi (Thoruni 1873), iii, 4, pp. 165–6, 471–2.
[47] Lahire, P. de, *Traite des roulettes*, 1706; see also Dingelday, F., 'Coniques', *Encyclopedie des sciences mathematiques*, iii (3), 158–9; Coolidge, J. L., *A history of the conic sections and quadratic surfaces*, p. 152, Oxford 1945.

small circle rolls without slipping along the inside of a larger circle (the arrangement for the generation of a roulette) then if the diameter of the larger circle is twice as great as the smaller; then:

1. The locus of a point on the circumference of the smaller circle is a straight line segment, a diameter of the larger circle.[48]
2. The locus of a point, which is not on the circumference but which is fixed with respect to the smaller circle, is an ellipse.

It is also clear that Kepler knew of part two of Lahire's theorem, as is to be expected of such a skilled mathematician. This point, now forgotten, should be given the emphasis it deserves, for it destroys the too often repeated false dichotomy[49] that writers on astronomy wish to place between the Ptolemaic-Copernican system of epicycles and the elliptical orbits of Kepler.

In the view of Boyer it was the second law of Kepler, the law of equal areas, that in the end brought about the abandonment of the Aristotelian circular motion in Newtonian celestial mechanics. Kepler personally did not make use of epicycles to generate the elliptical orbits, because such devices did *not agree with the natural causes that produce the ellipse*. But the ability of the epicycle as a generator of truly remarkable curves is something that modern astronomers too frequently overlook; some measure of that versatility is shown in Fig. 13. Boyer goes further and cautions against a 'specious emphasis upon intellectual *revolutions*' in the history of science and urges us to give credit to the obscure men who serve as indispensable links between the so-called 'giants'.

Copernicus was also aware that the Ptolemaic system produced an unusual oval figure for the centre of the lunar and mercurian epicycle. This is explained in Peurbach's *Theoricae novae Planetarum* (c. 1460),[50] which had been commented upon by his mentor Albert of Brudzew (Brudzewski) c. 1482. The drawing taken from Peurbach is reproduced as Fig 14 (see also Fig. 15). In Hartner's view this discovery is anticipated by the Arabian scholar Azarguiel,[51] whose full name is Ibrahim Ibn Yahya al-Naggash Abu Ishag Ibn al Zargali (c. 1100), and whose orbital construction for Mercury's orbit is shown in plate 2 (taken from vol. 3, p. 282, of *Libros Del Saber*). Whether Kepler's elliptical orbits owe anything to Peurbach or Azarguiel is for speculation, but the point is one which cannot fail to excite our imagination and show us that there is

[48] Part I of Lahire's theorem was anticipated by Nasir Eddin (1201–74) four hundred years before.
[49] See for example Rosen, E., *Three Copernican treatises*, p. 49, New York 1939 (also the Dover 1959 Edition). 'The ancient axiom of circularity was not shattered until Kepler demonstrated the ellipticity of the planetary orbits.'
[50] A beautiful copy of Peurbach's slim yet seminal work dated 1482 is to be seen in the British Museum. General Catalogue No. IA 20516.
[51] See Don Manuel Rico y Sinobas, Libros del Saber de Astronomica, for D. Alfonso X De Castile, 5 vols, Madrid 1863 (B.M. 1803 d.26).

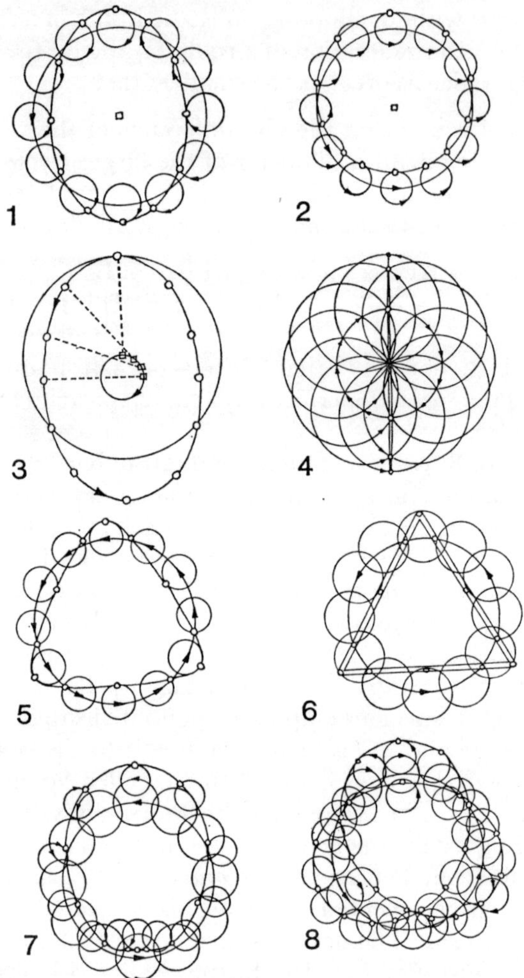

Fig. 13. (1) *Ellipse*: planet turns counter to revolution in the deferent. (2) *Circle*: planet turns in harmony with revolution in the deferent. (3) The same effect as (1), without the epicycle, using a deferent of radius equal to the epicycle. (4) *Rectilinear motion*: epicycle's radius approximates that of the deferent. (5) *Triangular figure*: the epicycle is controlled in speed. (6) The rectilinear limit of (5). (7) *Oviform*: figure using a second epicycle riding on the first. (8) A periodically repetitive configuration of same complexity. This selection of curves is after Norwood Russell Hanson.

seldom anything wholly new once it is brought to a fine clear focus by an intellect of unusual brilliance.

It has often been stated with cogency that it was not possible to choose between the Ptolemaic, the Copernican, the Tychonic nor the advanced Keplerian world systems on evidences then available from the behaviour of the solar system itself. Everything was a metaphysical argument and this is nowhere made more plain than by Santillana[52] when he reminds us that 'On Wolynski's count there were 2330 works published on astronomy between 1543 and 1687 . . . of those only 180 were Copernican' (see *Archivio Storico Italiano*, 1873, p. 12). In 1610 Galileo had seen and announced the satellites of Jupiter moving about their primary (*Sidereus Nuncius*, Venice 1610) and in the same year he saw the lunar phase changes of Venus,[53] all of which, contrary to the often held view, did not prove the Copernican world system but showed conclusively *only that Venus was placed in an heliocentric orbit* (Figs. 16 and 17), and this was compatible with the Tychonic system, which was geostatic with the planets circling about the Sun, which itself circled about the Earth.

Polanyi[54] has remarked with some force that it was fortunate that no experiment of the Michelson-Morley[55] type was carried out at that time, for its negative result would have served as decisive proof that the Earth was at rest.

Galileo can take posthumous comfort that the Inquisition did not have our sophisticated scientific tools to mislead them into unusual realms of 'truth'. We might here reflect that what Otto von Guericke[56] describes as 'De Systemate alio', that is the diurnal rotation of the Earth and its revolution about the Sun, had to wait respectively until 1851 and 1727[57] for conclusive evidence in support of these now accepted commonplaces. Galileo gave us facts with the acquisition of his Dutch

[52] Santillana, G. de, *The Crime of Galileo*, Chicago 1955, p. 164 n. See also: 'The impassioned defense of the Copernican system', by Paolo Antonio Foscarini in Thomas Salusbury's *Mathematical Collections* 1661–5, made available in English while the original Italian was utterly suppressed. (There is a Dawson reprint 1967.)

[53] The famous anagram *Haec immatura a me jam frustra leguntur*, o.y. appears in the Preface to Kepler's *Dioptrics* (Augsburg 1611) and is in effect a continuation of the *Siderius Nuncius*. When reconstituted the anagram reads, *Cynthiae figuras aemulatur mater amorum* (The Mother of the Loves rivals the phases of Cynthia).

[54] Polanyi, M., *Personal Knowledge*, London 1958, p. 152 n. (The point has also been made by G. J. Whitrow and H. Dingle in earlier writings.)

[55] *Michelson-Morley experiment*. An attempt to detect and measure the relative velocity of the earth and the ether by observations of interference fringes with a form of apparatus which can be rotated bodily into different orientations. The very small velocity indicated by the results of the experiment is much less than was expected, and was probably due to accidental causes.

It should be mentioned that Sir Edmund Whittaker was of opinion that when relativity theory had become generally accepted, the Michelson-Morley experiment was rediscussed with a much more complete understanding and exactitude. See E. Kohl, *Ann. d. Phys.*, *28*, pp. 259, 662 (1909).

E. Budde, *Phys. ZS.*, *12*, p. 979 (1911).

M. von Laue, *Ann. d. Phys.*, *33*, p. 186 (1910); *Phys. ZS*, *13*, p. 501 (1912).

A. Right, *Le Radium*, *11*, p. 321 (1919); *N. Cimento*, *18*, p. 91 (1919).

J. Villey, *Comptes Rendus*, *170*, p. 1175 (1920); *171*, p. 298 (1920).

E. H. Kennard and D. E. Richmond, *Phys. Rev.*, *19*, p. 572 (1922).

J. L. Synge, *Sci. Proc. Roy. Dub. Soc.*, *26*, p. 45 (1952); *Nature*, *170*, p. 244 (1952).

This confidence of Sir Edmund has not come to pass. See Professor H. Dingle's paper to the Hamburg meeting of the International Union for the History of Science. *Vistas in Astronomy*, *9*, 97 (1967); see also *Nature*, *216*, 119–24 (1967).

[56] Guericke, O. von, *Experimenta Nova*, 1672.

[57] Foucault's celebrated pendulum experiment at the Pantheon, Paris, was performed in 1851. In 1727, Bradley discovered the cause of the aberration of light due to the Earth's motion in orbit; the first stellar parallax came in 1837.

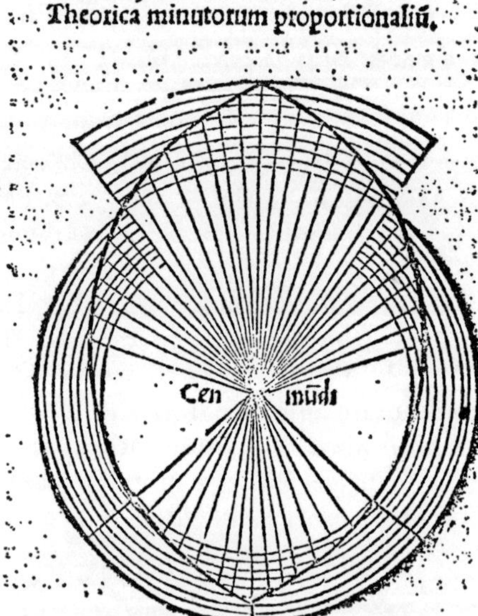

Fig. 14. The ovoid orbit of Mercury from Peurbach's *Theoricae noue planetarum*, 1482 (BM 1A 20516). According to Hartner, Peurbach is the first European author to speak of the curve of the centre of Mercury's epicycle in the words *Sexto ex dictis apparet manifeste centrum epicycli Mercurii propter motus supra dictos non ut in aliis planetis fit; circumferentiam deferentis circularem sed potuis figurae habentis similitudinem plana ovali periferiam describere.*

Fig. 15 (*opposite*). Mercury's orbit according to Ptolemy. Centre of epicycle $H_1 H_2 H_3 H_4$ appears uniform from E_1. Figures 1, 2, 3, 4, show centre of deferent A B C D in the course of half one tropical year. It revolves with uniform retrograde motion, while radius vector $E_1 H_1 E_1 H_2 E_1 H_3 E_1 H_4$ simultaneously carries round the centre of the epicycle with uniform direct motion. (*Right*) The ovoid form of the curve traced out by the centre of Mercury's epicycle H. Under certain conditions it approximates an ellipse.

56

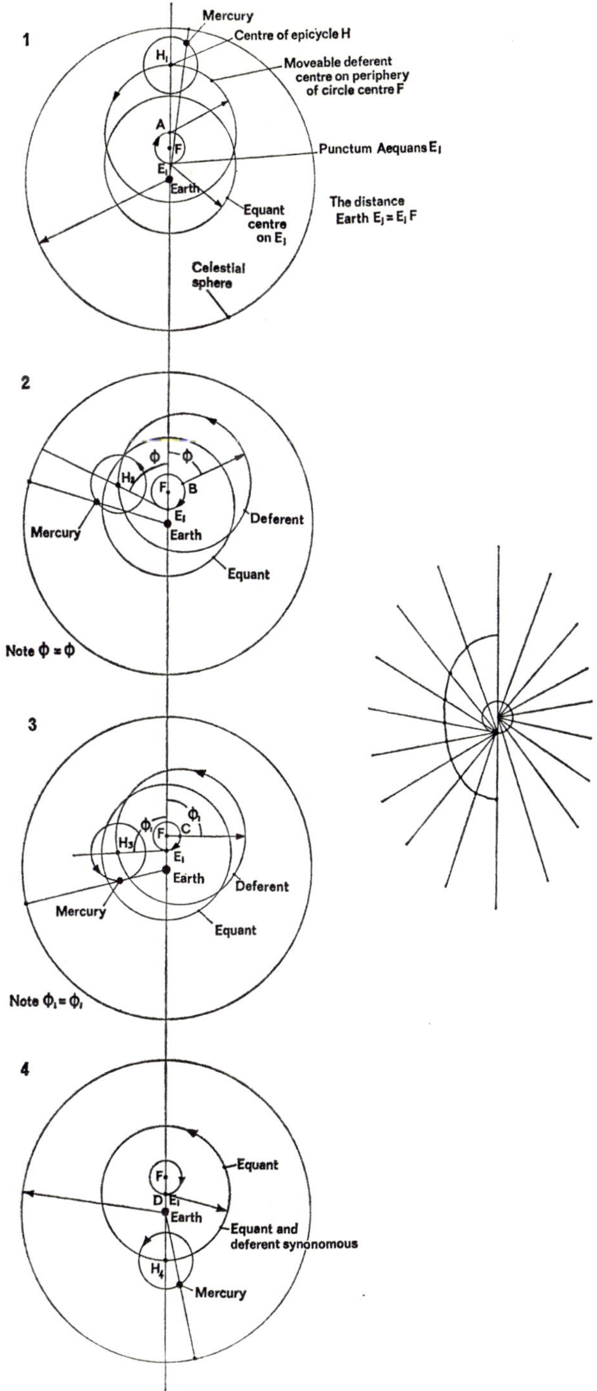

1
Mercury
Centre of epicycle H
H₁
Moveable deferent centre on periphery of circle centre F
A
F
E₁
Earth
Punctum Aequans E₁
The distance Earth E₁ = E₁ F
Equant centre on E₁
Celestial sphere

2
φ φ
H₂
F B
E₁
Earth
Mercury
Deferent
Equant
Note φ = φ

3
φ₁ φ₁
H₃ F C
E₁
Earth
Mercury
Deferent
Equant
Note φ₁ = φ₁

4
F
D E₁
Earth
Equant
Equant and deferent synonomous
H₄
Mercury

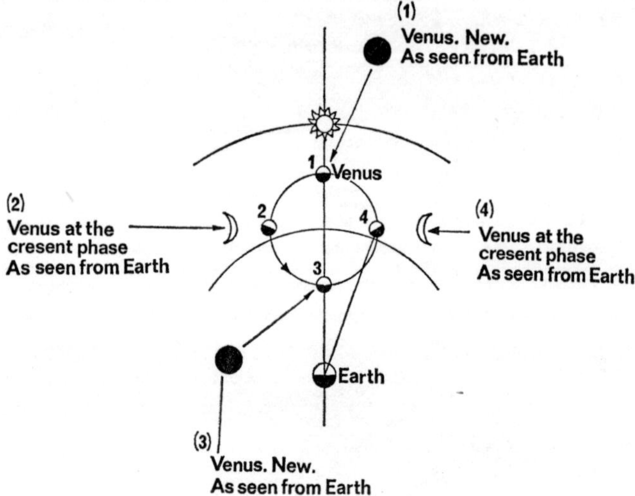

Fig. 16. Motion of Venus according to Ptolemy.

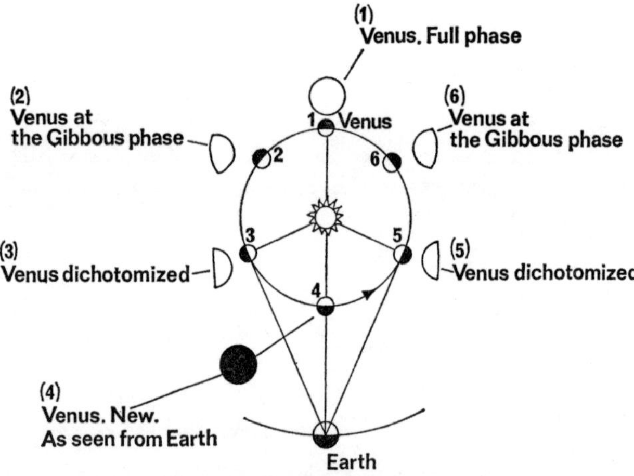

Fig. 17. Motion of Venus according to Copernicus.

perspective instrument,[58] but we should not fail to note his audacity in pointing it at the sky, his contributions to astronomy are of a wholly different quality from Kepler's. To the end of his life he failed to embrace the elliptical orbits, entertained an erroneous hypothesis of the

[58] *Dutch perspective instrument.* A flamboyant term for telescope—based on the invention of that instrument by the Dutch spectacle-maker Lippershey; see King, H. C., *The History of the Telescope*, London 1955, p. 30.
Rosen has shown (see Rosen, E., *The Naming of the Telescope*, New York 1947, p. 30) that the word *telescope* in place of Galileo's term *perspicillum* was first introduced in 1611 by the poet Demisiani at a banquet in honour of Galileo.

tides, considered Saturn to be 'triplicate' and lagged far behind Tycho in his understanding of comets. Yet his wonderful work on projectiles in his dialogue in the *New Sciences* lent potent aid to the solid establishment of celestial mechanics in the competent hands of Newton.

Galileo's work was confined to the familiar motions of falling bodies and projectiles near the Earth's surface; Newton took them out into the far flung spaces. The genesis of the universal law of gravitation which is Newton's greatest achievement is clearly set out in his own writings and commented upon by Cherry.[59]

Newton has left the following account of his early work on the dynamical theory of the solar system. After speaking of his discoveries of the binomial theorem and of the infinitesimal calculus, he proceeds:

'And the same year [1666] I began to think of gravity extending to yc orb of the Moon, and having found out how to estimate the force with wch a globe revolving within a sphere presses the surface of the sphere, from Kepler's Rule of the periodical times of the Planets being in a sesquialterate proportion of their distances from the centers of their Orbs I deduced that the forces wch keep the Planets in their Orbs must be reciprocally as the squares of their distances from the centers about wch they revolve; and thereby compared the force requisite to keep the Moon in her Orb with the force of gravity at the surface of the Earth, and found them answer pretty nearly. All this was in the two plague years of 1665 and 1666, for in those days I was in the prime of my age for invention.'

This summary may be amplified as follows: If a body moves with speed v in a circle of radius r, its acceleration is directed always towards the centre of the circle, and is of magnitude v^2/r. If the periodic time of a planet moving in a circle of radius r is T, its speed is $2\pi r/T$, so its acceleration is $4\pi^2 r/T^2$. From Kepler's third law, T^2 varies as r^3; hence the acceleration varies as $1/r^2$.

This suggests that the Sun attracts the planets with a force varying with the distance according to the inverse square law. Now the Moon is kept in its (roughly circular) orbit about the Earth by a force directed towards the Earth. Let us assume by analogy that this attractive force also follows the inverse square law, and calculate its accelerative effect at the Earth's surface. For the Moon,

$$r = 240,000 \times 5280 \text{ feet,}$$
$$T = 27\tfrac{1}{3} \times 86400 \text{ seconds}$$

approximately, so its acceleration is

$$\frac{4\pi^2 \times 240,000 \times 5280}{(27\tfrac{1}{3} \times 86400)^2} \text{ ft./sec.}^2$$

By the inverse square law the corresponding acceleration at the Earth's surface is 60^2 times this, since the Moon's distance is 60 times the Earth's radius; it comes out almost exactly to 32 ft./sec.^{2}, the familiar acceleration of gravity.

[59] Cherry, T. A., *Newton's Principia in 1687 and 1937*, Melbourne University Press, 1937.

The solar system

Here we have the historical origin of the famous law of gravitation: 'Every particle in the universe attracts every other particle with a force varying jointly as their masses and the inverse square of their distance part:

$$F = k m_1 m_2 / r^2.'$$

It is too frequently said that Newton's proof of the matter is complex and for this reason it is seldom seen. It is, however, too important to leave hidden in the *Principia* (under Section III, Proposition XI, Problem VI) and it is presented below in all its geometric elegance.[60]

If a body revolves in an ellipse; it is required to find the law of the centripetal force tending to the focus of the ellipse.

Let S be the focus of the ellipse. Draw SP cutting the diameter DK of the ellipse in E, and the ordinate QV in x; and complete the parallelogram QxPR. It is evident that EP is equal to the greater semiaxis AC: for drawing HI from the other focus H of the ellipse parallel to EC, because CS, CH are equal, ES, EI will be also equal; so that EP is the half-sum of PS, PI, that is (because of the parallels HI, PR, and the equal angles IPR, HPZ), of PS, PH, which taken together are equal to the whole axis 2AC. Draw QT perpendicular to SP, and putting L for the principal latus rectum of the ellipse (or for $\dfrac{2BC^2}{AC}$), we shall have

$$L \cdot QR : L \cdot PV = QR : PV = PE : PC = AC : PC,$$
also, $L \cdot PV : GV \cdot PV = L : GV$, and, $GV \cdot PV : QV^2 = PC^2 : CD^2$.

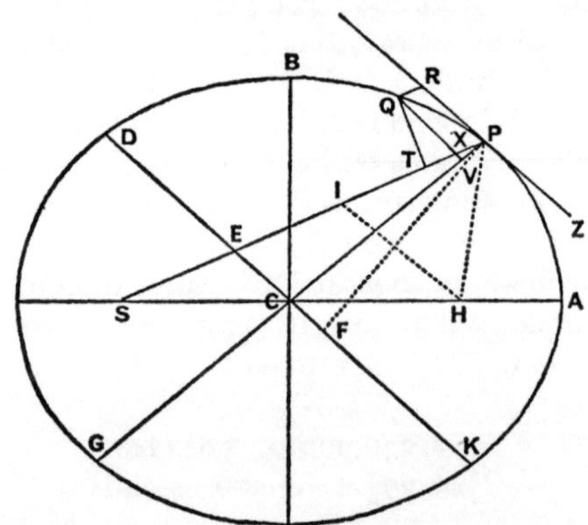

Now when the points P and Q coincide, $QV^2 = Qx^2$,[61] and Qx^2 or $QV^2 : QT^2 = EP^2 : PF^2 = CA^2 : PF^2$, and (since as is shown by the writers on the conic sections—all parallelograms circumscribed about any conjugate diameters of a given ellipse or hyperbola are equal among themselves) equals $CD^2 : CB^2$.

[60] It appears at pages 56 and 57 of Cajori's edition of the *Principia*, University of California Press, 1947.
[61] Cor. II, Lem VII states.

Multiplying together corresponding terms of the four proportions, and simplifying, we shall have

$$L \cdot QR : QT^2 = AC \cdot L \cdot PC^2 \cdot CD^2 : PC \cdot GV \cdot CD^2 \cdot CB^2 = 2PC : GV,$$

since $AC \cdot L = 2BC^2$. But the points Q and P coinciding, 2PC and GV are equal. And therefore the quantities $L \cdot QR$ and QT^2, proportional to these, will be also equal. Let those equals be multiplied by $\dfrac{SP^2}{QR}$, and $L \cdot SP^2$ will become equal to $\dfrac{SP^2 \cdot QT^2}{QR}$. And therefore the centripetal force is inversely as $L \cdot SP^2$, that is, inversely as the square of the distance SP.

From Newton's insight we obtain a synthesis of elliptical orbits and gravitational forces. From Kepler we obtain the basic laws of modern dynamical astronomy. It is, however, important to realize that we should not take the view expressed in many works devoted to the history of astronomy, namely, that Newton's success lay in the use of the scientific method that came in with Copernicus and was advanced by Brahe and Kepler and was something unknown to Ptolemy and the adherents to the older, now discredited, astronomy of classical times. A closer study of ancient astronomy will show that Ptolemy's work was very much scientific and that the rejection of a geocentric system for a heliocentric one is primarily based on a metaphysical revolution and not a scientific one. This profound point is ably demonstrated by Burtt and Kattsoff.[62] Similarly we should not, as is commonly held, ascribe elliptical orbits to Kepler in view of the theorem of Copernicus that is today incorrectly graced with the name of Lahire, indeed we should seek to widen our appreciation of the history of science in exactly those domains that are too frequently ignored, namely the domains of 'unknown' men and their works that form the indispensable connecting links between the great discoveries that are too often incorrectly presented as the mountain peaks and give the false impression of men of science jumping from one mountain top to the next. A powerful plea for the redress of this situation is made by no less a luminary than Sarton.[63]

[62] Burtt, *Metaphysical Foundations of Modern Physical Science*; Kattsoff, L. O., 'Ptolemy and Scientific Method', *Isis, 38*, 18.
[63] Sarton, G., 'A study of Early Scientific Textbooks', second Preface to *Isis, 38*, 137.

Chapter 3

Models of the solar system

In order that we might somehow understand what this proper movement of the sun is, certain very learned geometers have manufactured kinds of automats. They stacked a certain number of spheres one inside the other and the planets were, so to say, lodged within these. It is even said that Archimedes constructed such αὐτόματα of the celestial movements, that is, orreries which represent these movements to the eyes. . . .

This is the place to censure the perversity and quarrelsome disposition of Averroes and many other philosophers. They make fun of this doctrine which is put together with so much art because we cannot say that such mechanisms really exist in the heavens.

If only Averroes and the others would stop bringing confusion into established science. Why do they not show us laws of the celestial movements which are better adapted and through which we might set up exact computations? Since Averroes' arguments are extremely crude (prorsus βάναυσα), we need not repeat them here. Besides, geometers themselves never meant to claim that such models exist in the heavens. They only want to give an exact account of their movements.
MELANCHTHON

Over a long period of time the constitutive members of the solar system offer a remarkably complex anfractuosity to the Earth-bound eye of an assiduous observer as they make their paths backward and forward across the celestial vault. Attempts to bring these members within a clear mental system have been the burden of our two previous chapters and now we pause to notice how with consummate skill man has made for himself models of the system of the world in harmony with his insight into its mysteries.

In the tomb of Senmut[1] the ceiling is decorated with an astronomical device for easing the task for both the living and the dead of anticipating their wanderings. In the great Weingarten[2] Abbey on Lake Constance, sixteen miles north of Friedrichshafen, the brethren had constructed a static planetarium to show the future positions of the Sun, Moon and

---

[1] Pogo, A., 'The Astronomical Ceiling decoration in the tomb of Senmut', *Isis, 14*, 301 (1930).
[2] Bévenot, H. G., 'Weingarten Planetarium', *Isis, 8*, 300 (1926).

planets; while in the far West the Peruvian Quipu[3] is thought to have enabled its rude yet sagacious designer to keep track of the planets in their courses.

The idea of a model to represent the solar system is advanced by Plato in the tenth book of *The Republic* under the intriguing heading 'The spindle and whorl of necessity':

Another day's journey brought them to the place, and there, in the midst of the light, they saw the ends of the chains of heaven let down from above: for this light is the belt of heaven, and holds together the circle of the universe, like the under-girders of a trireme. From these ends is extended the spindle of Necessity, on which all the revolutions turn. The shaft and hook of this spindle are made of steel, and the whorl is made partly of steel and also partly of other materials. Now the whorl is in form like the whorl used on earth; and the description of it implied that there is one large hollow whorl which is quite scooped out, and into this is fitted another lesser one, and another, and another, and four others, making eight in all, like vessels which fit into one another; the whorls show their edges on the upper side, and on their lower side all together form one continuous whorl. This is pierced by the spindle, which is driven home through the centre of the eighth. The first and outermost whorl has the rim broadest, and the seven inner whorls are narrower, in the following proportions—the sixth is next to the first in size, the fourth next to the sixth; then comes the eighth; the seventh is fifth, the fifth is sixth, the third is seventh, last and eighth comes the second. The largest (or fixed stars) is spangled, and the seventh (or sun) is brightest; the eighth (or moon) coloured by the reflected light of the seventh; the second and fifth (Saturn and Mercury) are in colour like one another, and yellower than the preceding; the third (Venus) has the whitest light; the fourth (Mars) is reddish; the sixth (Jupiter) is in whiteness second. Now the whole spindle has the same motion; but, as the whole revolves in one direction, the seven inner circles move slowly in the other, and of these the swiftest is the eighth; next in swiftness are the seventh, sixth, and fifth, which move together; third in swiftness appeared to move according to the law of this reversed motion the fourth; the third appeared fourth and the second fifth. The spindle turns on the knees of Necessity; and on the upper surface of each cricle is a siren, who goes round with them, hymning a single tone or note. The eight together form one harmony; and round about, at equal intervals, there is another band, three in number, each sitting upon her throne.

We know that Archimedes[4] (287–212 B.C.) devised a model sphere to imitate the motions of the Sun, the Moon and the five planets, and through Cicero,[5] who saw and described it, we learn that it was in fact

---

[3] *Quipu.* A device of the ancient Peruvians and others for recording events, keeping accounts, sending messages, etc., consisting of cords or threads of various colours, knotted in various ways. The Peruvian quipu is described in the Treasury of Mathematics I, Midonick, H., Penguin Books, 1968, p. 145. See also Ascher, M. and Ascher, R., 'Code of Ancient Peruvian Knotted Cords (Quipus)', *Nature*, *222*, 529 (1969).

[4] Archimedes, *On Sphere Making*, see Dijksterhuis, E. J., *Archimedes* (Groningen 1938), ch. 1, p. 18.

[5] Cicero, *De Republica*, I, c. 14, pp. 21–2.

made and operated. That Archimedes was familiar with astronomy is readily understood when we recall that his father Pheidias was an astronomer of Syracuse; that the son's mechanical genius should lead him to construct the earliest known working model of the heavenly bodies need cause little surprise. Ovid[6] also makes reference to a miniature representation of the vast vault of heaven and Claudis Claudianus confirms it (A.D. *c.* 400) by saying that Jupiter must have laughed to see what he had created represented in a small glass sphere by the aged Syracusan. That this was a masterpiece of the instrument-makers art cannot be doubted, otherwise reference to it over such a long period would not have been sustained. An ornate celestial globe is said to have been made by none less than Eudoxus himself and the poet Aratus is thought to have based his celebrated poem the *Phaenomena* upon the constellations according to the system proposed by Eudoxus.

The earliest celestial sphere extant is the 26 in. (660 mm.) diameter, carved marble globe, the *Atlante Farnesiano* at Naples. Frederick II (1194–1250), the Roman Emperor, King of Sicily and Jerusalem and a member of the Hohenstaufen family, is said to have received a *tentorium*[7] from al Ashraf, the Sultan of Babylon, in which the Sun and the Moon go through their courses in their sure and correct periods and without error indicate the hours of night and day. In *c.* 1274 Petrus Philomena de Dacia[8] invented a new equatorium based on the simple analogue computers of Johannes Campanus known as *Theoricae Campani*. These computers were in fact geometric models of the Ptolemaic world system. The deferents and epicycles were represented by circular discs of parchment while the various lines of sight were represented by threads. These instruments, used with suitable tables, enabled the operator to arrive at the *aequatio centri* and the *aequatio argumenti* without recourse to the usual difficult computation. Dacia's improved instrument dealt with all the naked eye planets in a single instrument, whereas the older *Theoricae Compani* necessitated drawing some eighteen graduated circles. Undoubtedly this instrument was a stimulus to other constructors of the planetary systems. In the Middle Ages many astronomical clocks were made and one of the earliest recorded is that of Richard Wallingford (*c.* 1292–1335), who studied at Oxford and took the Benedictine habit at St Albans. His complex astronomical clock showing the motions of the Sun, the Moon and the stars also gave some indication of the ebb and flow of the tides. The clock aroused the greatest curiosity in *c.* 1320 and Edward III is said to have complained to Richard that he spent

<hr>

[6] Ovid, *Fasti*, VI, 277.
[7] See 'Chronica Regia Coloniensis Cont. N. Ann 1232', ed. G. Witz, p. 263 (Hanover 1880).
[8] See Zinner, E., 'Petrus de Dacia en middelalderlig dansk Astronom', *Nordisk Astronomisk Tidsskrift, 13*, 136 (1932).
Pedersen, O., *Vistas in Astronomy, 9*, 3 (1967).

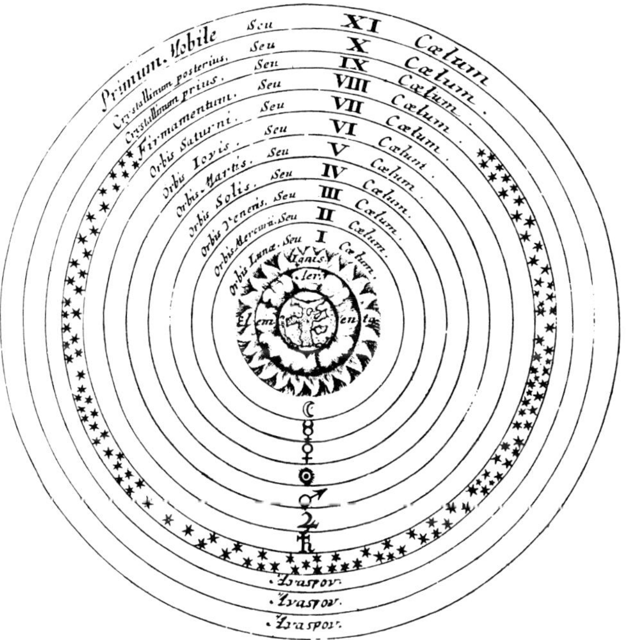

1. The geocentric system of the world as conceived by some of the Greek philosophers.

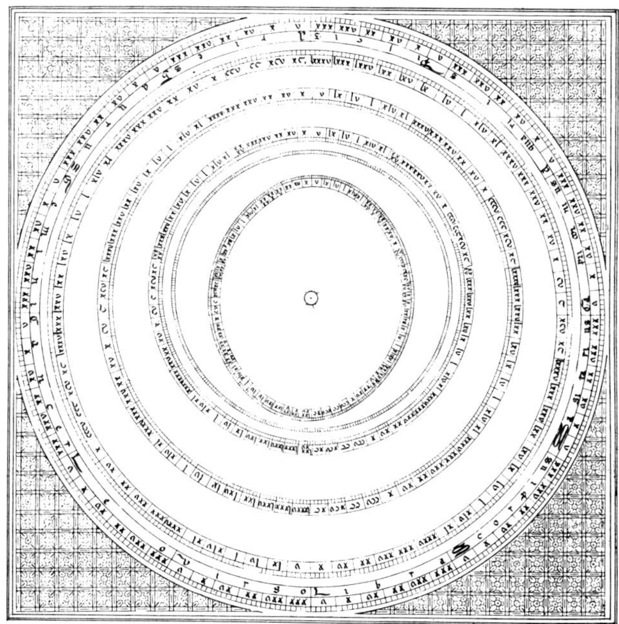

2. The circles of the Sun and Moon and the ovoid or elliptiform orbit of Mercury's epicycle according to Azarquiel (eleventh century).

NICOLAI CO

PERNICI TORINENSIS

DE REVOLVTIONIBVS ORBI-

um cœlestium, Libri VI.

Habes in hoc opere iam recens nato, & ædito,
studiose lector, Motus stellarum , tam fixarum,
quàm erraticarum, cum ex ueteribus, tum etiam
ex recentibus obseruationibus restitutos:& no-
uis insuper ac admirabilibus hypothesibus or-
natos. Habes etiam Tabulas expeditissimas , ex
quibus eosdem ad quoduis tempus quàm facilli
me calculare poteris. Igitur eme, lege, fruere.

Ἀγεωμέτρητος ἀδεὶς εἰσίτω.

Norimbergæ apud Ioh. Petreium,
Anno M. D. XLIII.

Axioma Astronomicum Motus cœlestis æqualis est et circularis.
vel ex æqualibus & circularibus compositus.

3. The title page of the *De revolutionibus* of Copernicus.

4. Cracow at the time of Copernicus. (From Schedel's World's Chronicle.)

AD SANCTIS-
SIMVM DOMINVM PAV.
LVM III. PONTIFICEM MAXIMVM,
Nicolai Copernici Præfatio in libros
Reuolutionum.

ATIS equidem, Sanctissime Pater, æ
stimare possum, futurum esse, ut simul
atq; quidam acceperint, me hisce meis li
bris,quos de Reuolutionibus sphærarũ
mundi scripsi,terræ globo tribuere quos
dam motus , statim me explodendum
cum tali opinione clamitent. Neq; enim
ita mihi mea placent , ut nõ perpendam,
quid alij de illis iudicaturi sint.Et quamuis sciam, hominis phi
losophi cogitationes esse remotas à iudicio uulgi , propterea
quòd illius studium sit ueritatem omnibus in rebus, quatenus
id à Deo rationi humanæ permissum est,inquirere,tamen alie
nas prorsus à rectitudine opiniones fugiendas censeo. Itaq; cũ
mecum ipse cogitarem, quàm absurdum ἀκρόαμα existimatu
ri essent illi,qui multorum seculorum iudicijs hanc opinionẽ
confirmatam norũt,quòd terra immobilis in medio cœli, tan
quam centrum illius posita sit, si ego contra assererem terrã
moueri,diũ mecum hæsi , an meos cõmentarios in eius motus
demonstrationem conscriptos in lucem darem, an uero satius
esset,Pythagoreorum & quorundam aliorum sequi exemplũ,
qui non per literas,sed per manus tradere soliti sunt mysteria
philosophiæ propinquis & amicis duntaxat. Sicut Lysidis ad
Hipparchum epistola testatur . Ac mihi quidem uidentur id
fecisse : non ut quidam arbitrantur ex quadam inuidentia
communicandarum doctrinarum,Sed ne res pulcherrimæ,&
multo studio magnorum uirorum inuestigatæ,ab illis contem
nerentur,quos aut piget ullis literis bonam operam impende
re , nisi quæstuosis,aut si exhortationibus & exemplo aliorum
ad liberale studium philosophiæ excitentur , tamen propter
stupidita

AD LECTOREM DE HYPO.
THESIBVS HVIVS OPERIS.

NON dubito , quin eruditi quidam, uulgata iam de
nouitate hypotheseon huius operis fama,quòd ter
ram mobilem,Solem uero in medio uniuersi ĩm
mobilẽ constituit,uehementer sint offensi, putẽtq;
disciplinas liberales recte iam olim constitutas , turbari nõ o
portere.Verum si rem exacte perpendere uolent,inueniẽt au
thorem huius operis,nihil quod reprehendi mereatur cõmi
sisse.Est enim Astronomi proprium,historiam motuum cœle
stium diligenti & artificiosa obseruatione colligere . Deinde
causas earundem,seu hypotheses,cum ueras assequi nulla ra
tione possit,qualescunq; excogitare & confingere,quibus sup
positis,ijsdem motus,ex Geometriæ principijs,tam in futuru,
quàm in præteritũ recte possint calculari.Horũ autẽ utruncq;
egregie præstitit hic artifex.Neq; enim necesse est , eas hypo
theses esse ueras,imò ne uerisimiles quidem,sed sufficit hoc u
num,si calculum obseruationibus congruentem exhibeant.ni
si forte quis Geometriæ & Optices usq;adeo sit ignarus, ut e
picyclium Veneris pro uerisimili habeat,seu in causa esse cre
dat,quod ea quadraginta partibus,& eo amplius, Solẽ inter
dum præcedat,interdũ sequatur.Quis enim nõ uidet, hoc po
sito,necessario sequi,diametrum stellæ in περιγείῳ plusq; qua
druplo,corpus autem ipsum plusq; sedecuplo,maiora, quàm
in ἀπογείῳ apparere,cui tamen omnis æui experientia refraga
tur? Sunt & alia in hac disciplina non minus absurda, quæ in
præsentiarum excutere,nihil est necesse.Satis enim patet,ap
parentiũ inæqualium motuũ causas,hanc arte penitus & sim
pliciter ignorare.Et si quas fingẽdo excogitat,ut certe quãplu
rimas excogitat , nequaquã tamen in hoc excogitat,ut ita esse
cuiquam persuadeat,sed tantum,ut calculum recte instituant.
Cum autem unus & eiusdem motus,uarie interdum hypothe
ses sese offerant(ut in motu Solis,eccentricitas,& epicyclium)
Astronomus eam potissimum arripiet,quæ compræhensu sit
quàm facillima.Philosophus fortasse,ueri similitudinem ma
gis re

---

stupiditatem ingenij inter philosophos,tanq; fuci inter apes
uersantur.Cum igitur hæc mecũ perpenderem, contemptus,
qui mihi propter nouitatem & absurditatẽ opinionis metuen
dus erat,propemodum impulerat me,ut institutum opus pror
sus intermitterem.

Verum amici me diu cunctantem atq; etiã reluctantem re
traxerũt, inter quos primus fuit Nicolaus Schonbergius Car
dinalis Capuanus,in omni genere doctrinarũ celebris. Proxi
mus illi uir mei amantissimus Tidemannus Gisius , episcopus
Culmensis,sacrarum ut est, & omnium bonarũ literarum stu
diosissimus.Is etenim sæpenumero me adhortatus est, & con
uitijs interdum additis efflagitauit,ut librum hunc æderem, &
in lucem tandem prodire sinerem,qui apud me pressus non in
nonum annũ solum,sed iam in quartum nouenniũ, latitasset.
Idem apud me egerunt alij non pauci uiri eminentissimi & do
ctissimi,adhortantes ut meam operam ad communem studio
sorum Mathematices utilitatem,propter conceptum metum,
conferre non recusarem diutius.Fore ut quanto absurdior plẽ
risq; nunc hæc mea doctrina de terræ motu uideretur, tanto
plus admirationis atq; gratiæ habitura esset,postq; per æditio
nem cõmentariorum meorum caliginem absurditatis sublatã
uiderent liquidissimis demonstrationibus. His igitur persua
soribus,eaq; spe adductus,tandem amicis permisi,ut æditionẽ
operis,quam diu à me petissent,facerent.

At nõ tam mirabitur fortasse Sanctitas tua,quòd has meas
lucubratiões ædere in lucem ausus sim,posteaq; tantum operæ
in illis elaborandis,mihi sumpsi,ut meas cogitationes de terræ
motu etiam literis cõmittere non dubitauerim, sed quod ma
gis ex me audire expectat,qui mihi in mentem uenerit,ut con
tra receptam opinionem Mathematicorum , ac propemodum
contra communem sensum,ausus fuerim imaginari aliquẽ mo
tum terræ.Itaq; nolo Sanctitatem tuã latere,me nihil aliud mo
uisse,ad cogitandum de alia ratione subducendorum motuum
sphærarum mundi,quàm quod intellexi,Mathematicos sibi
ipsis non constare in illis perquirendis.Primũ enim usq;adeo
incertisunt de motu Solis & Lunæ,ut nec uertentis anni perpe
iij   tuam

gis requiret,neuter tamen quicquam certi compræhẽdet, aut
tradet,nisi diuinitus illi reuelatum fuerit . Sinamus igitur &
has nouas hypotheses,inter ueteres,nihilo uerisimiliores inno
tescere,præsertim cum admirabiles simul,& faciles sint,ingen
temq; thesaurum, doctissimarum obseruationum secum ad
uehant.Neq; quisquam,quod ad hypotheses attinet,quicquã
certi ab Astronomia expectet,cum ipsa nihil tale præstare que
at,ne si in alium usum conficta pro ueris arripiat , stultior ab
hac disciplina discedat, quàm accesserit. Vale.

NICOLAVS SCHONBERGIVS CAR
dinalis Capuanus , Nicolao Copernico, S.

CVm mihi de uirtute tua,cõstanti omniũ sermone
ante annos aliquot allatũ esset,cœpi tum maiorem
in modũ te animo cõplecti , atq; gratulari etiã no
stris hominibus,apud q̃s tãta gloria florere.Intellexerã enim
te nõ modo ueterũ Mathematicorũ inuẽta egregie callere,sed
etiã nouã Mũdi rationẽ cõstituisse.Qua doceas terrã moueri:
Solem imũ mũdi , adeoq; mediũ locũ obtinere:Cœlũ octauũ
immotũ,atq; fixũ ppetuo manere:Lunã se unã cũ inclusis suæ
sphæræ elementis,inter Martis & Veneris cœlũ sitam,anni
uersario cursu circũ Solem cõuertere.Atq; de hac tota Astro
nomiæ ratione cõmentarios à te cõfectos esse, ac erraticarum
stellarũ motus calculis subductos in tabulas te cõtulisse,maxi
ma omniũ cum admiratione.Quamobrem uir doctissime,ni
si tibi molestus sum,te etiã atq; etiã oro uehementer ,ut hoc
tuũ inuentũ studiosis cõmunices,& tuas de mundi sphæra lu
cubrationes unã cũ Tabulis,& si quid habes præterea , q̃d ad
eandem rem pertineat, primo quoq; tempore ad me mittas.
Dedi autem negotiũ Theodorico à Reden,ut istic meis sum
ptibus omnia describantur,atq; ad me transferantur.Quod si
mihi morem in hac re gesseris,intelliges te cum homine no
minis tui studioso,& tantæ uirtuti satisfacere cupiente rem ha
buisse.Vale,Romæ,Calend.Nouembris,anno M.D.XXXVI.
ij

5. The Preface and Foreword to the De revolutionibus of Copernicus.

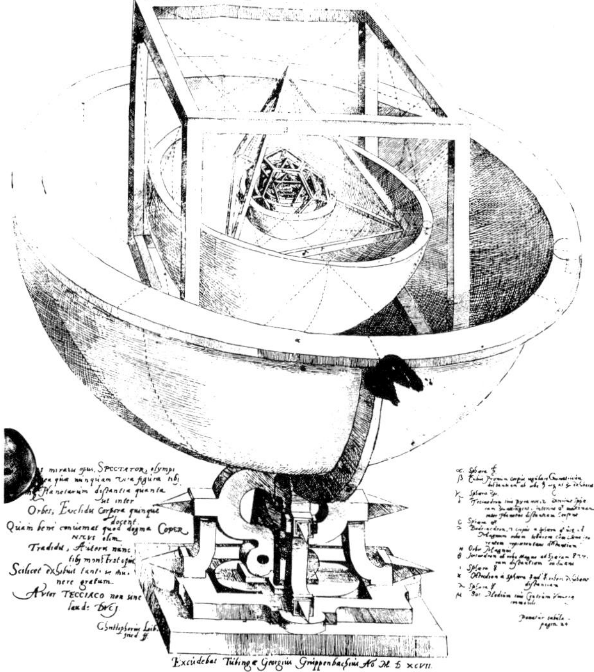

6. Kepler's explanation of the structure of the planetary system in which he uses the five regular polyhedra to fill the inter-planetary spaces. (From Kepler's *Harmonice Mundi*, 1619.)

7. Celestial globe of Atlante Farnesiano, *c.* 300 BC, National Museum, Naples.

8. Celestial globe by Martin Bylica of Olkusz, 1480. Diameter 400 mm., height 1350 mm., bronze. One of the largest medieval globes in existence it carries one astrolobe and two sundials. Many refinements enable it to be used as a type of orrery. Martin Bylica bequeathed it to the Jagellonian University of Cracow in 1493. (Nicholas Copernicus studied in Cracow and may have used the orrery.)

9. The Golden Jagellonian Globe, *c.* 1510. A clock in the shape of a globe encircled by an armillary sphere. The terrestrial globe is of copper, diameter 73 mm. On the South American continent three names are engraved, TERRA DE BRAZIL, MONDUS NOVUS, TERRA SANCTAE CRUCIS. The armillary sphere consists of three polar systems, the innermost representing the solar system of the Ptolemaic model and driven by clockwork.

10. Armillary sphere, bronze, height 375 mm., diameter 170 mm., *c.* 1570, signed 'P. Danfrie Fec'. Philippe Danfrie the elder was born in Cornwall.

11. English orrery, *c.* 1781. The mechanical parts are made of wood. There is a reference in the case to Georgium Sidus; the planet discovered by Herschel in 1781.

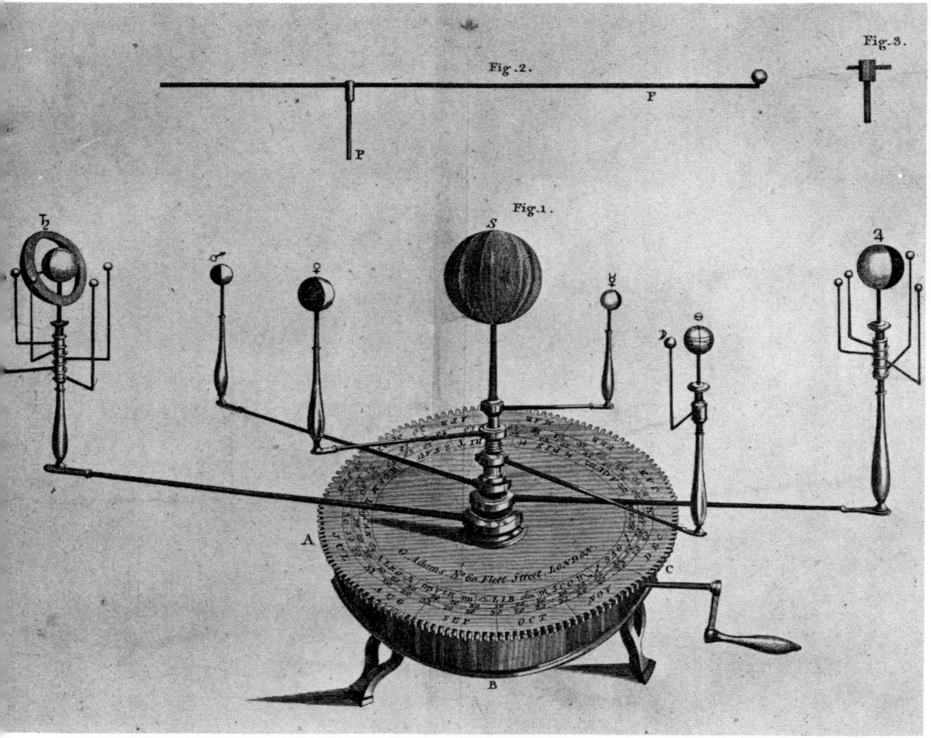

12. A typical orrery. (From a drawing *c.* 1793.)

13. A Tellurion, a **model of the** Earth/moon system, from **a** drawing *c.* 1793.

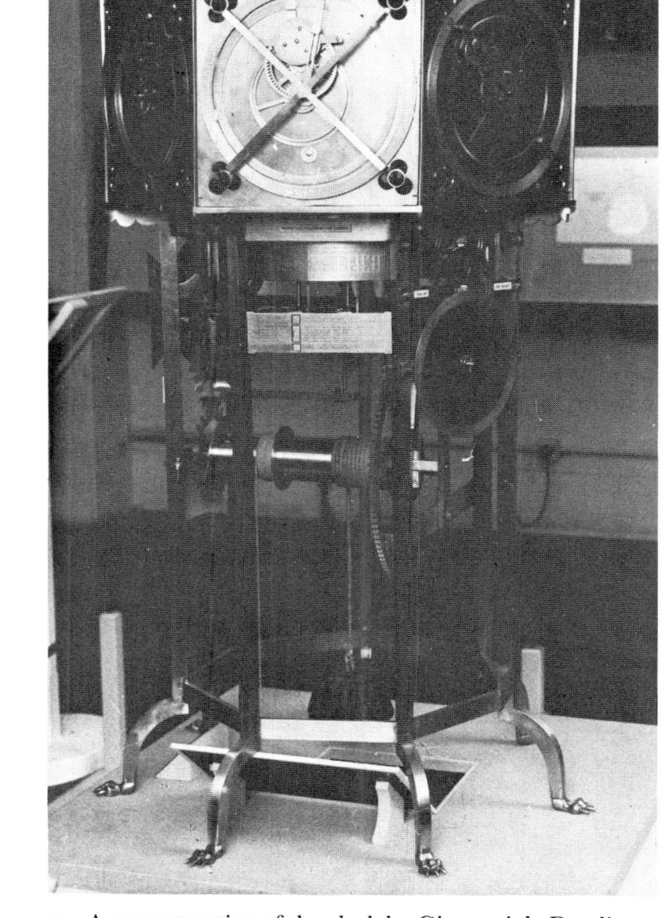

14. The famous astronomical clock of Strassburg Münster. It contains some fragments of the clock built by the mathematician Conrad Dasypodius in 1574.

15. A reconstruction of the clock by Giovanni de Dondi, c. 1364, made by Thwaites & Reed, London, from the instructions of the author's friend and colleague the late H. Alan Lloyd. The clock is now in the possession of the Smithsonian Institution, Washington, USA.

16. Rowley's tellurion, *c.* 1712, made for Charles Boyle, 4th Earl of Cork and Orrery, presented to the Royal United Service Institution, London, by the Earl's family. The instrument was restored in 1937 by Lieut-Commander R. T. Gould. All subsequent instruments of this type take their name from this; viz. an orrery.

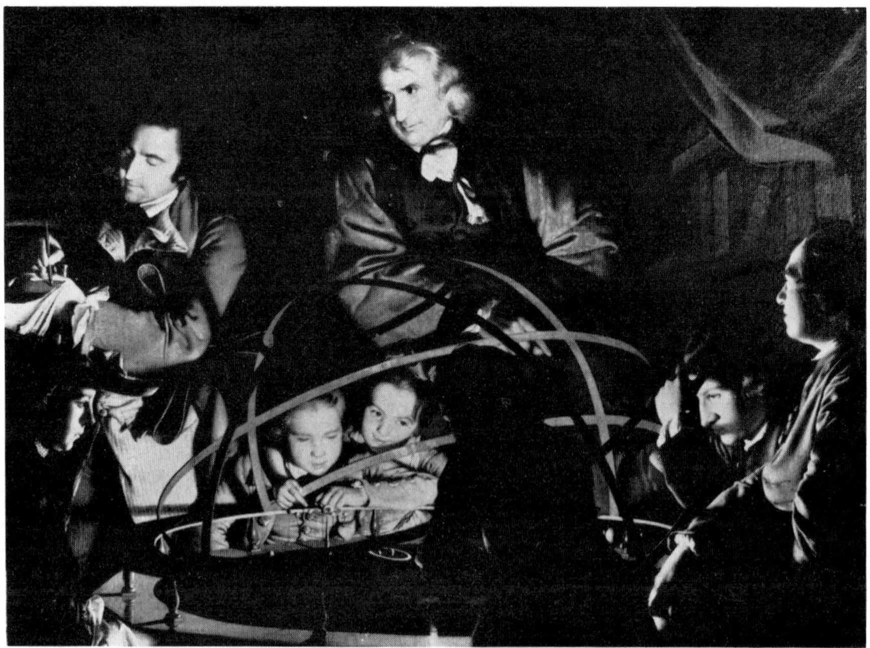

17. 'The Orrery', by Joseph Wright, A.R.A., 1766. 'A Philosopher giving a lecture on the Orrery in which a lamp is put in the place of the Sun.' This painting has become well known through the mezzotint by J. Boydell, and W. E. Pether, *c.* 1768.

18. Astronomical globe on the back of Pegasus. Made in Vienna in 1579 by Gerhard Emmozer.

19. An English tellurion signed, *W. & S. Jones, Fecerunt,
153 Holborn, London*. It stands on a folding tripod base.
The circular drum is nielloed with the months, mean and
sideral days, and the signs of the zodiac in concentric
rings. A crank for turning the mechanism has an ivory
handle connected with geared mechanism providing
movements of the Earth and the Sun. A dial indicates the
age and phases of the Moon. Height 40 cm. (Sold at

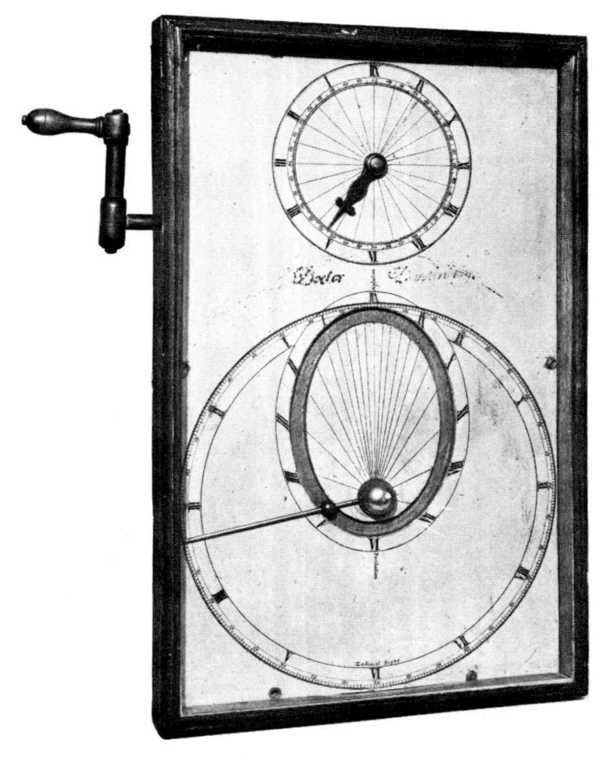

20. Newton's Orrery, 1860.

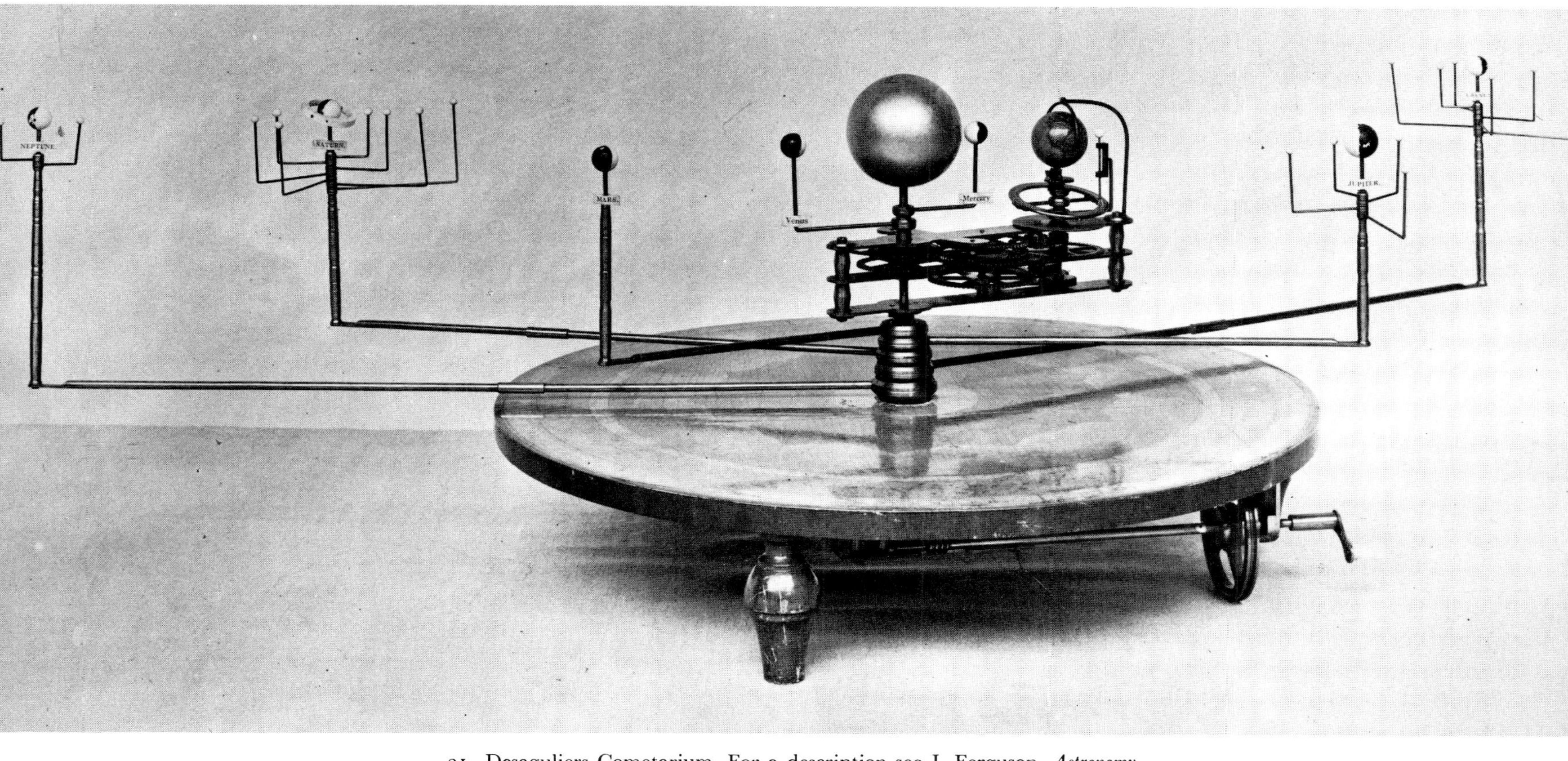

21. Desaguliers Cometarium. For a description see J. Ferguson, *Astronomy explained upon Sir Isaac Newton's principles*, London, 1770, p. 400, pl. IV.

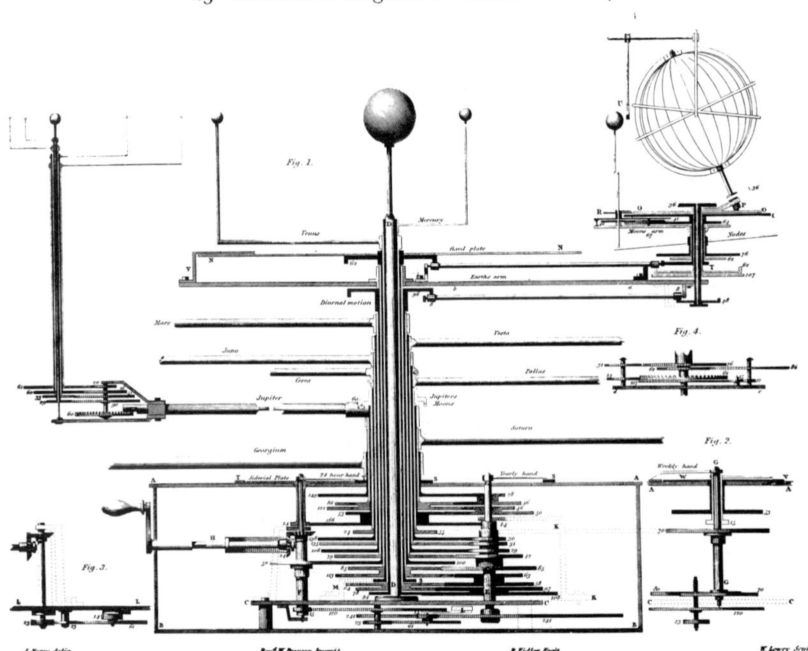

22. Rev. William Pearson's mean motion orrery (1813).

23. Schematic diagram of Pearson's orrery.

24. Armillary sphere of Santucci della Pomarance, made between 1588–93.

25. The Leyden 'Sphaera' by Stevin Tracey of Rotterdam. The brass rings of the armillary sphere are 1·5 m.
in diameter. It was presented to the University of Leyden
in 1710.

26. Orrery by T. Heath, *c.* 1757, height 1067 mm., diameter 940 mm.;
now in the possession of Harvard University, previously owned by H.
Blairman & Sons, London.

more money and time on his clock than on his church. Richard is said to have replied that others could build churches but not clocks.

In 1364, at about the same time that the great series of clocks for Strasbourg Cathedral were being constructed, Giovanni de Dondi produced his horological masterpiece—the Dondi clock, which was eventually lost, after many vicissitudes, in the wanton destruction of the Convent of San Yste by General Soult c. 1809. Fortunately we have the wonderful Dondi Manuscript preserved in the library of St Mark's at Venice and copies of it in the Ambrosia library of Milan, in Padua, Eton College[9] and the Bodleian Library.

The craftmanship and deep knowledge of the Ptolemaic system of the world which Dondi exhibits in his manuscripts are of an extremely high order of merit and we can but feel a deep sadness that the clock is lost to us. In recent years we are indebted to my late friend H. Alan Lloyd[10] for arranging for a modern reconstruction of the Dondi clock by English craftsmen at the famous old London firm of Thwaites & Reed of Bowling Green Lane, with engravings by Mr N. F. Fryer of the L.C.C. School of Arts and Crafts.[11]

A model of the Tychonic system of the world was made by the skilled hand of Jost Bürgi, the gifted imperial mechanic and watchmaker, who was born in Lichtensteig in Toggenburg (St Gall canton Switzerland). He worked for the Land-grave William of Hesse and delivered the model to Emperor Rudolph at Prague in c. 1593. This Tychonic planetarium was equipped with a gilded celestial globe. It is of interest to relate that this model was based on a design attributed to Nicolai Reymers, a native of Holstein, who claimed the Tychonic system of the world as his own creation and was sufficiently bold to introduce a rotating Earth.[12] Later Bürgi worked for Kepler at Prague.

The astronomical clock at Strasbourg contains a true planetarium dial showing the paths of the planets and has excited the human intellect more than possibly any other device of its kind. Ungerer[13] records that, according to Crommelin, in the course of the centuries roughly four hundred descriptions of this great clock have been published.

The Ptolemaic system continued to hold sway and the wide dissemination of these ideas through, *inter alia*, the writing of Peurbach c. 1450 may have given some impetus to the construction of the beautiful and

[9] The Eton College MS. is copied from one dated 1397.
[10] Lloyd, H. Alan, *Giovanni de Dondi's Horological Masterpiece 1364*, privately printed by the Author.
[11] The clock is now in the Smithsonian Institution of Washington, no one in the United Kingdom having shown sufficient interest to defray the cost of reconstruction.
[12] Dreyer, J. L. E., *Tycho Brahe*, pp. 183, 184, Dover Book.
Reymer's book is entitled *Nicolai Raymari Ursi Dithmarsi Fundamentum astronomicum* (Strasbourg 1588).
[13] Ungerer, A., *Les horloges astronomiques et monumentales les plus remarquables de l'antiquite jusqu'a nos jours* (Strasbourg 1931).

complex nested spheres of the Renaissance such as, for example, the Armillary Sphere[14] of Santucci della Pomarance (*c.* 1580) at Florence, which contains nine such spheres around a central Earth. The earliest surviving clock-driven model of the geocentric universe (*c.* 1680), in which a terrestrial globe by Vopell dated 1511 is surrounded by three spheres is at Cracow. In this model the celestial sphere revolves in sidereal time and a Solar image moves in the ecliptic, taking its drive from a gear with seventy-two teeth. The finest globes of this type were made by Bürgi and Georg Roll of Augsburg. One of Bürgi's globes[15] with a solar image moving in true solar time along a grooved ecliptic, is still extant at the Conservatoire des Arts de Metiers, Paris.

The first planetarium known to operate on a pseudo-Copernican[16] system of the world is that of the Dutch mathematician and astronomer Christiaan Huygens (1629–95). This elegant small instrument, which is preserved in fine condition at Leyden, was made by the clockmaker Johannes van Ceulen of the Hague, and described by Huygens in his *Opuscula postuma*, Leyden 1703.[17] The planetarium is in an octagonal case 65 cm. across and 17 cm. deep. The five planets run in circular tracks placed asymmetrically of the Sun (there is no moon) and are driven by a novel clockwork due to Huygens in which a large balance wheel and spring are used to regulate the movements.

A unique armillary sphere containing a clockwork pseudo-Copernican planetarium was made by Steven Tracery of Rotterdam and presented to the University of Leyden in 1710. According to Crommelin[18] it is the only planetarium known at that date to exhibit the four Galilean satellites of Jupiter and to have satellites that move about their primary in a realistic manner. This planetarium, and the one by Huygens previously referred to, are to be seen today in the Rijksmuseum Voor de Geschie-denis der Natuurwentenschappen at Leyden.

In the eighteenth century models to show the Earth's diurnal rotation and annual revolution about the Sun and obliquity of axis in causing the alternations of day and night and the succession of seasons (a tellurion) and more complex planetariums (orreries), were made for the edification of wealthy gentlemen and to adorn their libraries in company with the now ubiquitous terrestrial and celestial globes. The great Herschel speaks derogatorily of these as 'Those childish toys called orreries'.[19] We need not dismiss them, however, on this statement alone since a mind of his order, with its grasp of the system of the world, could well dispense with frivolous aids—for others a model gives a grasp of

[14] Only three such spheres are known; the other two are in the Library of the Escorial, Spain.
[15] See Mesnage, P., 'Un chef d'oeuvere de Jost Bürgi au Conservatorie des Arts de Metiers de Paris', *Journal Swisse Horlogerie* (July/Aug. 1943).
[16] A better term is heliocentric, since it does not employ epicycles: the planets move in circles.
[17] See *Oeuvres Completes de Christiaan Huygens*, XXI (1944), pp. 109–84, 579–652.
[18] Crommelin, C. A., 'Planetaria', *Antiquarian Horology* (March 1955).
[19] Herschel, *Astronomy*, VIII, p. 287.

detail not readily achieved by long drawn-out communion with the heavens, especially in a country notorious for its clear view of these wondrous regions only at times of singular inclemency. That the models remain 'childish' is true, nevertheless, when we consider the complex motions of the planets and no clockwork will ever suffice to show them in their true wanderings. No model at this date uses the Keplerian elliptical orbits but all offer an overall view of a system that could never be seen from outside its spatial ambit. We can close our brief survey by pointing to one or two examples of orreries which attempt to come to terms with the system of the Sun.

The original orrery is thought to be that of George Graham and Thomas Tompion (now at Oxford),[20] built in 1709 and copied by John Rowley for his patron Charles Boyle, 4th Earl of Orrery, the great nephew of the famous Robert Boyle. It was the Earl's instrument which gives us the genesis of the term 'orrery'. This instrument is not a model of the planetary system but only a model of the Earth/Moon system and is in effect a complex tellurion. The first American made orrery was constructed by Rittenhouse of Pennsylvania and is mounted vertically in a case having the form of a triptych. One outer case shows the movements of Jupiter's satellites, the middle case shows the Sun and the planets and the other outer case contains models for demonstrating eclipse phenomena of the Sun, Earth and Moon. The Rittenhouse orrery is a proud possession of the library at Princeton University.

A large pseudo-Copernican planetarium, on the scale of $1 : 10^{12}$ made without appeal to any other designer of these complex instruments, was made by the untutored Eise Eisinga and set into the ceiling of his house in Leeuwarden of West Friesland c. 1774–81. The planets move in their correct periods about the Sun, Saturn taking some $29\frac{1}{2}$ years to complete a circuit. Everything is driven by ingenious cartrate peg gearings made from wood.

A large planetarium embodying true Keplerian orbits was made by F. Meyer of the Zeiss Works in a cylindrical room of 36 ft. diameter, 9 ft. high at the Deutsches Museum of Munich in c. 1920, but it is now dismantled.

One of the most beautiful of the eighteenth-century orreries is that of Thomas Heath, whose workshops were believed to have been in the Strand, London. It shows planets out to Uranus, and in the view of Maddison[21] it should be dated c. 1781, but Adams believes that Heath died in 1775. Clearly the Uranus system of the model is a later addition. The orrery is of exquisite workmanship. It is supported by a parcel, gilt, mahogany pedestal elaborately carved with acanthus foliage and with

[20] See Gabb, G. H., and Sherwood Taylor, F., 'An Early Orrery', by Thomas Tompion and George Graham, *The Connoisseur* (Sept. 1948), pp. 24, 25.
[21] Maddison, F., 'An Eighteenth Century Orrery', by Thomas Heath, *The Connoisseur, 141*, No. 569, 163 (1958).

four lion-paw feet. The twelve-sided case enclosing the mechanism is just over 3 ft. wide and it has a removable glazed cover. The mechanism is of brass, the scales are silvered and the bodies of the planets of bone or ivory. The Sun is an engraved brass sphere and the Earth a miniature globe. Twelve ebony panels decorated with ormolu mounts represent the signs of the Zodiac.

A modern orrery of complex design was exhibited at the South Bank Exhibition in London during the Festival of Britain, 1951. It was made by Sondes Place Research Institute and is now in the Science Museum at South Kensington. The scale is based on a logarithmic scaling down of the solar system and then a reduction, by a simple factor, to bring Pluto's orbit to a maximum diameter of 16 ft. (488 cms.). The same treatment is given to the sizes of the Sun and the planets. The speeds of the revolutions of the planets about the Sun are scaled up linearly, as are the planetary rotational periods on their individual axes. The planetary bodies move in concentric circles[22] about the Sun, and the inclinations of the orbits are simplified. The power is supplied from a $1\frac{1}{2}$ h.p. electric motor. The ecliptic plane is tilted, for the convenience of the spectator, at 45° to the horizontal. This advanced orrery embodying the latest skills is now in error. Uranus is shown with six moons, Neptune with three, the rotation of Venus is in the wrong direction and the tenth moon of Saturn is wanting. A small table orrery (without any wheelwork) that I designed in 1950 shows the salient features of the system of the Sun and was used with the Sondes Place orrery in a series of lectures given over the television network of the BBC by Professor R. A. Lyttleton, entitled The Modern Universe (November 1955). That these instruments continue to excite the mind is supported by the spasmodic issue of both British and United States Patent specifications over the past fifty years.[23]

Today these models are outmoded by the optical planetarium instruments designed by Dr W. Bauersfeld,[24] but these revert to a 'true' mimic heaven and the extra terrestrial standpoint of the observer is lost.

---

[22] It is of interest to note that the Huygens planetarium of *c.* 1680 attempts to cope with the planets' variable speeds in their Keplerian orbits by using eccentric circles having their centres displaced by a calculated amount from the Sun.

[23] U.S. Patent Specification 3 286 374 1966
   U.K. Patent Specification 884 026 1961
                       7 264 1914
                         399 1913
                 3 180
                 3 179
                 4 785   1912
                 7 487
             22 823
             23 047
             15 446 1911

[24] See Werner, H., *From the Aratus Globe to the Zeiss Planetarium*, translated by Degenhardt, A. H. (Stuttgart 1957).

A LIST OF SOME WORKS DIRECTED TO ORRERIES

CHRISTIAAN HUYGENS. 'Descriptio Automati Planetarrii', with 4 engraved plates. In his *Opuscula postuma*, Leyden, 1703.

WILLIAM STUKELEY'S drawing of Stephen Hales's planetary machine, ca. 1705. Reproduced (from original in Bodleian Library, Oxford) in A. E. Clark-Kennedy, *Stephen Hales*, Cambridge, 1929.

R. T. GOULD. 'The Original Orrery Restored: An Early 18th-Century Mechanical Model of the Solar System', with coloured illustrations. In *The Illustrated London News*, vol. CXCI (December 18, 1937), pp. 1102–1103, 1126.

JOHN HARRIS. *Astronomical Dialogues between a Gentleman and a Lady: wherein the Doctrine of the Sphere, Uses of the Globes, and the Elements of Astronomy and Geography are explain'd. In a Pleasant, Easy, and Familiar Way. With a Description of the famous Instrument, call'd the Orrery.* Second Edition. London, 1725. (First edition, 1719). With engraving of 'The Orrery. Made by Mr. John Rowley M^r of Mechanicks to his Maj^ty.'

J. T. DESAGULIERS. *Lectures of Experimental Philosophy . . . To which is added, a Description of Mr. Rowley's Machine, called the Orrery, which represents the Motion of the Moon about the Earth, Venus and Mercury about the Sun, according to the Copernican System.* London, 1719.

EDMUND STONE. *The Construction and Principal Uses of Mathematical Instruments. Translated from the French of M. [Nicolas] Bion . . . To which are Added, The Construction and Uses of such Instruments as are omitted by M. Bion; particularly of those invented or improved by the English.* London, 1732. With engraving of the Orrery.

'The Great Orrery Four Feet in Diameter Made by Tho: Wright, Mathematical Instrument-maker to his Majesty for the Royal Academy at Portsmouth.' Engraving. In Joseph Harris, *The Description and Use of the Globes, and the Orrery*, Third Edition, London, 1734.

'The Orrery, made by James Ferguson', 1745. Engraving. In James Ferguson, *Astronomy explained upon Sir Isaac Newton's Principles*, Philadelphia, 1806. (First edition, London, 1761.) Wheel-work of Ferguson's orrery. Engraving. Plate VII in James Ferguson, *Select Mechanical Exercises: Shewing how to construct different Clocks, Orreries, and Sun-Dials, on Plain and Easy Principles*, Third Edition, London, 1790.

'Four Wheel'd Orrery', by James Ferguson. Engraving in his *A Dissertation upon the Phaenomena of the Harvest Moon. Also, The Description and Use of a New Four-Wheel'd Orrery*, London, 1747.

DAVID JENNINGS. *An Introduction to the Use of the Globes and the Orrery.* Third Edition. London, 1766.

DAVID RITTENHOUSE. 'Description of a New Orrery', March 27, 1767. In William Barton, ed., *Memoirs of the Life of David Rittenhouse*, Philadelphia, 1813, pp. 199–202.

Orrery by Thomas Heath & Wing, ca. 1770, at All Souls College, Oxford. Photographs from R. T. Gunther, *Early Science in Oxford*, Vol. II, Oxford, 1923.

'The New Portable Orrery, Invented and Made by William Jones.' Engraving in his *The Description and Use of a New Portable Orrery*, Sixth Edition, London, 1812. (First edition, 1782.)

ANTIDE JANVIER. *Des Revolutions des corps célestes par le Mécanisme des Rouages.* Paris, 1812. With plates of planetary machines by Janvier, and by Huygens.

ABRAHAM REES, ed. *The Cyclopaedia, or Universal Dictionary of Arts, Sciences, and Literature.* Philadelphia, n.d. (First edition, London, 1819–1820.) With plates of orreries by Janvier, William Pearson and others.

DAVID BREWSTER, ed. *The Edinburgh Encyclopaedia.* Philadelphia, 1832. (First edition, Edinburgh, 1830.) With plates of orreries.

ASA SMITH. *Smith's Illustrated Astronomy, designed for the Use of the Public or Common Schools in the United States.* New York, 1848. With plate of commonly used type of 19th-century orrery.

ISAAC GREENWOOD. *Explanatory Lectures on the Orrery, Armillary Sphere, Globes and other Machines, Instruments, and Schemes made use of by Astronomers: Accompanied With a Great Variety of Physical Experiments and Curious Remarks.* Boston, 1734.

JOSEPH POPE. 'Mr Joseph Pope's description of an Orrery of his construction', November 12, 1794. In *Memoirs of the American Academy of Arts and Sciences* [Boston], Vol. II, Part II, pp. 43–45, Charlestown, 1804.

RICE, HOWARD, C., Jr. *The Rittenhouse Orrery.* Princeton's eighteenth-century Planetarium 1764–1954.

See also COHEN, I. BERNARD, *Some Early Tools of American Science.* An account of the Early Scientific Instruments and Mineralogical & Biological Collection in Harvard University, 1950.

Chapter 4

The telescopic appearances
of the components as seen
from the Earth and
Earth-launched space craft

Appearance, wait for me a little,
Let me see what you are, and what you represent.
EPICTETUS

The inferior planets Mercury and Venus, and the trans-Saturnian planets Uranus, Neptune and Pluto, are very difficult objects for telescopic observation from the Earth. Much has been written about the faint markings to be seen on Mercury and Venus, but it is difficult to be certain of its reliability and little has any real lasting value to astronomy. I offer no photographs of the trans-Saturnian planets, few are available and those that are have no aesthetic content. An optical measurement which is not as simple as it might appear gives the apparent diameter of a planet. Turbulence in the earth's atmosphere, distortions of the eye and the telescope, the finite width of the micrometer wire, and darkening of the planet's limb by scattering in its atmosphere lead to uncertainties of the order of $\pm$ 0·1″ in the apparent semidiameter of a planet—amounting to uncertainties of the order of $\pm$ 20 km. in the radius of a terrestrial planet or $\pm$ 200 km. in the radius of a major planet.

The difficulty of observing Mercury, especially in high latitudes, is too well known to require any more than a line. The best conditions are obtained when the planet's elongations occur at the same time as it enjoys a high declination.

The planet is always much brighter in the gibbous phase than in the crescentic. The measurement of its disc is made difficult by the phase effect, because of which the diameter is taken as coincident with the line of the cusps. The Flagstaff Observatory made extensive measurements in 1898 and the planet's diameter was found to be 3400[1] miles in place of the more usually accepted 3000 miles. The angular diameter is, how-

[1] Sharonov gives 5140 km. (3187 miles) on a weighted mean angular value of 7″·09 $\pm$ 0″·05 or 0·403 Earth diameters and criticizes the value of 6″·68, so often quoted in popular works that is based on Leverrier's values solely from transit observations.

ever, generally given in seconds of arc and these have been tabulated by Sharonov,[2] but he does not include the Flagstaff measurements.

The observing of Venus is also difficult. When the planet has a large angular diameter it is in the crescentic phase, and as soon as the phase is favourable to the detection of surface features its distance from the Earth is great.

There is a phase effect[3] known as the Schröter Effect,[4] which refers to the subjective observation of the planet, when the observed phase is found not to agree with the theoretical phase.

The determination of the angular diameter is again tabulated by Sharonov. From the available data he thinks the true diameter of Venus lies between 12,000 km. (7440 miles) and 13,000 km. (8060 miles). A critical analysis by Rabe gave an angular diameter of 8"·70 for the weighted mean value and this corresponds to a linear diameter of 12,620 km. (7824 miles) or 0·989 Earth diameter.

The planet Mars exhibits clear surface markings. The Equatorial diameter in seconds of arc is tabulated by Sharonov.

It is to be made clear that the planet can satisfactorily be observed only for a few months before and after opposition, and these periods occur at intervals of two years two months. In these few months the planet makes about one sixth part of its revolution about the Sun, and some fifteen years are needed to follow one cycle of its seasonal fluctuation. The great eccentricity of its orbit puts the planet in favourable and unfavourable distances from Earth, indeed, the distance varies between 63 and 35 million miles.

Huygens is the first to have produced drawings of the planet's surface features c. 1659.

A map of the main surface features is shown (pl. 49). The polar caps are clearly seen in even a modest glass. The Mariner 6 and 7 close-up photographs of Mars[5] show that the polar caps have a ragged edge, already noticed from telescopic observations, and this is thought to be due to differences in the height of the cap. Radar measurements suggest that regions of the Martian equator differ in height by as much as 8·3 miles. Collinson[6] has given a detailed description of the surface markings.

Jupiter's disc is one of the most rewarding for telescopic study and may justly be called a very beautiful object. A modest magnification of 45 diameters increases the size of the disc to that equal to the naked-eye view of the Moon.

[2] Sharonov, V. V., *The Nature of the Planets* (Moscow 1958).
[3] See Warner, B., *The Phase Anomaly of Venus*, J.B.A.A., *73*, 65 (1962).
[4] *Schroter Effect*. A term advanced by J. Hedley Robinson, see J.B.A.A., *81*, 181 (1971).
[5] Mariner 6 and Mariner 7 flew past the Planet Mars on 1969 July 31 and August 5 respectively; see Miles H., J.B.A.A., *80*, 107 (1970). Mariner 9 went into orbit about Mars, arriving 1971 Nov. 12, and a descent capsule of the Soviet spacecraft Mars Three made a soft landing on the planet 1971 Dec. 2.
[6] Collison, E. H., 'The Planet Mars', B.A.A. reprint no. 2 (1956).

The disc of the planet is crossed by a multiplicity of bands parallel to the planet's equator and these are divided into dark belts and light zones. In the South Tropical Zone lies the famous Great Red Spot. For a detailed understanding of the planetary surface see Peek, B. M., *The Planet Jupiter*, London 1958.

Saturn without doubt is the most beautiful of the planets. The disc is sometimes marked with belts and spots, but the features are difficult to detect, especially owing to the brightness and shadows of the rings.

The changes, in the surface features of the planets that exhibit them, are best studied over a protracted period with the same instrument, provided it has an appreciable aperture, and the seeing is good. The student may care to consult the beautiful photographs recorded over some 56 years by Slipher[7] with the 24 in. Lowell refractor and the beautiful drawings[8] recorded over some thirty-five years with refractors of at least 29 cm. aperture at the French and North African stations of the Plateau de Revard (N. 45°40′55″ E. 3°38′23″ 1550 m. altitude). Sétif (N. 36°1′10″ E. 3°4′25″ 1113 m. altitude), Laghouat (N. 3°48′2″ E. 0°32′32″ 735 m. altitude), Massegros (N. 44°18′2″ E. 0°50′ 900 m. altitude), and Toury (N. 48°11′ W. 0°22′ 136 m. altitude). More recent studies of the changes in planetary surfaces are recorded in the section reports published in the *Journal of the British Astronomical Association*.

[7] Slipher, E. C., 'A photographic study of the Brighter Planets' (1964), Lowell Observatory Flagstaff Arizona and National Geographic Society Washington, D.C. Congress Cat. No. 64–18807.
[8] *Observations des Surfaces Planetaires Observatoire Jarry Desloges*, 10 vols (Paris 1908–46), (R.A.S. Library QB601).

Chapter 5

The planets and their satellites

*Now go, write it before them
in a table and note it in a book.*
ISAIAH

MERCURY

Mercury is the smallest major planet and the nearest to the Sun. Its symbol is ☿. Its distance in terms of the Earth's distance from the Sun (the astronomical unit[1] a.u.) is 0·387 a.u. For long it was thought that the rotation of Mercury[2] was similar to that of the Moon, in having its period of rotation equal to that of its orbital revolution. This was based on Schiaparelli's researches of 1882. It is now known that the period of rotation is 58·7 days and that the period of revolution about the Sun (the orbital sidereal period) is 87·969 days. The resolution of this difficult problem was achieved in 1965 by observers at Cornell University, using its 1000 ft. diameter radio-radar telescope at Arecibo Ionospheric Observatory, Puerto Rico. There are reasons of dynamics for expecting a rotational period of $\frac{2}{3}$ of the orbital period, that is 58·646 days. The long standing puzzle of why the rotational period should have been thought to equal the orbital period is now seen to lie in the fact that three synodic periods offer the same face of the planet to the view of the observer on Earth, and the planet is then at the same phase, combined with the fact that after this interval the planet is also again at favourable apperances, and all this with a complete absence of any visibly distinctive markings on the planet's surfaces to guide astronomers. These facts taken together not unnaturally led to the persistent erroneous view that Mercury always turned the same face to the Sun. The discovery[3] of the

---

[1] The astronomical unit is 149,600,000 km. or 92,957,209 miles.
[2] Schröter estimated the rotational period to be 24 h. 5m. 30s. The famous *Handbook of the BAA* recorded it 'incorrectly' up until 1967 as do all astronomical textbooks prior to that date. e.g. Barlow & Bryan, *Elementary Astronomy*, 5th ed., 1946, give the rotational and revolutionary periods as equal at 88 days. Even books published in 1970 copy the wrong figures, for example *Handbook for Planet Observers*, Roth, Faber & Faber (1970).
[3] See *Nature*, *206*, 1240 (1965), *208*, 375, 575 (1965); *Science*, *150*, 1717 (1965), Smithsonian Institution Astrophysical Observatory, Special Report No. 188R (1965 Oct. 13).

true state of affairs is an achievement of the first rank for the large sophisticated instrument and its operators at Arecibo.

The mass of Mercury is so small that its value was long in doubt. Leverrier[4] put the ratio of the mass of Mercury to that of the Sun as 1:3,000,000. It is now given as 1:6,200,000, which is approximately one half of the original estimated value. The latest report on the point from the Lincoln Laboratory[5] gives the mass as 0·055 of the Earth's mass and a mean density of 5·42 with 60 per cent of the mass within the core.

The most interesting phenomenon connected with Mercury is that of the occasional transit over the disc of the Sun at the time of inferior conjunction. Kepler[6] remarked on this and skilfully predicted in 1629 from the Rudolphine Tables the transit of 1671 November 7, witnessed by Gassendi at Paris in that year.

These phenomena occur only at the time when the planet is near one of its nodes, and the Sun and the Earth come into a satisfactory alignment, thus they occur in May or November. The average interval between these events is $7\frac{2}{3}$ years. The transits are generally repeated in a cycle of 46 years,[7] during which eight occur in May and six in November. The transit[8] for 1970 occurred on May 9, 8 h UT (i.e. Universal Time): others will occur as follows:

that of 1973 on Nov. 10 11 h UT
1986 ,, Nov. 13 4 h ,,
1993 ,, Nov. 6 4 h ,,
1999 ,, Nov. 15 22 h ,,
2003 ,, May 7 8 h ,,
2006 ,, Nov. 8 22 h ,,
2016 ,, May 9 15 h ,,

A perplexing problem is offered by the secular motion of the perihelion of Mercury's orbit. In 1845 Leverrier found that this motion, when derived from observations of the transits, was greater by 35″ of arc per century than it should be from the gravitational influences of all the other planets, according to the accepted Newtonian gravitational theory. In 1859 S. Newcomb evaluted it at 41·2″ ± 2·1″/century. In 1947 G. M. Clemence evaluated it as 42·6″ ± 0·9″/century. The difficulties were

4 See Grant, R., *History of Physical Astronomy*, p. 125 (1852).
5 See *The Observatory*, *89*, 48 (1969 April).
6 Kepler, *Supplement to Vitellionem*.
7 Other periods also exist of 7; 13 and 33 years but the 7 year period does not arise in May transits. The cause lies in the near-commensurabilities of the planetary periods.
29 sidereal periods of Mercury = 7 sidereal years − 5·69 d.
54 sidereal periods of Mercury = 13 sidereal years + 2·01 d.
137 sidereal periods of Mercury = 33 sidereal years − 1·67 d.
191 sidereal periods of Mercury = 46 sidereal years + 0·34 d.
8 A list of transits from 1605 to 2108 is given in 1970 *Yearbook of Astronomy*; see also Porter, J. G., *J.B.A.A.*, *80*, 182, (1970). Proctor gives a useful list in his *Old and New Astronomy* (1892).

such that in the late nineteenth century Asaph Hall suggested a modification of Newtonian gravitation from

$$\int = \frac{mm_1}{d^2} \text{ to } \int = \frac{mm_1}{d^{2 \cdot 00000016}}$$

to account for the discrepancy, but this was so arbitrary and repugnant to physicists as to be rejected. On the view that a missing mass would answer the problem best, Leverrier showed that an inferior planet, in the plane of the ecliptic, at a distance of one half of Mercury's distance from the Sun, would satisfy the equations.

No intra-Mercurial planets have, however, ever been found (despite the claims by Lescarbault),[9] and now it is established that the shift of the perihelion point is satisfied by Einstein's law of gravitation. This point was confirmed by Madge G. Adam at the discussion of the present state of relativity 1962[10] from which it appears that the predicted excess of the advance of the perihelion over the Newtonian value is

$$12\pi^2 \, a^2 \, c^{-2} \, T^{-2} \, (1 - e^2)^{-1} \text{ of a}$$

revolution per revolution (i.e. 43″/century)
where T is the period

    a   the semi major axis of the ellipse
    e   the eccentricity
    c   the velocity of light.

More recently Dicke has formulated a scalar-tensor gravitational theory as a rival to general relativity, and Roxburgh[11] has enlarged and criticized Dicke's[12] view, based on measurements of the Sun's body, which give a difference in the equatorial and solar radii of 35 km. This distortion of the Sun's body from a sphere is said to produce a perihelion advance of the planets. For Mercury the scalar-tensor theory of gravitation gives 39″ of arc per century due to relativistic effects and the difference of 4″ (43″ − 39″) is accredited to the solar oblateness[13] and con-

---

[9] See Gould, R. T., *Oddities*, 2nd ed. (London 1944), ch. X, The Planet Vulcan, pp. 194–203.
[10] See *Proc. Royal Soc.*, A 270, p. 298, 1962.
    Mercury has at all times afforded much occupation to astronomers; because to observe that planet is a matter of considerable difficulty. The great Copernicus died without ever having seen it, and therefore could believe only in its existence. The celebrated Maestlinus, the tutor of the immortal Kepler, used to say, that this planet was calculated only to expose astronomers to the danger of losing their reputation; so that, when he knew of any one employed in tracing out its intricate course, he would advise him to employ his time on some better subject. Riccioli calls Mercury a false deceitful star (*sidus dolosum*) the eternal torment of astronomers, which eluded them as much as the terrestrial Mercury did the alchymists. Lalande, that respectable veteran of astronomy, after being forty-six years engaged with this celestial rebel, pronounced its course *une orbite inextricable*. From *The Percy Anecdotes*, 6, 144 (1826).
[11] Roxburgh, I. W., 'Implications of the Oblateness of the Sun', *Nature* 1077 (1967).
[12] Dicke, R. H., *Nature*, 202, 432 (1964).
[13] See p. 146.

sequently the perihelion advance[14] would appear to be better catered for in the scalar-tensor theory.

Radio waves in the 3·45 and 3·75 cms. wavelength band have been detected by Howard *et al.*,[15] to be emitted from Mercury.

VENUS

Venus is the second major planet in order of distance from the Sun. Its mean distance from the Sun is 0·723 a.u.; it has a mass of 0·815 times that of the Earth, and a mean density of 5·23[16] It closely resembles the Earth in size. Its symbol is ♀.

At inferior conjunction Venus approaches the Earth more closely than any other major planet, but at that position it is practically invisible to the eye of an observer on Earth, since it comes between the light of the bright solar disc and the Earth.

To the eye it appears to oscillate from one side of the Sun to the other, the complete period of oscillation being 1·6 years. At the greatest elongation it is about 45° from the Sun. When east of the Sun it is an evening 'star' and when west a morning 'star'. It is often of exceptional brilliance and has been thought by many to be the star of Bethlehem.[17] To the ancients it was known as two separate bodies, Hesperus and Phosphorus.

The rotation of the planet has, as has that of Mercury, posed a problem of severe obtuseness.

In 1726 Bianchini inferred a period of 24 days. The great Herschel could not, even with his superior instruments, detect any significant detail on its surface to yield a value. Schiaparelli, after exhaustive studies, thought it was a case in harmony with that of the Moon and the Earth, in that Venus always presented the same face to the Sun. Later researches led astronomers to favour a period of revolution about the Sun of 224 days and a period of axial rotation of between 40 to 225 days.[18]

[14] It should be noted that the observed effect is 120 times greater than 43″/century and that the residual value of 43″ is only obtained after the subtraction of the general precession of the equinoxes and the planetary perturbations. See Dicke, R. H., *The Theoretical Signficance of Experimental Relativity*, 1964, p. 27.

[15] Howard, W. E., Barret, A. H., Haddock, F. T., *Astron. J.*, *66*, 287, (1961); *Ap. J.*, *136*, 995 (1962).

[16] See *The Observatory*, *89*, 48 (1969 April).

[17] Matthew 2⁹ records that the star *went before them and stood over* where the young child lay. Astronomically the evidence is slender indeed. That a star should go before a traveller—that is to say share in his terrestrial wandering—is counter to all experience; stars and planets are notorously so remote that they stand fixed and unconcerned with the perambulations of men. That the star physically went before the Wise Men and stood over the place where the young child lay is not a natural event; it lies in the sphere of the miraculous and is consequently beyond the recall of either positional or mathematical astronomy. The position of the Sun, Moon and the five naked eye planets is known to a fine fraction of a degree for the period 601 B.C. to A.D. 1649 from Tuckerman's Tables. See Vols. 56 and 59 of the Memoirs of the American Philosophical Society. Vol. 56 covers the period 601 B.C. to A.D. 1. Vol. 59 covers the period A.D. 2 to A.D. 1649, and was computed on an I.B.M. 7094 computer in 12 hours of computer time.

[18] See B.A.A. Handbook for 1953. Barlow & Bryan, *Elementary Mathematical Astronomy*, revised by H. Spencer Jones, 5th Edition (1946).

This fiction was held by all competent authorities until the spring of 1964, when the results of the radar investigations of Venus by the 1000 ft. diameter Arecibo telescope at Puerto Rico, were published at the I.A.U. meeting in Hamburg.[19] The rotation is, to everyone's surprise, *retrograde* with a period of 247 ± 5 days.[20] The revolution about the Sun is effected in 224·701 days and the direction in the orbit is direct. Venus suffers a small advance of its perihelion, as does Mercury, but it is very small, the observed residual is 8″·4 ± 4·8.

The question of whether Venus has any moons has periodically excited astronomers, but none has ever been discovered.

Venus transits[21] the solar disc in the same manner as Mercury and these events occur in pairs, eight years apart. After such a pair there is an interval of more than a century before another pair, and these take place at the opposite node of the orbit. A period of 243 years gives an extremely close approximation to recurrences of circumstances. The transits, according to Newcomb, are as follows:

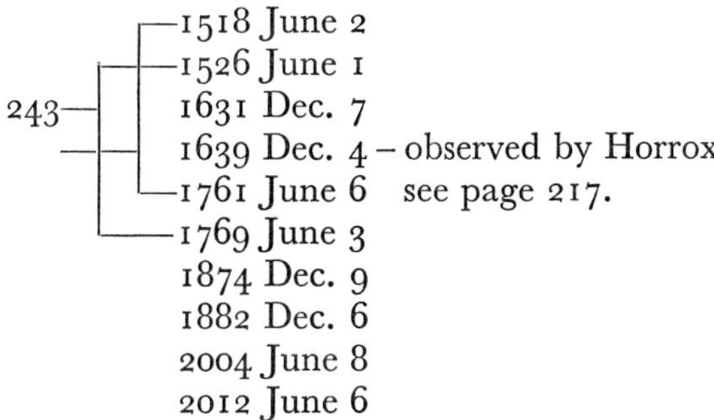

1874 Dec. 9
1882 Dec. 6
2004 June 8
2012 June 6

The emission of radio waves from Venus has been investigated by *inter alia* Roberts[22] and Lynn *et al.*[23] It is thought that the emission comes from the deep layers of the atmosphere of Venus. The data suggests the the radiation belts of Venus do not contribute any important part to the generation of these waves.

EARTH

The Earth is treated in the chapter devoted to the Earth/Moon system (see chapter 10).

[19] See *Sky and Telescope, 28,* 4.73, 341 (1964). Carpenter, R. L., Studies of Venus by Centimeter Wave Radar, *Astron J. 69,* 2.
[20] Firsoff believes this rotation to be erroneous in that the radar method measures not the rotation of the planet's surface but a belt of plasma moving counter to the planet. Firsoff, V. A., *J.B.A.A., 80,* 303 (1970). In 1961 Boyer and Camichel discovered the retrograde sense of the rotation from the clouds; see *Nature, 227,* 477 (1970), *Ann Astrophys 24,* 531 (1961).
[21] See Porter, J. G., 'Transits of Mercury and Venus', *J.B.A.A., 80,* 182 (1970). (See p. 216).
[22] Roberts J. A., *Planet and Space Science, 11,* 221 (1963).
[23] Lynn, V. L., Meeks, N. L., Schigian, M. P., *Astron. J., 68,* 284 (1963).

MARS

Mars is the fourth planet in the order of distance from the Sun, and is the first of the superior planets, being further from the Sun than is the Earth. Its symbol is ♂.

The planet revolves about the Sun in 687 days approximately (686·98 d), and rotates on its axis in 24 hours 37 minutes 23 seconds. Its rotation has not been in doubt since Cassini in Italy brought the planet under observation in 1666.[24] The eccentricity[25] of its orbit, 0·0933773 ($\frac{1}{11}$ th), is greater than that of any other superior planet. At perihelic oppositions it is markedly nearer to the Earth than at aphelic oppositions, the distances varying from 35×10^6 to 63×10^6 miles. The mean interval between oppositions is 2 years 49·5 days but owing to the eccentricity the actual excess over two years ranges from 36 days to more than 75 days. From a close study of the orbits of Mars and the

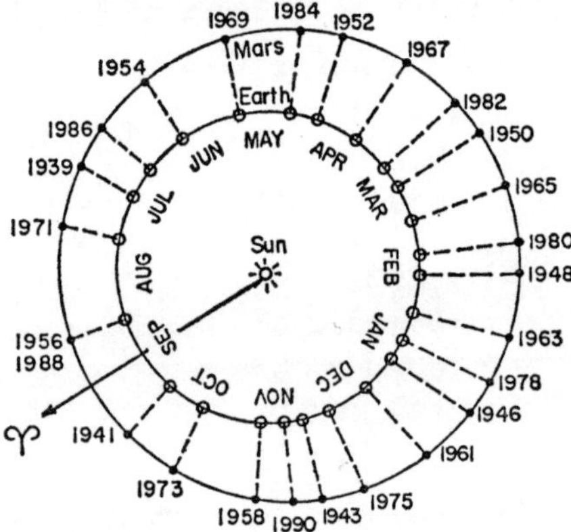

Fig. 19. Oppositions of Mars from the Earth, 1939–1990. The relative distances are shown by the lines joining the orbits.

Earth, it can be seen that the points of opposition travel round the orbits in about 16 years, so that oppositions at perihelion, when Mars is at its nearest, occur at intervals of 15 or 17 years (Fig. 19).

The great controversy over the Martian surface[26] features is now

[24] Hooke studied the planet in the same year and communicated a note to the Royal Society (*Phil. Trans.* No. 14) but he was doubtful concerning the exact period and he placed it between 12 and 24 hours. Cassini gave it as 24 h. 40 m. Miraldi confirmed it in 1704/1719 at 24 h. 39 m. (*Mem. Acad. des Sciences*, 1720.) The first to see the Martian spots was Fontana in 1636.

[25] Eccentricity expresses the shape of an elliptical orbit and is defined by the ratio of the distance between the foci to the longest diameter. All ellipses have eccentricities between zero and one. The eccentricity of a circle is zero, that of a parabola is one and that of an hyperbola greater than one.

[26] See Schiaparelli, '*Osservazioni astronomiche a fisiche sull asse di rotazione e sulla topografia del planeta Marte*', (Rome *c.* 1877). Lowell, P., *Mars and its Canals* (1906); *Mars as the Abode of Life* (1909).

terminated with the close approach of the Mariner spacecraft[27] and the receipt on Earth of pictures of the surface of the planet, disclosing none of the legendary canals.

Further details of the planet's surface may be expected in 1971 and 1974 with the planned flights of Mariner orbiting spacecraft.[28]

The orbit of Mars lies outside that of the Earth, and Mars is never seen from Earth at the crescentic phase, but at the quadratures it is appreciably gibbous.

Mars has two small moons (or satellites), Phobos and Deimos, which revolve rapidly about the parent planet. These have been used to provide data for an accurate calculation of the oblateness of the planet. The best value so far leads one to believe that the polar radius is less than the equatorial radius by about 17 km.[29] The satellites are discussed in greater detail below (See page 91).

Mars is known to emit radio waves in the 3 cm. wavelength band. The first observations of this phenomenon were made by Mayer *et al.*[30]

JUPITER

Jupiter is the largest planet of the Sun's family and his size is so great that it exceeds the collective mass of all the other planets in the proportion of five to two. His symbol is ♃. The body of the planet is oblate to such a degree that its ovoid shape is readily visible to the eye, when aided by a good glass. Needless to say, the speed of rotation of the planet is high.

The orbit of Jupiter has a mean distance from the Sun of 483×10^6 miles. The eccentricity of his orbit is 0·048, about $\frac{1}{20}$ th, so that his maximum and minimum distances vary considerably between about 504×10^6 and 462×10^6 miles. His orbital period of revolution about the Sun is 11·86 years (4332·59 days) and the Jovian day, the shortest for any planet, is dispensed in the remarkable time of 9 h. 50 m. 30 s. The rotation of this great planet was inferred by Kepler, and the rotation of the outer surface markings in the atmosphere of the planet, witnessed by Hooke,[31] and later recorded by Cassini in *c.* 1665. The rotation is today considered in two separate rotational periods.[32] System I for the equatorial region of the visible atmosphere within 10° of the equator, rotating in 9 h. 50 m. 30 s., and System II rotating in 9 h. 55 m. 41 s. Radio methods have led to the use of a System III, with a period in the polar regions of 9 h. 55 m. 41–12 s.

Jupiter has twelve satellites and these disport themselves about the

[27] The best pictures of the Martian surface have been obtained by Mariner 6 at an altitude above the planet of 2,150 miles covering an area of view of some 560 miles by 430 miles. Elevations of 15 km. have been detected by radar; see *Nature*, *223*, 562 (1969); *224*, 751 (1969).
[28] See *Nature*, 222, 1023 (1969); *234*, 67, 335 (1971); *New Scientist 52*, 11 (1971).
[29] *Nature*, *224*, 751 (1969).
[30] Mayer, C. H., McCullogh, T. P., and Sloanaber, R. M., *Ap. J.*, *127*, 11 (1958).
[31] *Phil. Trans. of Roy Soc.* No. 1.
[32] See Peek, B. M., *The Planet Jupiter* (London 1958).

body of the great planet. Their movements are considered in greater detail below. The four large satellites in the equatorial plane are of a special interest to the historian of science, since they are the first celestial bodies to have been discovered with the telescope, in the skilled hands of Galileo (1610).

Jupiter is known to emit continuous radio waves at a wavelength of 3·15 cm. and this was first observed by Mayer *et al.*[33, 34] These waves may come from the region of Jupiter's atmosphere adjacent to the visible cloud layer. In more recent years Jupiter has been found to emit sporadic radio waves.

Burns[35] thinks that the sporadic emissions are associated with the magnetic properties of the two large moons Io and Europa. Other authorities[36] associate the decameter radio signals with Io, which travels in the analogue of the inner Van Allen radiation belt, surrounding Jupiter, and gives rise to a plasma wake, which streams ahead of it by about 130,000 km. and stimulates the radio emission from the Jovian atmosphere. The radio signals are said to be circularly polarized.[37, 38] Jupiter's magnetosphere is discussed by Gledhill[39] and the Jovian continuum radiation based on the data gathered by Radio Astronomy Explorer satellite (RAE–1) launched on 1968 July 4 into a 5860 km. retrograde orbit circular in shape and inclined at 59° with a period of 3 h. 44 m. is given by Weber and Stone.[40]

The outstanding feature of the transient phenomena in the ever changing patterns of the Jovian atmosphere, visible from Earth, is the Great Red Spot of the southern hemisphere. It has an elliptical shape having its long axis in zenocentric latitude 22°S. It occupies 30° of longitude and 10° of latitude. The period of the spot has varied over the past 100 years and its range lies between the limits of 9 h. 55 m. 3 s. and 9 h. 55 m. 44 s. The average period is 9 h. 55 m. 38 s. The spot has perplexed astronomers for many years.[41] A recent suggestion of considerable interest has been presented by Hide.[42] He points out that the theory

[33] Mayer, C. H., McCullogh, T. P., and Sloanaker, R. M., *Ap. J.*, *127*, 1; *Proc. I.R.E.*, *46*, 260; *Ap. J.*, *127*, 11 (1958).
[34] Mayer, C. H., *Astron. J.*, *64*, 43 (1959).
[35] Burns, J. A., *Science*, *159*, 971 (1968).
[36] See Marshall, L., and Libby, W. F., 'Stimulation of Jupiter radio emission by Io', *Nature*, *214*, 126–8 (1967); see also McCullick, P. M., *Planetary* and *Space Science*, *19*, 1297 (1971).
[37] See *Nature*, *224*, 1011 (1969).
[38] For a detailed consideration of the radio emission of Jupiter as understood up until 1965 see Zheleznyakov, V. V., *Radio Emission of the Sun and Planets*, translated by Massey, H. S. H. (Pergamon Press 1970). More recently Kemp *et al.* have discovered circular polarization of scattered light from Jupiter. See *Nature*, *231*, 169 (1971); *232*, 165 (1971).
[39] Gledhill, J. A., *Nature*, *214*, 155 (1967).
[40] Weber, R. R., and Stone, R. G., *Nature*, *227*, 591 (1970).
[41] Phillips, T. E. R., 'Jupiter', 14th ed. of *Encyclopaedia Britannica* (1929).
[42] Hide, R., 'Origin of Jupiter's Great Red Spot', *Nature*, *190*, 895 (1961). See also Reese, E. J., *Icarus*, *14*, 343 (1971), for a summary of the 1970 observations with the 61 cm. telescope of New Mexico State Observatory. The spot dimensions are given as 27,800 km. by 13,800 km. For the recent theory that the spot is a Cartesian diver see *Icarus*, *14*, 319 (1971).

of hydrodynamics suggests that Jupiter rotates so rapidly, that the effect on the general circulation of the atmosphere of quite a shallow 'topographical feature' of the 'solid planet' will be attenuated very slowly with height, and thus the 'feature' will make its presence manifest at the level of heavy cloud at the top of the atmosphere. Specifically, the atmospheric flow cannot surmount the 'feature' if h, the height of the 'feature' above (or depth below) the general level, exceeds:

$$h \equiv ad\ (U/L\ \Omega)$$

where a is a number of order unity, U is a characteristic wind speed in the vicinity of the 'feature', L is a typical horizontal dimension of the 'feature', d is the depth of the Jovian atmosphere and Ω is the angular speed of rotation of the solid planet. When this criterion is satisfied, there will be a stagnant column of fluid stretching, not necessarily in a vertical direction, from the 'topographical feature' at the bottom of the atmosphere all the way to the top of the atmosphere. This column of fluid will partake of the motion of the 'topographical feature' and virtually no mixing will occur between the fluid within the column and that without. Effects of this kind were predicted and demonstrated long ago by Sir Geoffrey Taylor.[43]

If L is 25,000 km., d is 12,000 km., U is 2 m/sec. and Ω is $1\cdot5 \times 10^{-4}$ rad./sec., according to the equation, h is 6 km.

The equation shows that h is inversely proportional to L, and suggests a possible explanation of the uniqueness of the Great Red Spot, for if there were no 'topographical feature' having L in excess of say 40,000 km. and h nowhere substantially exceeding the critical value for the Spot, other 'topographical features' would only produce spots on those occasions when U has a sufficiently low value.

This appears to be borne out by the occasional appearance of short-lived spots, slightly smaller than the more permanent Great Red Spot.

SATURN

Saturn is the sixth major planet in the order of distance from the Sun and the third superior planet. His symbol is ♄. It was the most distant planet known up until 1781, when Uranus was discovered by the elder Herschel. Its period of revolution about the Sun is $29\frac{1}{2}$ years or to be more precise 10,759 days. The interval between oppositions is from 12 to 13 days in excess of a year ($378\cdot09$ d).

The period of axial rotation is very short 10 h. 14 m. and this was first given by Herschel[44] after a long series of observations between 1793 and 1794 involving some 100 rotations of the planet. The figure of the

---

[43] Taylor, G. I., *Proc. Roy. Soc.*, A *104*, 213 (1923).
[44] Herschel estimated the rotation to be 10 h. 16 m. *Phil. Trans. of Roy. Soc.*, p. 62 (1794).

planet is markedly elliptical and this also was measured with care by Herschel, using a filar micrometer fitted to his reflector of 20 ft. focal length.[45] The polar diameter is 107,700 km. and the equatorial 119,300 km. The planet is surrounded by a spectacular ring system and ten satellites, forming the most beautiful planetary system in the Sun's family and a never-to-be forgotten sight in a good glass.

The ring system caused endless difficulty to early investigators.[46] Galileo saw the planet triple, and was 'unfortunate' enough to see it later at that part of its great orbit about the Sun when the Earth dweller sees the ring plane edge on and thus it disappears from view; this happens approximately every fifteen years.[47] Galileo could not explain this phenomenon: 'What', he remarks, 'is to be said concerning so strange a metamorphosis? Are the two lesser stars (the outer parts of the rings) consumed after the manner of the solar spots? Have they vanished and suddenly fled? Has Saturn perhaps devoured his own children? Or were the appearances indeed illusion or fraud with which the glasses have so long deceived me—I do not know what to say in a case so surprising, so unlooked for and so novel.'[48]

It was left to Huygens to explain the mystery in the famous logogriph

aaaaaaa ccccc d. eeeee g. h. iiiiiii llll
mm nnnnnnnnn oooo pp q rr s ttttt uuuuu

which when correctly resolved was seen to stand for:
'annulo cingitur tenui plano nusquam cohaerente ad eclipticam inclinato'—or in English: 'the planet is surrounded by a slender flat ring everywhere distinct from its surface and inclined to the ecliptic.' Nothing could be more succinct or pleasingly expressed.

Huygens believed the rings to be thin, flat and solid, and this view even survived the discovery by Cassini in 1675[49] of the division within them which has ever afterwards carried his name. It was the revelation of the semi-transparent crepe ring that caused a thorough revision of the facts, and it was Clerk Maxwell, who announced the answer in his Adams Prize Essay of 1857, in which he confirmed the speculative suggestion of the Cassinis and Thomas Wright, by showing that a system of rings could be stable only if it consisted of discrete particles. In 1867 Kirkwood showed that the divisions in the rings are due to perturbations of the particles by Saturn's larger satellites, principally Mimas. The explanation lies in the art of celestial dynamics, from which it appears that a satellite sweeps clean a zone in the myriad of particles in the ring system at a distance corresponding to a simple fraction of its period.

[45] *Phil. Trans. of Roy. Soc.*, Part I, p. 17 (1790).
[46] Shapley, D., 'Pre-Huygenian Observations of Saturn's Rings', *Isis, 40*, pp. 12–17.
[47] This point is explained in greater detail below.
[48] *Systema Saturnium* (Hagae 1659).
[49] *Anc. Mém. Acad. des Sciences*, Tome X, p. 583.

The Cassini division is at a distance that corresponds to particles having a period of revolution equal to half the period of Mimas, one-third that of Enceladus, one-quarter that of Tethys and one-sixth that of Dione. The inner ring and the crepe ring have similar disturbances from the commensurabilities of these periods, *mutatis-mutandis*.

For a more full treatment of this interesting subject, which goes back to Kepler's third law $p^2 \propto a^3$, and for the influence of the satellites, Rhea and Titan, on the ring system, see Antoniadi's[50] studies of the subject. The dimensions of the ring system are as follows:

		Miles	Ratio	km.
Ring A	outer	169,199	1·0000	272,300
	inner	148,880	0·8801	239,600
Ring B	outer	145,525	0·8599	234,200
	inner	112,530	0·6650	181,100
Ring C	inner	92,771	0·5486	149,300
Saturn	equatorial diameter	75,100	0·4440	120,862
	polar diameter	67,200	0·3973	108,148

A continuous emission of radio waves from Saturn in the 4–4·5 cm. wave band has been detected by Drake and Ewen.[51] It is thought to come from the regions of the atmosphere adjacent to the cloud layer.

URANUS

Uranus is the seventh major planet in order from the Sun, at a distance of 19·182 a.u., and is the first of the trans-Saturnian planets. His symbol is ♅. It was the first major planet to be added to man's knowledge of the Sun's family by telescopic observation. The planet was not known to the ancients. It was 'discovered' by the elder Herschel, who thought it to be a comet, on 1781 March 13.[52]

Strange to relate the planet is visible to the unaided human eye as a sixth magnitude star; provided one knows where to look for it. It was seen long before its true nature was known. Flamsteed had observed it as a fixed star and Lemonnier also made eight observations of it in 1768–9. It was also seen, and its position accurately recorded, by Tobias Mayer[53] using a quadrant and pendulum clock at the beginning of 1756. It is said to have had its position measured on twenty-three occasions before 1781, all the observers taking it to be a sixth magnitude star.[54]

[50] Antoniadi, *L'Astronomie*, *44*, 163. See also *J.B.A.A.*, *25*, 114 (1915) *J.B.A.A.*, *26*, 63 (1916).
[51] Drake, F. D., Ewen, H. T., *Proc. I.R.E.*, *46*, 53 (1958).
[52] The new planet's true nature was made known by Lexell. Laplace communicated it to the Academy of Sciences in 1783 Jan. The name Uranus was proposed by Bode; See Herschel, 'Account of a comet', *Phil. Trans. Roy. Soc.*, *71*, p. 472.
[53] Forbes, E. G., 'The Life and Work of Tobias Mayer', *Quat. J.R.A.S.*, *8*, No. 3 (September 1967).
[54] *Sky and Telescope*, *37* 14 (1969).

Grant remarks that its discovery marks the commencement of the long series of brilliant discoveries and sublime speculations, which adorned the astronomical career of Sir William Herschel—and that it was the legitimate reward that might be expected eventually to crown the exertions of an astronomer, who continued from night to night, with unwearied enthusiasm, to explore the heavens with optical appliances, which owed their exquisite character solely to the resources of his own genius.

The orbit has a large eccentricity (0·047) so that the actual distance from Uranus to the Sun varies by 270×10^6 km. (168×10^6 miles). Five moons revolve about the planet in a system of remarkable 'regularity' as judged by the smallness of the orbital eccentricities and inclinations; yet strange to relate this beautiful system has its plane of symmetry inclined $97° 53'$ to the plane of its orbit about the Sun. In consequence the revolution of the satellites and the planet's rotation, which are in the same sense as all the other planets are seen in the 'retrograde'[55] sense, when viewed from the plane of the ecliptic. The rotation[56] on the axis is fast, the day on Uranus is dispensed in 10 h. 49 m. The revolution about the Sun is slow taking 84 years, or more accurately 30,685 days.

On 1949 March 25 the planet completed its second sidereal revolution round the Sun since its discovery.[57]

NEPTUNE

Neptune is the eighth planet in distance from the Sun. It has a mean distance of 30·057900 astronomical units and revolves in its great orbit about the Sun in about 165 years (60,190 days). His symbol is Ψ. His period of axial rotation was finally settled by spectroscopic investigations in 1928 and is made in about 16 hours (15 h. 48m.). Neptune has two satellites, the inner one of which (Triton) moves in a retrograde orbit, that is, in the retrograde sense with respect to all other planetary motions about the Sun. It was originally assumed from this, that the rotation of the parent planet would be found to be retrograde but this is not so, and the anomaly remains.

The plane of Neptune's orbit crosses that of the 'outermost' planet Pluto and conditions can arise, when, for example, Pluto comes to perihelion, under which Neptune may take pride of place in being the outermost body of the Sun's family. These conditions arise according to Taylor[58] in the years 1970 to 1980 and Neptune will hold this distinction for some twenty years thereafter.

Neptune will always retain a special fascination for the student of the

[55] Herschel, *Phil. Trans. Roy. Soc.*, p. 47 (1798).
[56] Lowell & Slipher detected it by spectroscopy in 1912.
[57] Steavenson, W. H., *J.B.A.A.*, *59*, pp. 34–5 (1949).
[58] Taylor, Gordon E., *J.B.A.A.*, *80*, 13 (1969).

solar system in regard to its truly remarkable discovery brought about by the mathematical skill and deep insight of two able mathematicians, each a master of the planetary orbits and in complete ignorance of each other's complex and substantially identical researches.

I speak of John Couch Adams (1819–92) and Urbain Jean Joseph Leverrier (1811–77). Neptune is the only planet to have been discovered by the solution of the inverse problem of perturbations, based on the anomalous movements of another planet (Uranus) from the elements derived (those of Bouvard) for it. The story has been told with consummate skill and care by Smart,[59] and anyone interested in the history of astronomy may turn to his writings for the facts, painstakingly gathered over many years of research. The facts, may, however, be restated with advantage.

The planet was discovered by Galle of Berlin from advice supplied by Leverrier, following a rapid search of the heavens on 1846 September 23. The final solution to the position of the hypothetical planet had, unknown to Leverrier and to the French nation, been anticipated by the young English astronomer and mathematician—Adams, who had arrived at a solution of this inordinately complex problem[60] towards the end of 1843. He communicated his solution to Airy, the Astronomer Royal, in 1845 October and a refined solution on 1846 September 2. Leverrier had also enlisted Airy's aid since the English astronomer had at his command the best optical telescopes for immediate use in settling the question of the hypothetical planet's existence and position. It subsequently transpired that Airy had kept Adams's[61] paper in his desk for eight months with an ineptitude past belief and that he was guilty of showing some discourtesy to this gifted junior who had sought his help.

Airy, with the trust of both mathematicians, now saw the urgency of the situation, and Challis at Cambridge was asked to look for the planet. Again with shocking ineptitude Challis found it on three separate occasions in July to September of 1846, all prior to Galle's discovery; but without recognizing it as the planet. All this only came on an unsuspecting world when Sir John Herschel communicated Adams' researches to the *Athenaeum*. The subsequent controversy, the disbelief of the French, the shocking treatments given to Adams by the Royal Society and the Royal Astronomical Society, are only made palatable, even today, by the gentlemanly stand of Adams, who never once reproached his unappreciative countrymen or showed any of the irascibility of the French. Today everyone is prepared to grant the

[59] Smart, W. M., 'John Couch Adams and the discovery of Neptune', *Occasional Notes of the R.A.S.*, No. 11 (1947 Aug.); *The Observatory*, 66, 378 (1946).
[60] Airy, The Astronomer Royal, informed Bouvard in 1837 'I cannot conjecture what is the cause of the errors, (of Uranus' position) but I am inclined in the first instance to ascribe them to some error in the perturbations. . . . If it be the effect of any unseen body it *will be nearly impossible* ever to find out its place.'
[61] From contemporary pen portraits of Adams and Airy.

discovery to both Adams and Leverrier but the loss to England will never be easily equated with the shabby behaviour of those, who, most of all, should have encouraged one of her most brilliant and gifted sons.[62]

According to Smart, Arago acclaimed Leverrier's discovery of Neptune in 1846 as 'one of the noblest titles of his country to the gratitude and admiration of posterity'. Polanyi says 'no contribution to knowledge could be more useless than was the discovery of this remote planet', and points out that Sir R. S. Ball comments on the fact that Lalande would have discovered Neptune in 1795 if only he had believed what he saw on the 8th and 10th of May in that year. . . . But had he done so how lamentable would have been the loss to science. The discovery of Neptune would then merely have been an accidental reward to a laborious worker instead of being one of the most glorious achievements in the loftiest departments of human reason. (See Ball, Sir R. S., *The Story of the Heavens*, London 1891, p. 288; and Polanyi, M., *Personal Knowledge*, London 1958, p. 182). Now Lyttleton enters the lists with an *ex post facto* analysis, showing clearly that both Adams and Leverrier made incorrect assumptions and failed completely to specify the elements of the unknown planet. He gives the following comparison table

VALUES FOUND BY ADAMS AND LEVERRIER FOR NEPTUNE
COMPARED WITH ACTUAL VALUES

	Adams	Leverrier	Actual
Mean distance from sun, a'	37·2	36·2	30·07
Eccentricity, e'	0·121	0·108	0·0086
Longitude of perihelion, w'	299°	285°	44°
Sun/mass of Neptune	6670	9350	19300
Longitude at discovery	330°·9	327°·4	328°·4

and proceeds to show that the planetary positions and the cosmic machinery were, at the date of discovery, kind to the two illustrious but misguided mathematicians. He then advances in 1968 a subtle route that they may have taken in 1846 (see Lyttleton, R. A., *The Mysteries of the Solar System*, Oxford 1968, also Vistas in Astronomy, 3, 25 (1960)) to have achieved a more satisfactory answer. He can hardly appeal for any greater success than posterity has continued to accord to their genius— but we should not fail to note that provided one succeeds in scientific analysis to enrich the world of appearances, the underlying 'truth' of the analysis is totally irrelevant to our culture.

[62] A contemporary account of this affair, often neglected, is that to be found in Professor R. Grant's famous *History of Physical Astronomy*, 1852, Appendix III, entitled 'Reflections on certain circumstances connected with the discovery of the planet Neptune'.

PLUTO

Pluto, ♇, the last major planet to be discovered, was found by photographic techniques[63] at the Lowell Observatory by Clyde Tombaugh in 1930, a few degrees from the position predicted[64] by Lowell and Pickering many years previously. It is a faint object of fifteenth magnitude. Its mean distance from the Sun is 39·52 a.u.; and its orbit is so large that it requires 250·6 years to make one revolution around the Sun.

The Plutonian day is 6d·390 and the inclination of the orbit to the plane of the Earth's orbit is large, being 17°, with the largest eccentricity (0·2502358) of any of the Sun's major planets. It is this that places part of its orbit within that of the orbit of Neptune. There is no opportunity for a collision however, owing to the high inclination of 17°. Cohen and Hubbard[65] show that the orbit is stable and that no instability results from the fact that the perihelion distance of Pluto is less than the radius of the orbit of Neptune.

The diameter of the planet has long been in doubt, but a size has now been put on it from a close occultation of a faint star on 1965 April 28/29. Two figures have been derived giving the diameter as <5800 km. (3600 miles) and <6800 km. (<4225 miles). The planet is calculated to have a mass of 0·14 times that of the Earth and an average density of 1·4 times that of the Earth. It is in simple terms a planet about the size of Mars with a density twice as great.[66]

THE SATELLITES OF THE PLANETS

No satellites of the inferior planets, Mercury and Venus, are known. It was long held by many that, according to one theory[67] of the origin of the solar system, Mercury could only have a satellite by capture of an asteroid since its periodicity of rotation was (according to then current ideas) equal to that of its revolution about the Sun, but now that this dynamical situation has been shown to be *untrue*, then the satellite hypothesis also must be discarded.

A similar line of invalid argument was adduced for Venus. It is said that any satellites of Venus brighter than visual magnitude 12 would have been discovered by now, and it is estimated that no satellite of Venus larger than 200 km. in diameter can exist.

---

[63] Grosser, M., has given an extensive appraisal of the subject; see his 'The search for a planet beyond Neptune', *Isis*, *55*, 163 (1964). A good account is to be found in *The Griffith Observer*, *16*, No. 1 (Jan. 1952).
[64] The first recorded mention of a ninth planet is due to Dr T. Hussey, a British amateur astronomer in a conversation with Bouvard; see Airy, G. B., *Monthly Notices of the R.A.S.*, *7*, 124 (1846); see also Flammarion, C., 'There is a planet beyond Neptune', *L'Astronomie*, p. 660, (Paris 1884).
[65] *The Observatory*, *85*, 43 (1965).
[66] *Sky and Telescope* (Oct. 1965), 213; (Feb. 1969), 71.
[67] Kuiper's Theory (see Chapter 14).

Planet and Satellite	Discoverer	Date	Instrument	Mean distance from Primary astronomical units	Mean distance	
					(Equatorial radius of planet = 1)	km.
MARS						
I Phobos	A. Hall	1877	26″ refractor, Washington	0·0000625	2·82	9,380
II Deimos				0·0001571	7·06	23,500
JUPITER						
V	E. E. Barnard	1892	36″ Lick refractor	0·0012099	2·531	180,500
I Io	Marius/Galileo			0·0028193	5·911	421,600
II Europa	—	1610	optic tube	0·0044857	9·405	670,800
III Ganymede	—	—	—	0·0071552	15·003	1,070,000
IV Callisto	—	—	—	0·0125845	26·388	1,882,000
VI	C. D. Perrine	1904	36″ Crossley reflector	0·076723	160·8	11,470,000
X	S. B. Nicholson	1938	36″ Crossley reflector	0·078345	166	11,850,000
VII	C. D. Perrine	1905	36″ Crossley reflector	0·078455	165	11,800,000
XII	S. B. Nicholson	1951	100″ Hootver reflector	0·142	297	21,200,000
XI	—	1938	—	0·151	316	22,600,000
VIII	P. J. Melotte	1908	30″ Greenwich reflector	0·157	329	23,500,000
IX	S. B. Nicholson	1914	36″ Crossley reflector	0·158	332	23,700,000
SATURN						
X Janus	Texerau/A. Dollfus	1966/7	82/43″ reflector	0·00106	2·5	149,500
I Mimas	W. Herchel	1789	48″ reflector	0·0012406	3·07	185,400
II Enceladus	—	—	—	0·0015916	3·95	237,900
III Tethys	G. D. Cassini	1684	100′	0·0019703	4·88	294,500
IV Dione	—		136′	0·0025235	6·25	377,200
V Rhea	—		34′	0·0035241	8·73	526,700
VI Titan	C. Huygens	1655	12′	0·0081660	20·2	1,221,000
VII Hyperion	Bond & Lassell	1848	15″/24″ Harvard, Liverpool	0·0099115	24·53	1,479,300
VIII Iapetus	G. D. Cassini	1671	17′	0·023798	59·01	3,558 400
IX Phoebe	W. H. Pickering	1898	24″ Bruce	0·086375	215	12,945,500
URANUS						
V Miranda	G. R. Kuiper	1948	82″ MacDonald reflector	0·000872	5·10	123,000
I Ariel	W. Lassell	1851	24″ reflector, Liverpool	0·0012820	7·938	191,700
II Umbriel	—	—	—	0·0017860	11·06	267,000
III Titania	W. Herschel	1787	18·7″ reflector	0·0029308	18·14	438,000
IV Oberon	—	—	—	0·0039187	24·26	585,960
NEPTUNE						
I Triton	W. Lassell	1846	24″ reflector, Liverpool	0·0023747	15·5	353,400
II Wereid	G. R. Kuiper	1949	82″ MacDonald reflector, Texas	0·0371797	245	5,560 000

Italic figures refer to satellites with retrograde orbits.

Mean distance from primary angular at mean opposition distance (' ")	Sidereal period of revolution (d h m)	Sidereal period days	Mean synodic period (d h m s)	Mass ratio: satellite/planet	Diameter km.	Orbital inclination to planet's equator proper plane	Orbital eccentricity
24·6	0 7 39	0·318910	7 39 26·6		13	118°	0·0210
1 01·8	1 6 17	1·262441	1 06 21 15·7		8	1146°	0·0028
59·4	0 11 57	0·448179	11 57 27·6		200	0°24'	0·003
2 18·4	1 18 27	1·769138	1 18 28 35·9	$3·8 \times 10^{-5}$	3240	0°2'	0·0000
3 40·1	3 13 14	3·551181	3 13 17 58·7	$2·5 \times 10^{-5}$	2830	0°28'	0·003
5 51·2	7 3 43	7·154553	7 03 59 35·9	$8·2 \times 10^{-5}$	4900	0°11'	0·0015
10 17·6	16 16 32	16·689018	16 18 05 06·9	$5·1 \times 10^{-5}$	4570	0°15'	0·0075
62 45	250	250·57	265 22 43		100	27°38'	0·15798
64 05	260	263·55	275 17 09		20	29°01'	0·13029
64 10	260	259·65	276 04 56		30	24°46'	0·20719
116	625	631·1	551		20	147°	0·16870
123	692	692·5	597		20	164°	0·20678
129	739	738·9	635		20	145°	0·378
130	758	758	645		20	153°	0·275
25·3	0 17 55	0·7490	17 59		800	0°0'	0·0
30·0	0 22 37	0·942422	22 37 12·4	$6·7 \times 10^{-8}$	500	1°31'	0·0201
38·4	1 8 53	1·370218	1 08 53 21·9	$1·5 \times 10^{-7}$	600	0°01'	0·00444
47·6	1 21 18	1·887802	1 21 18 54·8	$1·1 \times 10^{-6}$	1000	1°06'	0·0000
1 01·0	2 17 41	2·736916	22 17 42 09·7	$1·8 \times 10^{-6}$	1000	0°01'	0·00221
1 25·1	4 12 25	4·517503	4 12 27 56·2	$4·0 \times 10^{-6}$	1350	0°21'	0·00098
3 17·3	15 22 41	15·945452	15 23 15 31·5	$2·5 \times 10^{-4}$	4950	0°20'	0·2890
3 59·4	22 6 38	21·276665	21 07 39 05·7		500	0°26'	0·1042
9 35	79 7 56	79·33082	79 22 04 59	$2·5 \times 10^{-6}$	1100	14°43'	0·02828
34 51	550 28	550·45	523 13		200	150°	0·1659
9·9	1 24 50	1·414	1 09 55 31		300	0°	0
14·5	2 12 29	2·520383	2 12 29 39·0		800	0°	0·0028
20·3	4 3 28	4·144183	4 03 28 25·8		600	0°	0·0035
33·2	8 16 56	8·705876	8 17 00 01·2		1100	0°	0·0024
44·5	13 11 7	13·463262	13 11 15 36·5		1000	0°	0·0007
16·9	5 21 3	5·876833	5 21 03 29·8	$1·3 \times 10^{-3}$	3700	159°57'	0·000
4 23·9	500	359·881	362 01		300	27°48'	0·749

The solar system

The Earth's Moon is the largest satellite relative to its parent planet and its mass in relation to the parent is greater than that to be found in any other system (0·0123 of the Earth's mass). Furthermore its orbit is everywhere concave to the Sun, and we propose to treat the Moon separately and deal with it and the Earth as a double planet—the Earth/Moon system. This approach is even more valid today, now that the latest researches appear to show that the Moon never formed part of the Earth and that it is in some 'sympathy' with earthquakes on Earth which may, according to Kozyrev[68] trigger the venting of gases from lunar craters.

Excluding the Moon there are thirty-one satellites known and the first of these, the four large moons of Jupiter, were discovered by Galileo in Padua in 1610. The latest to be found, the inner moon of Saturn, was discovered by Dollfus in Meudon and recorded by Texereau in Texas in 1966/7.

All but six of these bodies revolve in direct orbits about their primaries. The six that possess irregularities are the outer four satellites of Jupiter, the outermost of the ten satellites of Saturn and the innermost satellite of the system of Neptune. The outer seven satellites of Jupiter are strongly perturbed by the Sun, and Hunter[69] has shown, by the application of de Vogelaere's method and by using the electronic computer of the University of Glasgow, that retrograde orbits may generally be more stable and that the orbit of Jupiter VIII (Iapetus) is certainly more stable in the retrograde sense than in the direct.

The thirty-one satellites known should not, it seems, be considered to be the totality of the satellite component of the Sun's system. The arguments for the presence of undiscovered satellites have been marshalled by Bowell and Wilson.[70] They postulate two new satellites for Jupiter, one for Saturn and one for Uranus. Their main figures are as follows:

Planet	Period of satellite				Mean distance from primary in equatorial planetary radii
Jupiter	0d.	20h.	40m.	47s.	3·66
	0d.	19h.	59m.	24s.	3·58
Saturn	0d.	14h.	45m.	03s.	2·31
Uranus	1d.	16h.	16m.	08s.	9·03

Porter[71] has explained that in most cases the satellites are so close to the primary, and so distant from the Sun and the other planets, that perturbations by these bodies can be ignored. Mutual perturbations are, however, important in the case of the larger satellites, and this is

[68] See *Nature*, 222, 404 (1969).
[69] Hunter, R. B., *The Observatory* (1967 Aug.), 151; also *Monthly Notices of the R.A.S.*, 136, 245 (1967).
[70] Bowell, E. L. G., and Wilson, L., *Nature*, 216, 669 (1967).
[71] Porter, J. G., 'The Satellites of the Planets', *J.B.A.A.*, 70, 33 (1959).

particularly true of the four great moons of Jupiter and the effect of Titan on the neighbouring satellites of Saturn. In addition, the motion of close satellites is always modified by disturbances due to the oblateness of the parent planet. The effect in such cases, on a satellite having direct motion about its primary, is to cause the line of nodes to regress, while the apse line (the line joining pericentre to apocentre) advances. In the case of our own Moon, the nodes regress in a period of 18·59 years, while the apse line advances through one revolution in 8·85 years.

A similar effect is seen to take place in other satellite systems; thus the two satellites of Mars show a regression of their nodes when referred to the Martian equator.

Laplace has shown that there exists, for any satellite orbit, a fixed plane such that the components of the disturbing force, perpendicular to this plane, balance each other, leaving no resultant force at right angles. Referred to this plane, the inclination of the satellite orbit is virtually constant, while the nodes will regress steadily. The Laplacian plane lies between the plane of the satellite orbit and the plane of the planet's equator and it passes through their intersection. In most cases the only significant disturbing force arises from the oblateness of the planet's body, and the Laplacian plane then coincides with the planet's equatorial plane.[72] It should not be forgotten that the visual analysis of the positions of the satellites is one of immense difficulty, owing to the great distances that separate them from the Earth and to their small size. At the distance of Neptune (30 a.u.) a tenth of a second of arc represents 1350 miles and the scope for inaccuracy is greatly increased from what it would be in the vicinity of the Moon or Mars.

Something of the scale can be seen in reverse if we give the size of our Sun, as it would be seen from the exterior planets. It is well known that to our eyes on Earth the great ball of the Sun exhibits a disc of but half a degree of arc (31′40″). Seen from Mars his disc would appear to have a diameter of 21′11″, from Jupiter 6′11″, from Saturn 3′27″ and from Neptune he would exhibit a disc having the very small diameter of 0′44″·25 of arc.

THE SATELLITES OF MARS

Mars has two extremely faint satellites, Phobos and Deimos, and these were found from a deliberate search by Asaph Hall[73] using the superb 26 in. refractor, built by Alvan Clark for the US Naval Observatory, at the close opposition of the planet in 1877. Mädler (1794–1874) had made a search some years before from Berlin but his $3\frac{3}{4}$ in. refractor was

[72] See Brower, D., Clemence, G. M., 'Orbits and Masses of Planets and Satellites', *Vistas in Astronomy*, vol. I, ch. III, pp. 31–94.
[73] Hall, A., 'The Satellites of Mars', Appendix, Washington Observations (1875).

not optically equal to the task. In 1862–4 d'Arrest also made a search from Copenhagen but without success. The difficult optical task and the need for skilled observation was dispensed with by Swift, however, who discovered them with that priceless instrument—the naked intellect.[74]

From investigations of the secular changes of the planet's orbital elements given by Sharpless in 1945, Kerr and Whipple in 1954 and Kuiper in 1956, Porter[75] records that Phobos is approaching Mars, while Deimos is receding.

These elements appear to be in some doubt and the secular acceleration of Phobos is thought to be much less than that previously specified.[76] The angular velocity of Phobos increases by about 1 per cent in $2 \cdot 2 \times 10^6$ years and this implies that the satellite will collide with its primary in 35 to 40×10^6 years from the present day. To use Asquith's famous response—we shall have to wait and see.

The inclination of the twin orbits is in the case of this system referred to the Laplacian plane. Phobos with its fast period of revolution, which is one third the rotational period of the parent planet, will appear from the Martian surface to rise in the *west* and set in the *east* at least twice[77] each Martian day. Deimos, in contradistinction, will move but slowly in the Martian sky,[78] owing to its period of revolution being almost equal to the rotational period of the parent planet. Both satellites will suffer eclipses in the considerable shadow cone of Mars.

The moons of Mars are difficult objects to observe with the telescope. Deimos has been seen with apertures of eight and ten inches and M. du Martheray has reported seeing them with a glass of but five inches diameter by making use of an occulting bar, with which to obscure the bright disc of the planet. It should be remembered that Deimos rides at a distance of 14,600 miles above the parent planet but Phobos is close in to it at a distance of 3700 miles, not very much outside the 'Roche Limit'.[79]

Phobos is reported[80] from the Mariner 6 and 7 missions to be elongated, approximately $22 \cdot 5$ by $17 \cdot 5$ km. and to have a geometric albedo of about $0 \cdot 065$, the lowest in the solar system.

[74] The Martian Moons were predicted by Swift in his discussion of the skills of the astronomers of Laputa, *Gulliver's Travels* (1727). For an analysis of the matter see Nicolson, M., and Mohler, N. M., 'The Scientific Background of Swift's Voyage to Laputa', *Annals of Science* (1937), pp. 229–334; see also *The Observatory*, *66*, 289 (1946).

[75] Porter, J. G., *J.B.A.A.*, *70*, 33 (1960).

[76] See Wilkins, G. A., 'Motion of Phobos', *Nature*, *224*, 789 (1969); see also *The Observatory*, *90*, 38 (1970).

[77] Phobos takes about $4\frac{1}{2}$ hours to move across the Martian sky.

[78] Deimos will take about 60 hours to rise and set.

[79] Roche Limit (or Roche's Limit). A critical value, according to the researches of Roche, for the radius of the orbit of the secondary star, in a binary system, in terms of that of the primary, to ensure stability (see footnote 95 on page 100).

[80] *Nature*, *226*, 316 (1970).

THE SATELLITES OF JUPITER[81]

Twelve satellites of Jupiter are known (Fig. 20). The most conspicuous of these are the four moons discovered by Galileo[82] and often referred to as 'the Galilean satellites' in honour of their illustrious discoverer. They form with their fifth companion, discovered by Barnard using the 36 in. Lick refractor in 1892, a system that is intrinsically separate from the other satellites of the Jovian system, in that they all move substantially

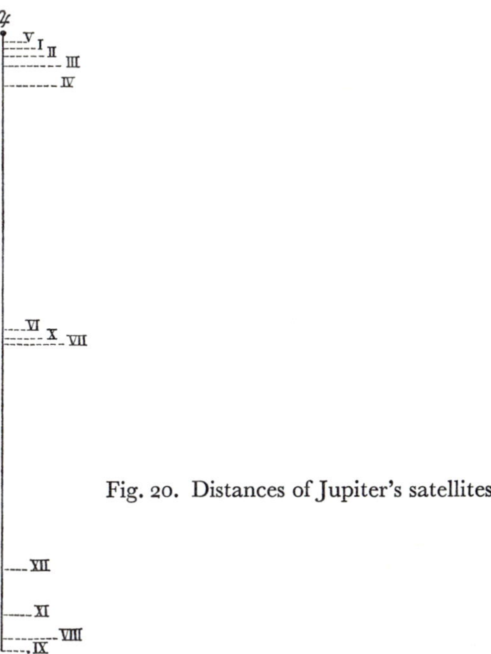

Fig. 20. Distances of Jupiter's satellites.

in his equatorial plane. The four big moons J_I, J_{II}, J_{III} and J_{IV} (or to give them their names: Io, Europa, Ganymede and Callisto) are a delight to the observer on Earth as they wander from side to side of the vast body of the parent planet displaying, as the conditions allow, the phenomena of multiple occultations, eclipses[83] and transits across the face of Jupiter.

[81] See Peek, B. M., *The Planet Jupiter* (London 1958), pp. 255–68, and *Icarus*, *15*, 174, (1971).
[82] Historical research shows that Simon Marius astronomer at the court of the Margrave of Ansbach *c.* 1605, was an independent observer of the four large satellites of Jupiter a few weeks before Galileo made his discovery; see *J.B.A.A.*, *41*, 165, 415. Borelli in his astronomical work *Theorica mediceorum planetarum ex causis physicis deducta* (Florence 1666) was the first to consider the influence of attraction on the satellites of Jupiter.
[83] The accurate timing of the mutual eclipses of the satellites of Jupiter enabled Ole Roemer (1644–1710) in 1675 to show that light has a finite velocity. The summated interval between the eclipse phenomena was variable by some 1000 seconds and this was explained by the time taken for light to pass across the Earth's orbit about the Sun; now known to be 186×10^6 miles, hence the speed of light is seen to be 186×10^3 miles/second.

The solar system

Porter[84] has had much experience in computing the motions of the Galilean Satellites. He states that:

The theory of the satellites of Jupiter involves the difficulty of their mutual perturbations. They form a tightly-bound system, so that it is impossible to deal with one satellite at a time; the theory must include all four satellites. Thus the mean motions of J_I, J_{II} and J_{III} are almost in the ratio $4 : 2 : 1$, and Laplace showed that the mean motion of J_I plus twice that of J_{III} is equal to three times that of J_{II}. Expressed in terms of their mean longitudes (L),

$$L_1 - 3L_2 + 2L_3 = 180°$$

This equation must be satisfied; if any satellite is disturbed, as for instance by J_{IV}, the motions adjust themselves to this equality once more. As a result, the three satellites cannot all be in conjunction or opposition at the same time. This is at once obvious from the equation, since it is impossible for L_1, L_2 and L_3 to be equal. It will be noticed that J_{IV} does not share in this commensurability.

The complexity of the motion of these four large moons may be gathered by reference to the number of terms used in Sampson's theory. It is usual in this class of work to expand the equations for the true longitude of the satellite in its orbit into a series which includes all appreciable terms. Of these the largest (and the only one normally included in an undisturbed orbit of small eccentricity) is the equation of the centre.

The complexity of these expressions, therefore, increases considerably, and the following table shows the number of terms used by Sampson:

	Number in longitude	Number in latitude	Number in distance	Total number
J_I	28	6	9	43
J_{II}	37	10	7	54
J_{III}	35	8	8	51
J_{IV}	22	9	6	37

(It might be noted that in Brown's lunar theory, where much greater accuracy is called for, there are 1179 terms in longitude alone.) Sampson's[85] tables are not easy to use, and for most purposes they give times of satellite phenomena to an unnecessary precision. The tables include all terms having coefficients of $1''$ or more, and give positions to $0°·00001$. For prediction purposes the simpler method developed by Andoyer* (1915) is used; this employs only the main terms of Sampson's Tables, giving times to $0^m·1$ for eclipses and to the nearest minute for other phenomena; positions of the satellites in their orbits can be obtained to $0°·001$, the necessary data being published in the *Connaissance des Temps* each year.

Porter then refers to the work of Enso Mora, who calculated the times

---

[84] *J.B.A.A.*, *70*, 46 (1960).
[85] Sampson, R. A., *Tables of the Four Great Satellites of Jupiter* (London 1910).
* Andoyer, *Bulletin Astronomique*, *32*, 177 (1915).

at which Jupiter is seen from the Earth without his four large satellites, owing to their being rendered invisible from *inter alia* eclipse or occultation. The next occasion is on 1980 April 9, but the event is not observable from the United Kingdom. Mora refers, according to Porter, to the cycles of these recurring phenomena as not being useful, and he appeals to but one, that of 48 years less 24 days, which is twice the period of 24 years less 37 days quoted by Flammarion. Porter observes that

cycles of this kind are obviously of less value than those which, like the Saros,[86] refer only to the heliocentric positions of two bodies; in this case we are dealing with geocentric positions of six bodies. It is, however, interesting to compare the appearance of the satellites at intervals of 23·8984 years, and some of the phenomena in Mora's list can be fitted to this cycle. The earliest reference to such recurrent phenomena seems to be Bradley's discovery in 1735 of the great inequality of J_{II}, with a period of 437·6 days, after which J_I, J_{II} and J_{III} return to the same positions with respect to Jupiter's shadow. It will be noticed that the period of J_{IV} is not commensurate with this figure, but Mora's cycle of 17,507·54 days contains 9892 synodic periods of J_I, 4926 of J_{II}, 2443 of J_{III}, 1045 of J_{IV}, and is also equal to 40 periods of the great inequality of J_{II}.

If the rapid motion of the inner satellites of Jupiter was a surprise to Galileo, one wonders what he would have thought of the fifth satellite. This tiny body is only about 100,000 miles from Jupiter, so that it travels round at half the distance of the Moon from the Earth in 12 hours; it is therefore the swiftest-moving satellite and travels 25 times as fast as our Moon. Its speed about Jupiter is 16·4 miles/sec or about 1000 miles a minute, which is not far short of the speed of the Earth round the Sun. As a result of its proximity to the oblate planet, the line of nodes regresses rapidly at the rate of $2°·478$ per day. It is from this value that we derive the figure of 1/15 for the compression of Jupiter.

The surface temperatures of the Galilean satellites have been examined by Richardson and Shum[87] for over a lunation of each satellite and their results are more detailed than those obtained by Murray[88] *et al*, who gave only the observed integrated brightness temperatures of each of the four moons, as follows:

J_I 135°K. J_{II} 140°K. J_{III} 156°K.   and J_{IV} 168°K.

The seven outermost satellites of Jupiter are small in comparison with the Galilean satellites and they have all been discovered photographically since 1904. They are faint objects and as seen from Earth have apparent magnitudes of between 14·7 and 19·0 when Jupiter is at opposition.

[86] *Saros*. The Babylonian name for the number 3600, and hence for a period of 3600 years. Adopted by modern astronomers as the name of the cycle of 18 years and 10⅓ days, in which solar and lunar eclipses repeat themselves.
[87] Richardson, P. D., Shum, Y. M., 'Surface Temperatures of the Galilean Satellites of Jupiter', *Nature*, *220*, 897 (1968).
[88] Murray *et al*, *Astrophys. J.*, *137*, 986 (1964).

The solar system

The mean distances of these satellites from Jupiter are large, compared with their distances from the Sun (about 1 in 80), and the mass of the Sun is so much greater than that of Jupiter (1047 : 1) that the orbits are very strongly perturbed by the Sun. It is clear, from even a cursory study of their elements, that they fall into three very distinct groups:

1. Jvɪ, Jx and Jvɪɪ moving in direct orbits with similar elements at 163 Jupiter radii.
2. Jxɪɪ moving in a retrograde orbit at a distinct inclination and lower eccentricity than those of group 3 at 297 Jupiter radii.
3. Jxɪ, Jvɪɪɪ and Jɪx moving in retrograde orbits with similar elements at 325 Jupiter radii.

The orbits of the retrograde satellites Jvɪɪɪ, Jɪx and Jxɪ are not even approximately elliptic, yet as pointed out by Roy *et al*[89] they have approximately equal mean motions,* and if the Sun is considered as a satellite of Jupiter, the ratio of its mean motion to those of the retrograde satellites is within the ratio of 1 to 7 and, as seen from the table below, these satellites are nearly commensurable in mean motion with the Sun, the ratio being 1 to 6.

Indeed, the Sun's mean motion about Jupiter is almost exactly $\frac{1}{6}$ of the mean motions of Jvɪɪɪ, Jɪx and Jxɪ. If we consider the other satellites the Sun's mean motion about Jupiter is nearly $\frac{1}{7}$ of the mean motion of Jxɪɪ, while the satellites Jvɪ, Jvɪɪ and Jx form a group whose mean motions are nearly commensurable with the Sun's at $\frac{1}{17}$, and once again their average mean motion is almost exactly commensurable with that of the Sun.

	$\dfrac{n_2}{n_1}$	$\dfrac{n_2}{n_1} - \dfrac{1}{6}$
Sun Jvɪɪɪ	0·17055	+ 0·00388
Sun Jɪx	0·17196	+ 0·00529
Sun Jxɪ	0·15982	− 0·00684
Sun and mean of Jvɪɪɪ, Jɪx, Jxɪ	0·16725	+ 0·00059

[89] Roy, A. E., Ovenden, M. W., 'On the occurrence of commensurable mean motions in the solar system', *Monthly Notices of the R.A.S.*, *114*, 232 (1954), *115*, 296 (1955); see also Hunter, R. B., 'Motions of satellites and asteroids under the influence of Jupiter and the Sun', *Monthly Notices of the R.A.S.*, *136*, 41 (1967).

* The *mean motion* (n) of a planet is the average angular velocity, i.e., the average angle swept out by the radius vector r in unit time. If P is the sidereal period of the planet, $n = 2\pi/P$ but for computing purposes, the mean motion is usually expressed in degrees per day, so that $n = 360°/P$ where P is measured in days.

	$\dfrac{n_2}{n_1}$	$\dfrac{n_2}{n_1} - \dfrac{1}{17}$
Sun J_{VI}	0·05784	— 0·00098
Sun J_{VII}	0·06002	+ 0·00120
Sun J_X	0·05867	— 0·00015
Sun and mean of J_{VI}, J_{VII}, J_X	0·05883	— 0·00001
Sun J_{XII}	0·14564	+ 0·00278

where n_1 n_2 are the mean motions of the two bodies. It would appear that these commensurabilities are directed toward an intrinsic stability of the system. One should also note that the stability is enhanced in retrograde orbits by the fact that the satellite is travelling in a direction opposite to that of the disturbing body, the Sun, so the perturbations are applied but for a short time.

THE RINGS AND MAJOR SATELLITES OF SATURN

The rings of Saturn are properly considered within the scope of a chapter on the satellites, since they are firmly thought to be made up of a multitudinous array of small individual particles, each behaving as a true satellite to the parent planet and revolving in the equatorial plane. Beyond the rings lie ten large satellites S_{III}, S_{IV}, S_V, S_{VI} and S_{VIII}, having diameters in excess of 1000 km. and the others being much smaller, with diameters in the range 200 km. to 600 km. (Fig. 21, page 98.)

Titan (S_{VI}) is a large satellite having a diameter of 4950 km. Titan is a large and massive moon exceeded in size and mass only by Jupiter's Ganymede (5000 km.). It is a little larger than the planet Mercury and is remarkable as the only known satellite to possess an atmosphere, postulated from molecular bands of methane detected by refined spectroscopic studies.

The rings of Saturn were shown to be of a particulate nature by the mathematical investigations of Clerk Maxwell;[90] Hagihara[91] concludes that the stability of the ring satellite system is strengthened by the presence of the commensurability relation in their mean motions, that is to say, the occurrence of the gaps in the rings at distances at which the periods of the particles would be commensurate with the periods of the inner satellites.

[90] Maxwell, J. C., *Adams' Prize Essay* (1857).
[91] Hagihara, Y., *Stability in Celestial Mechanics*, Jubilee publication (Tokyo 1957).

It is well known that the Great Cassini division in the rings corresponds to the orbits of particles with a period of $11\cdot304$ hours, that is one half of the period of Mimas (S_I) ($0\cdot942$ d., i.e. $22\cdot608$ h.) and one quarter that of Tethys (S_{III}) ($1\cdot887802$ d, i.e. $45\cdot307$ h.). It was subsequently shown by Goldsborough[92] that the gap between the parent planet and the rings is caused by the action of all the satellites, and that Mimas (S_I) is mainly responsible for the Cassini division. The formation of the Cassini division

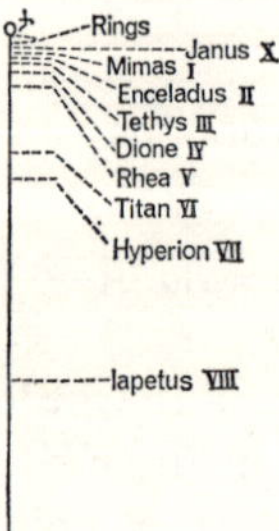

Fig. 21. Section through Saturn's satellite system.

due to the perturbations of Mimas upon the ring particles is not completely satisfactory. It having been shown by Walsh and Zimmerman (*Nature, 230*, 233, 1971) that the orbit in which a body would have exactly half the period of Mimas lies well away from the centre of the Cassini division and that a satellite such as Mimas would possibly produce a gap only a fraction of the present width of the division—that is to say about 3000 km., indeed calculations show that Mimas may produce a gap of but some 90 km. width. Undoubtedly the complex nature of Saturn's system suggests a complicated early history and it is postulated that Mimas may at one time have been a much larger body than it is today. The diaphanous nature of the Crepe ring is due primarily to the action of Dione, S_{IV} and Rhea, S_V.

It can be seen from the elements of the satellites, that Titan revolves about Saturn in approximately sixteen Earth days and that the periods of

[92] Goldsborough, G. R., *Phil. Trans. Roy. Soc.*, A222, 101 (1921); *A244*, 1 (1951).

the other satellites extend over the range 0·749 d. for Janus to 550·45 days for Phoebe. D'Arrest (1822–75) has shown that in 465·75 of our days—

Mimas	completes 494 orbits	
Enceladus	completes 340 orbits	
Tethys	completes 247 orbits	$(2 \times 247 = 494)$
Dione	completes 170 orbits	$(2 \times 170 = 340)$

and that the following approximations hold:

4 orbital periods of Titan are equal to 3 orbital periods of Hyperion.

5 orbital periods of Titan are equal to 1 orbital period of Japetus.

5 orbital periods of Rhea are equal to 1 orbital period of Hyperion.

Alexander[93] has given a detailed account of the discovery of what are now known as the Kirkwood gaps in the rings:

In 1866 the American astronomer, Daniel Kirkwood, propounded an interesting theory to explain why it was that at certain mean distances from the Sun there was a notable scarcity of minor planets, whereas at distances a little less or greater these small bodies were fairly plentiful. He pointed out that the gaps occurred at mean distances where, if planetoids had been revolving round the Sun, their revolution periods would be simple fractions of Jupiter's period. They would therefore pass fairly near to the great planet once in every few revolutions and be perturbed, and the cumulative effect of the repeated perturbations would force such minor planets into new orbits. The gaps among the planetoids are conspicuous at distances where their orbital periods would be $\frac{1}{3}$, $\frac{2}{5}$, $\frac{3}{7}$, $\frac{1}{2}$, and $\frac{3}{5}$ of Jupiter's period, and there is a very large group of them with a period just under half that of Jupiter.

In the following year (1867) Kirkwood applied his theory to the particles of Saturn's rings. In this case the perturbing bodies would be the inner satellites Mimas, Enceladus, Tethys and Dione, which, revolving round Saturn comparatively near to, and almost in the plane of, the rings, and being of great mass in comparison with the tiny ring particles, could easily produce a very disturbing effect on them. For example, a ring particle having a revolution period equal to half that of Mimas, about a third that of Enceladus, a quarter that of Tethys, and a sixth that of Dione, would be perturbed by one or other of those satellites once or twice a day, and would very soon be forced into a different orbit. As a matter of fact a particle with such a period (11·3 hours) would be situated within Cassini's division, and so Kirkwood's theory gives a very attractive explanation of the existence of that prominent gap in the ring system, and also of the brightness of the outer part of ring B, where there may be a congestion of particles forced out of the division, with periods a little less than half that of Mimas, just as the very numerous Hecuba group of minor planets is found with periods just under half that of Jupiter.

According to Antoniadi (L'Astronomie, vol. 44, p. 163) Meyer extended Kirkwood's theory about the Cassini division to the next two satellites, suggesting that a particle in the division would have a period $\frac{1}{9}$ that of Rhea and $\frac{1}{33}$ that of Titan, but it seems evident that the perturbations by Mimas are

<hr>

[93] Alexander, A. F., o'D. *The Planet Saturn* (London 1962).

the preponderating factor. In 1871 Kirkwood extended his theory to explain the existence of the marking a little outside the middle of ring A known as Encke's division, which though not a gap in the ring probably indicates a sparser distribution of particles (Proc. Amer. Phil. Soc., vol. 12, p. 163).

It will be noticed that Kirkwood applied his theory only to Cassini's and Encke's divisions, but it has since been extended, especially by Lowell, to other places in the ring system where particles would have periods that are simple fractions of that of Mimas or that of Enceladus. Two other 'Kirkwood gaps' have been suggested in the middle and outer part of ring A; three somewhat outside and three inside the middle of ring B; and also one at the junction of rings B and C. The extension of Kirkwood's theory therefore supplied observers with a new motive for seeking faint markings that might indicate a certain sparseness of particles, on the bright rings; the former expectation of ring gaps based on Laplace's theory had been removed by the work of Clerk Maxwell.

There is today some doubt about the reality of the minor gaps. Kuiper[94] has examined the rings using the 200 in. Hale telescope, using a power of 1170, and he has come to the conclusion that *only one division exists*, viz., the Cassini division, the width of which is one third that of ring A. The other divisions are either minor intensity ripples or are non-existent. Even the famous Encke division is now in doubt and is thought to be a pronounced ripple at the position where Ring A changes its intensity abruptly.

The latest results by Franklin in South Africa are thought to indicate that the particles are only a few metres in size, and this leads some to speculate that this vast assemblage of small material may be the disruption of a satellite that came within the Roche limit[95] and was not able to re-coalesce into a single body.

Kuiper speaks with authority on the structure of the rings.

He is of opinion that only very wide limits to the size of the particles

[94] Kuiper, G. P., *Trans. I.A.U.*, *9*, 255.

[95] Edouard Roche was professor of mathematics at Montpellier. (See *Mem de l'Acad des Sci. . . . de Montpellier*.) He calculated that within a circle of 2·44 times the radius of the planet, no fluid satellite with a density equal to Saturn could survive undamaged owing to its disruption by tidal forces. Jeffreys (MN. *107*, 260, 1947) has extended Roche's argument to *solid* satellites.

The Roche limit for Saturn can be derived in a very simple way. Consider two particles of mass m and radius r, at a mean distance d from the centre of Saturn. They are taken to be in mutual contact and aligned along a projected radius vector of Saturn, so that one is at distance $(d - r)$ and the other at distance $(d + r)$ from the planet. These particles can remain as a stable gravitational configuration as long as their mutual attractive force is greater than the difference in the forces on each by Saturn. Saturn's mass (M) exerts a force on the first particle of $GMm/(d-r)^2$, the difference being

$$GMm \frac{1}{(d-r)^2} - \frac{1}{(d+r)^2} \approx GMm \frac{4r}{d^3}$$

The mutual gravitational attraction for the particles is $Gm^2/4r^2$ so that the stability criterion, Roche's limit, is

$$\frac{Gm^2}{4r^2} = \frac{4GMmr}{d^3} \quad \text{or} \quad d = \left(\frac{16Mr^3}{m}\right)^{\frac{1}{3}}$$

Taking the density of the particles to be approximately 1 gives a value of d of $\sim 1\cdot5 \times 10^5$ km. The rings of Saturn, which extend to a distance of $1\cdot3 \times 10^5$ km., are thus within the Roche limit for instability.

of the rings can be set optically, and he puts them as larger than 1–2 microns. From evidences obtained from the infra red spectrum and from the albedo it is thought that the particles are mostly snow crystals.

If one further examines the probable effects of differential rotation in the rapidly revolving Ring, one is led to suppose that the snow particles are a few centimetres in size, and somewhat cylindrical in shape, with the axes orientated at right angles to the plane of the Ring. This raises the question whether such small particles could have survived in a vacuum through long periods of time, exposed as they are to solar radiation. The average temperature of the sunlit flakes will be 60–70°K.; but most particles will be only partially illuminated and be colder. At 70°K. the vapour pressure of water vapour is about $10^{-25.6}$ mm. Hg, from which the evaporation rate is found to be roughly 10^{-6} molecules per cm.^{2} per sec. The evaporation during geologic time would therefore be about 10^{11} molecules per cm.^{2}, a layer much less than one molecule thick. Thermal evaporation losses will, therefore, have been negligible for the ring particles. Losses due to photo-dissociation by solar ultra violet light will probably have been larger, though λ3060 of OH has not been found in the spectrum of the ring and its surroundings. At higher temperatures thermal evaporation rapidly increases; at 100°K. it would, during geologic time, have amounted to a layer of ice about 1 cm. thick.

The rings are known to disappear from the view of an observer on Earth at regular intervals and, indeed, Galileo was eventually faced with this phenomenon. The rings are in the equatorial plane of Saturn and tilted at 26° 44′ to the orbital plane, which is itself tilted at 2° 29′ 22″·1 to the ecliptic, and they are presented edgewise to the Earth on two occasions during each revolution of Saturn about the Sun (10,759·20 days). Porter[96] shows that three conditions for a disappearance of the rings may arise.

(a) When the Earth is in the plane of the rings, which are too thin to be seen in the edgewise presentation.

(b) When the Sun is in the plane of the rings, so that then their surface is not illuminated and is invisible.

(c) When the Sun and the Earth are on opposite sides of the ring plane and an observer on the Earth is then constrained to look toward the dark side of the rings.

Some details of these phenomena have been explained by Hepburn[97] and Goodman.[98] It is seen that there may, under certain conditions, be either one or three passages of the ring plane through the Earth. This is

[96] Porter, J. G., *J.B.A.A.*, 70, 52 (1960).
[97] Hepburn, P. H., *J.B.A.A.*, 30, 158, 240 (1920).
[98] Goodman, J. W., 'The Edgewise Presentation of Saturn's Rings', *Sky and Telescope*, p. 128 (Sept. 1965).

The solar system

made exceptionally clear by Hepburn's diagram, Figs. 22 and 23, which convention is also adopted by Goodman.

The sizes of the rings of Saturn are given below in a table taken from

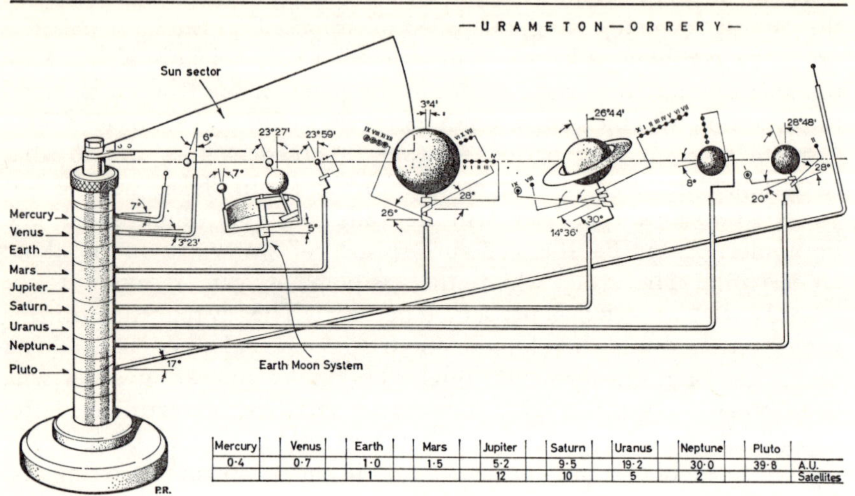

Mercury	Venus	Earth	Mars	Jupiter	Saturn	Uranus	Neptune	Pluto	
0·4	0·7	1·0	1·5	5·2	9·5	19·2	30·0	39·8	A.U.
		1		12	10	5	2		Satellites

	km.	miles	*Lyot's divisions*	Position in ring	*Kirkwood's gaps*	Position in ring
RING A						
outer	272,300	169,199	*RING A* (outer edge)	(0)		(0)
					$\frac{3}{7}$ E	10
			Outer division	20	—	
			Inner division:			
			outer edge	36		
			minima ⎫	40	$\frac{3}{5}$ M	39
			of ⎬ {	62	$\frac{2}{5}$ E	57
			light ⎭	80		
			inner edge	80		
inner	239,600	148,880	(inner edge of ring)	(100)		
			CASSINI'S DIVISION		$\frac{1}{2}$ M, $\frac{1}{3}$ E	
RING B						
outer	234,200	145,525	*RING B* (outer edge)	(0)		(0)
			Outer division:			
			outer edge	11		
			darkest part	17	$\frac{5}{11}$ M	24
			—	—	$\frac{4}{9}$ M	24
			inner edge	39		
			Middle division:	42	$\frac{3}{7}$ M	41
			—	—	$\frac{2}{5}$ M	57
			Inner division:			
			outer edge	70		
			darkest ⎫	80	$\frac{3}{8}$ M	76
			parts ⎬ {	89	$\frac{1}{4}$ E	89
			inner edge	96		
inner	181,100	112,530	(inner edge of ring)	(100)		(100)
RING C			DIVISION BETWEEN		$\frac{1}{3}$ M	
inner	149,300	92,771	RINGS B AND C			

Note: E represents Enceladus; M represents Mimas.

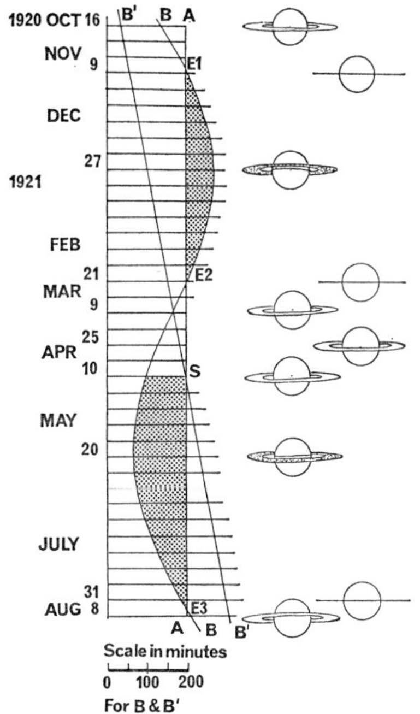

Fig. 22. Diagram of the varying elevations of Earth and Sun in relation to the ring-plane and the ring-phases of Saturn in 1920–1 (see Hepburn, P. H., *J.B.A.A. 30*, 241), based on Barnard's drawings of 1907.

AA ring plane of Saturn (left—south; right—north).

BB, B¹B¹ are taken from the Nautical Almanac and give respectively the saturnicentric latitudes of the Earth and the Sun refered to the ring plane.

BB crosses AA three times at E₁ E₂ E₃ representing three passages of the Earth through the plane of the rings extended.

Note: For a detailed paper see Turner, H. H., MN. *68*, part 6, April 1968.

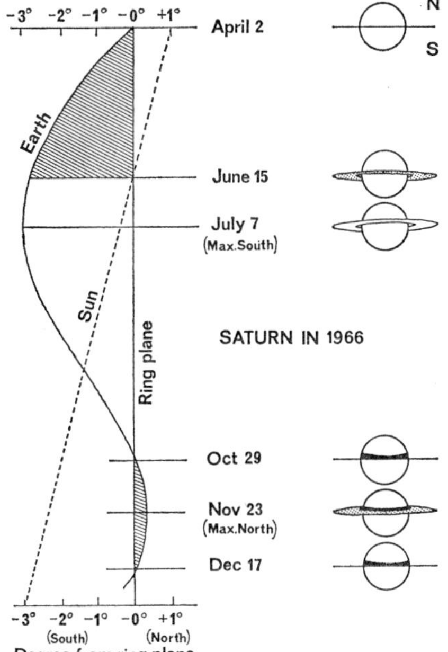

Fig. 23. Saturn's rings and their shadow as seen from the Earth in 1966 (after J. P. Hamilton).

the *Handbook of the British Astronomical Association* with minor additions for ease of comparison. The divisions of the rings are reproduced from the standard works of Alexander.[99]

Of the ten satellites the inner six (S_X, S_I, S_{II}, S_{III}, S_{IV}, S_V) move in direct orbits of almost circular form close to the plane of the rings. At a greater distance from the body of the planet, but much in harmony with the inner six moons, move Titan and Hyperion (S_{VI}, S_{VII}). Titan is nearly half a million miles outside the orbit of Rhea, the outermost of the six (S_V).

Titan and Hyperion form a 'close' pair at 0·008 and 0·009 astronomical units from their primary. Hyperion is, however, a satellite of unusual interest, in that it illustrates a particular case of the problem of three bodies. Porter[100] explains that the apse line moves in a *retrograde* direction at the rate of 18°·56 every year, and recalls Newcomb's[101] investigations, from which it was concluded that this was caused by the close commensurability of the motions of Titan and Hyperion.

The mean motion of Titan is 22°·57702 and that of Hyperion 16°·91999. Thus

$$3T - 4H = + 0°·051$$

Porter goes on to explain the significance of this angle.

Suppose Titan and Hyperion are in conjunction at a given time, then after three revolutions of Hyperion they will again be in conjunction at almost the same point. There is an interval of 63·6377 days between the conjunctions, and in this time Titan will have moved through 1436°·75 and Hyperion through 1076°·75. These angles are 3°·25 short of 4 revolutions of Titan and 3 of Hyperion; hence the point of conjunction retrogrades through 3°·25 in 63·6377 days—which is 0°·051 per day, or 18°·56 per year. The line of apses is forced to follow the same law and as a result, the conjunctions of these two satellites always occur at the same place in the orbit, and this coincides with the aposaturnium of Hyperion, that is, the point of greatest distance from Saturn.

The commensurabilities of the inner satellites have already been referred to, but it is necessary now to show the very well known relationship between S_I, S_{II}, S_{III} and S_{IV}. Mimas, Enceladus, Tethys and Dione.

If S_I is the mean motion of satellite I, the following rule can be stated:

$$5\,S_I - 10\,S_{II} + S_{III} + 4\,S_{IV} = 0°·509.$$

The mean motions for the inner satellites are:

[99] Alexander, A. F. O'd., *J.B.A.A.*, *64*, 28 (1954); also *The Planet Saturn*, 408 (London 1964).
[100] Porter, J. G., *J.B.A.A.*, *70*, 55, 56 (1960).
[101] Newcomb, S., *A.P. of A.E.*, III, p. 3 (1884).

Mimas	$381°\cdot994444$
Enceladus	$262°\cdot73194$
Tethys	$190°\cdot69795$
Dione	$131°\cdot53497$

Roy and Ovenden[102] have given further commensurabilities and these are shown below:

Satellite	$\dfrac{n_2}{n_1}$	$\dfrac{n_2}{n_1} - \dfrac{1}{2}$
S_{III} S_I	$0\cdot499\ 216$	$-\ 0\cdot000\ 784$

		$\dfrac{n_2}{n_1} - \dfrac{1}{2}$
S_{IV} S_{II}	$0\cdot500\ 643$	$+\ 0\cdot000\ 643$

		$\dfrac{n_2}{n_1} - \dfrac{2}{3}$
S_{II} S_I	$0\cdot687\ 790$	$+\ 0\cdot021\ 123$

		$\dfrac{n_2}{n_1} - \dfrac{5}{7}$
S_{III} S_{II}	$0\cdot725\ 827$	$+\ 0\cdot011\ 541$

		$\dfrac{n_2}{n_1} - \dfrac{1}{3}$
S_{IV} S_I	$0\cdot344\ 337$	$+\ 0\cdot011\ 004$

		$\dfrac{n_2}{n_1} - \dfrac{2}{3}$
S_{IV} S_{III}	$0\cdot689\ 756$	$+\ 0\cdot023\ 089$

		$\dfrac{n_2}{n_1} - \dfrac{3}{4}$
S_{VI} S_{VII}	$0\cdot749\ 434$	$-\ 0\cdot000\ 566$

		$\dfrac{n_2}{n_1} - \dfrac{2}{7}$
S_{VI} S_V	$0\cdot283\ 310$	$-\ 0\cdot002\ 404$

		$\dfrac{n_2}{n_1} - \dfrac{1}{5}$
S_{VIII} S_{VI}	$0\cdot201\ 001$	$+\ 0\cdot001\ 001$

[102] Roy, A. E., and Ovenden, M. W., *Monthly Notices of the R.A.S.*, *114*, 323 (1954).

		$\dfrac{n_2}{n_1} = \dfrac{3}{5}$
S_V S_{IV}	0·605 847	$+$ 0·005 847
		$\dfrac{n_2}{n_1} = \dfrac{3}{7}$
S_V S_{III}	0·417 887	$-$ 0·010 684
		$\dfrac{n_2}{n_1} = \dfrac{1}{5}$
S_I S_V	0·208 616	$+$ 0·008 616
		$\dfrac{n_2}{n_1} = \dfrac{2}{7}$
S_{II} S_V	0·303 313	$+$ 0·017 599
		$\dfrac{n_2}{n_1} = \dfrac{1}{6}$
S_{IV} S_{VI}	0·171 643	$+$ 0·004 976
		$\dfrac{n_2}{n_1} = \dfrac{1}{5}$
S_V S_{VII}	0·212 322	$+$ 0·012 322
		$\dfrac{n_2}{n_1} = \dfrac{2}{9}$
S_{VII} S_{VIII}	0·268 203	$-$ 0·017 511

It is shown further that:

NUMBER OF MEAN MOTION RATIOS n_2/n_1 WITHIN A LIMIT ϵ OF AN INTEGER RATIO

ϵ	0·0119	0·0089	0·0059	0·0030	0·0015
Number observed	33	26	20	12	6
Chance	17	13	9	3	2

Iapetus and Phoebe, the two outermost satellites, are at great distances from Saturn (0·024 and 0·086 a.u. respectively). Iapetus is in a direct orbit inclined at 14°·7 to the equatorial plane of Saturn, whereas Phoebe is in a retrograde orbit having the great inclination of 150° to the equatorial plane. Phoebe is the only satellite to have been discovered from the Southern Hemisphere. It was found by Pickering using the superb Bruce doublet of the Harvard Southern Station in the mountain observatory of Arequipa in Peru.

No chapter on the satellites is today complete without a special word on the discovery of Janus (S_X), which is the very latest of the Sun's family to be found. Dollfus[103] explains that he was led to look for the new satellite from perturbations caused to the ring system. The discovery was made with the 43 in. reflector at the high altitude Pic du Midi observatory at the exact time when the Earth passed through the plane of Saturn's Rings, i.e. on 1966 December 17. The examination of the photographic plates delayed the announcement of the discovery until New Years's Eve 1967. The first confirmation came on 3 January, when R. L. Walker reported images of the satellite on plates taken on 18 December with the 61 in. reflector at Flagstaff Arizona. Later Dollfus was able to detect the satellite on four photographs taken by the French astronomer, J. Texereau on 1966 October 29 with the 82 in. reflector of the McDonald Observatory.

THE SATELLITES OF URANUS

Five satellites revolve about Uranus in the equatorial plane, which is tilted at such an angle to the ecliptic plane that, when referred to it, the satellites are placed in retrograde orbits. If the regularity of a system is equated with the smallness of the orbital eccentricities and inclinations to the orbital plane then the regularity of this unusual system is not excelled in any part of the solar system; and yet we find the plane of symmetry inclined at the truly remarkable value of $98°$ to the orbital plane of the planet about the Sun. Clearly this extreme obliquity has had no noticeable effect on the wonderful symmetry of the satellite orbits.

The system may conveniently be subdivided into the massive Titania –Oberon pair (approx. 1000 km. diam.), the intermediate Ariel–Umbriel pair (approx. 700 km. diam.), and Miranda (300 km. diam.). The distances of the satellites from the parent planet are such that they obey the Bode relationship and according to Roy[104] they also fit the distance relation devised by Miss Blagg.[105]

Remarkable commensurabilities are found to exist within this system of satellites and we are indebted to D'Arrest[106] for pointing out that all four satellites cannot be in conjunction except at one determined point. (He was only aware of four.) Today the commensurabilities are usually expressed

$$n_5 - 3n_1 + 2n_2 = 0°\!\cdot\!079$$
$$n_1 - n_2 - 2n_3 + n_4 = 0°\!\cdot\!0034$$

For ninety-seven years only four satellites were known in the system of

[103] Dollfus, A., 'The discovery of Janus', *Sky and Telescope*, p. 136 (Sept. 1967).
[104] Roy, A. E., *J.B.A.A.*, *63*, 212 (1953).
[105] See chapter 12, p. 223.
[106] D'Arrest, *Astron. Nachr.*, *79*, 23 (1872).

The solar system

Uranus and then in 1948 Kuiper discovered Miranda (Uv), close in to
the primary; indeed, it is less than ten seconds of arc away from it as seen
from Earth. Its actual distance is but 66,000 miles above the surface of
the planet, and at the great distance of Uranus from the Sun (19·18 a.u.),
the discovery speaks much for the optical quality of the 86 in. reflector
of the McDonald Observatory and the conditions for good photography
in the skies of Texas.

THE SATELLITES OF NEPTUNE

Two satellites revolve about Neptune. The inner one, Triton, is large
(3700 km. diam.) and the outer, Nereid, is small (300 km. diam.). The
inner has a retrograde orbit and the outer a direct orbit. The disparity
in the orbits is even more marked when we note that Nereid is at the
great distance of 0·037 a.u. from the primary, whereas Triton is close in
at 0·002 a.u. Further, the eccentricity of Triton's orbit is zero, whereas
Nereid has an elliptic orbit of high eccentricity, viz. 0·7493. In contrast
to the beautifully regular system of satellites moving about Uranus,
Neptune presents us with a system that is small yet highly irregular in
its behaviour. The confirmation of Triton's retrograde orbit was not
made until long after its discovery.[107] The conditions here are quite
alien to the system of Uranus, where the revolutions of the satellites are,
strictly speaking, in harmony with the main motions of the solar system
but 'retrograde' by virtue of the high inclination of the orbit to the
ecliptic. In contradistinction the motion of Neptune's satellite Triton is
retrograde to the rotation of the parent planet and to the revolution of
the planet about the Sun. In consequence we have at the outermost
boundary of the solar system a satellite body moving in an anomalous
manner, such as that seen in the systems of Saturn and Jupiter by their
wayward satellites Phoebe, (Six) and Jxii, Jxi, J$viii$ and Jix.

We should not fail to note that Nereid's eccentricity is the greatest of
any satellite and this enables it to vary its distance from its primary
within the truly astonishing limits of 867,000 miles (1,395,306 km.) to
over 6×10^6 miles ($9·7 \times 10^6$ km.).

---

[107] The satellite was discovered in 1846 by Lassel but not recognized as such until his observations of 8 and 9 July 1847. The motion, whether direct or retrograde, was still in doubt as
late as 1852 (see Professor Grant's *History of Physical Astronomy*, p. 287) and 1858 (see Sir John
F. W. Herschel's *Outlines of Astronomy*, p. 369).

William Lassel (1799–1880) was a Lancashire brewer and keen amateur astronomer.
Grant reminds us of the citation at the award of the gold medal to Lassel by the Astronomical
Society: The simple facts are, that Mr. Lassel cast his own mirror, polished it by machinery
of his own contrivance, mounted it equatorially in his own fashion, and placed it in an
observatory of his own engineering. A private man, of no large means, in a bad climate, and
with little leisure, he has anticipated, or rivalled, by the work of his own hands, the contrivance of his own brain, and the outlay of his own pocket, the magnificent refractors with
which the Emperor of Russia and the citizens of Boston have endowed the observatories of
Pulkowa and the Western Cambridge.'—*Month. Proc. Ast. Soc., Feb.* 1849.

A useful table of the Sun/Planet Mass ratios given by Kaula and based on the work of Wilkins,[108] Ash *et al.*[109] and Anderson[110] is given below.

SUN PLANET MASS RATIOS

	I.A.U. adopted	Improved Solutions	
	(Wilkins)	(Ash et al.)	(Anderson)
Mercury	6 000 000	6 021 000 ± 53 000	6 005 000 ± 18 000
Venus	408 000	408 250 120	408 522·6 0·6
Earth + moon	329 390	328 900 60	328 900 1·8
Mars	3 093 500	3 111 000 ± 9 000	3 098 600 600
Jupiter	1 047·355		1 047·44 0·02
Saturn	3 501·6		3 499·1 0·4
Uranus	22 869		22 930 6
Neptune	19 314		19 070 21
Pluto	360 000		400 000 ± 40 000

[108] Wilkins, G. A., *Quart. J. Roy. Astron.*, *5*, 23 (1964); *6*, 70 (1965).
[109] Ash, M. E., Shapiro, I. I., and Smith, W. B., *Astron. J.*, *72*, 338 (1967).
[110] Anderson, J. D., *J.P.L. Tech. Rep.*, *32–816*, 298 (1967).

MODEL OF THE PLANETS AND THEIR SATELLITES

TABLE OF PLANETARY DATA

	Sun	Mercury	Venus	Earth	Moon	Mars	Jupiter	Saturn	Uranus	Neptune	Pluto
YEAR OF DISCOVERY	—	—	—	—	—	—	—	—	1781	1846	1930
Mean distance from Sun (millions of miles)		36·0	67·2	92·9	—	141·5	483·3	886·1	1783	2793	3666
Mean distance from Sun (astronomical units)		0·387	0·723	1·00	—	1·524	5·202	9·539	19·182	30·057	39·520
Equatorial dia. km.	1392000	4880	12112	12756	3476	6790	142800	119300	47100	48400	<5800
Equatorial dia. miles	864947	3032	7526	7926	2160	4219	83732	74129	29266	27837	<3604
Polar dia. km.	—	—	—	12714	—	6750	133500	107700	43800	44000	—
Polar dia. miles				7900		4194	82953	66922	27216	23740	
Diameter. Earth = 1	109·12	0·379	0·965	1	0·2725	0·532	11·197	9·355	3·70	3·51	0·47
Mass. Earth = 1	33×10^4	0·055	0·815	1	0·0123	0·108	318	95	14·52	17·25	0·14
Mass. Sun = 1	1	$1·6 \times 10^{-7}$	$2·4 \times 10^{-6}$	3×10^{-6}	—	$3·2 \times 10^{-7}$	$9·5 \times 10^{-4}$	$2·9 \times 10^{-4}$	$4·4 \times 10^{-5}$	$5·2 \times 10^{-5}$	—
Mean density. Earth = 1	0·26	∼1	∼0·9	1	—	0·7	0·24	0·12	0·2	0·29	—
Density. Water = 1	1·409	5·41	4·99	5·517	3·342	3·94	1·330	0·706	1·70	2·26	5·50
Volume. Earth = 1	1303600	0·055	0·902	1	0·0203	0·150	1318·9	743·7	47·1	53·7	0·10
Number of Moons	—	—	—	1	—	2	12	10	5	2	—
Sidereal period of axial rotation	25d·38	59d ± 5	**247d** ± **5**	23h 56m 04s		24h 37m 23s		10h14m		15h 48m	
Sidereal period of axial rotation					27d·322		9h50m 30s		**10h49m**		6d·39
Mean Synodic period days	—	115·88	583·92	—	—	779·94	398·88	378·09	369·66	367·49	
Sidereal period of revolution days		87·969	224·701	365·256	—	686·980	4332·59	10759·20	30685·0	60190	250·6 years
Inclination of equator to ecliptic	7°15′				1·°32′						
Inclination of orbit to ecliptic		7°	3°23′	—	—	1°50′	1°18′	2°29′	0°46′	1°46′	17°
Escape Velocity kms^{-1}	617·5	4·22	10·27	11·18	2·38	5·02	60·22	36·24	22·40	24·90	5·10
Orbital eccentricity. e.	—	0·2056	0·0067	0·0167	—	0·0933	0·048	0·0556	0·0472	0·0085	0·2502

Bold figures refer to retrograde rotations.

Chapter 6

The atmospheres of the planets

—the earth, seems to me a sterile promontory,
this most excellent canopy, the air, look you,
this brave o'er-hanging firmament,
this majestical roof fretted with golden fire,
why, it appeareth nothing to me but a foul
and pestilent congregation of vapours.
SHAKESPEARE

Up until recent years *all* our information of the atmospheres of the planets in the solar system with the one exception of our own Earth has been based upon spectroscopic data gathered on the Earth. Now we have some evidence from a direct sampling of the atmosphere of Venus and from a flight in close proximity to the Martian surface. Still, our knowledge is extremely small concerning the atmospheres of our neighbouring worlds. Spectroscopic studies of remote planetary atmospheres in space made on the Earth are difficult to interpret owing to the absorption of light by the Earth's atmosphere and the high content of water vapour and oxygen in it, which have a swamping effect upon the detection of small amounts of these materials in the planetary atmospheres under examination.

It can be shown from purely academic considerations that certain gases may be lost from a planet. It is well known, from the kinetic theory of gases, that the mean speed V of a gas molecule is a function of the temperature of the gas T and the mass m of the molecule under consideration and that the said mean speed may be given in a somewhat inaccurate[1] form by the simplified expression:

$$1/2 \ mv^2 = 3/2 \ kT$$

which may be rewritten

$$V = \sqrt{\frac{3kT}{m}}$$

---

[1] Most simplifications lead to incomplete and thus inaccurate descriptions. The escape flux is based on the particles whose velocity vector lies in the outbound hemisphere with a magnitude equal to or greater than the velocity of escape. The mathematics and the many assumptions made concerning the density of the planet's surface and the isothermal nature of the atmosphere are complex. See *Nature Physical Science*, *232*, 101 (1971).

27. Orrery by Francois Du Commun, *c.* 1813. The sphere was painted by Girardet.

28. The house of Eisinga called 'De Ooyevaar' (The Stork) after the stork chiselled in the gable.

29. The orrery of Eise Eisinga in the ceiling of his house at Franeker, Friesland, Netherlands.

30. English orrery by Tho. Tompion & Geo. Graham, London, *c.* 1710.

31. Front view of David Rittenhouse Orrery, built for the college of New Jersey; restored in 1954.

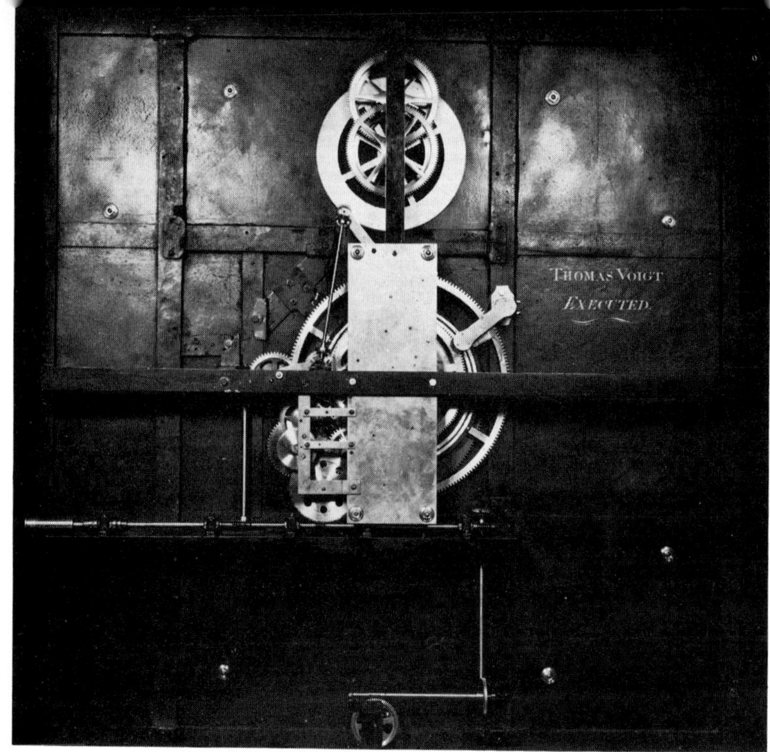

32. Rittenhouse Orrery (rear view).

33. Joseph Pope's Orrery, *c.* 1787. This orrery is one of the largest known it is 1990 mm. in diameter. Now in the Collection of Historical Scientific Instruments, Harvard University. (Bought by public lottery in 1788.)

34. Orrery of Le Verrier, *c.* 1847, height 108 cm., diameter 78 cm. It demonstrates the movements of the planet Neptune.

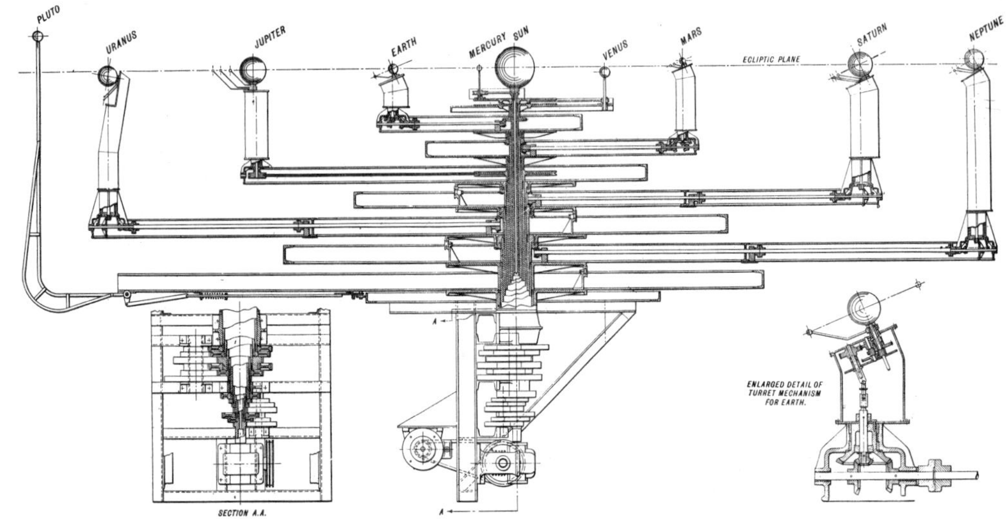

35. Mechanised solar system model made by the Sondes Place Research Institute and displayed at the Festival of Britain in 1951.

36. Urameton Orrery, 1955, designed and made by the author. Loaned to the BBC to illustrate Professor Lyttleton's television lecture 'The Modern Universe', 1955, November 22.

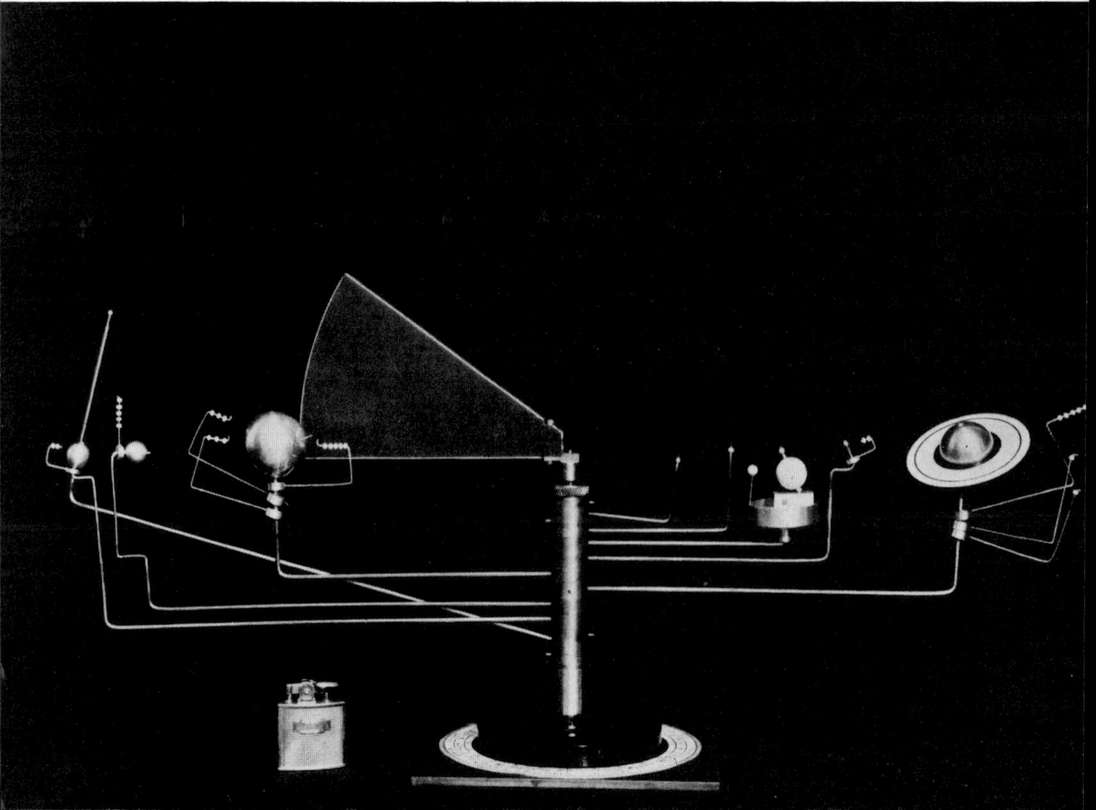

37. The 100 in. (254 cm.) 'Hooker' reflecting telescope of the Mount Wilson Observatory, U.S.A.

38. The Jodrell Bank 250 ft. (7620 cm.) radio telescope, Cheshire, England. (See B. Lovell, *The Story of Jodrell Bank*, 1968.)

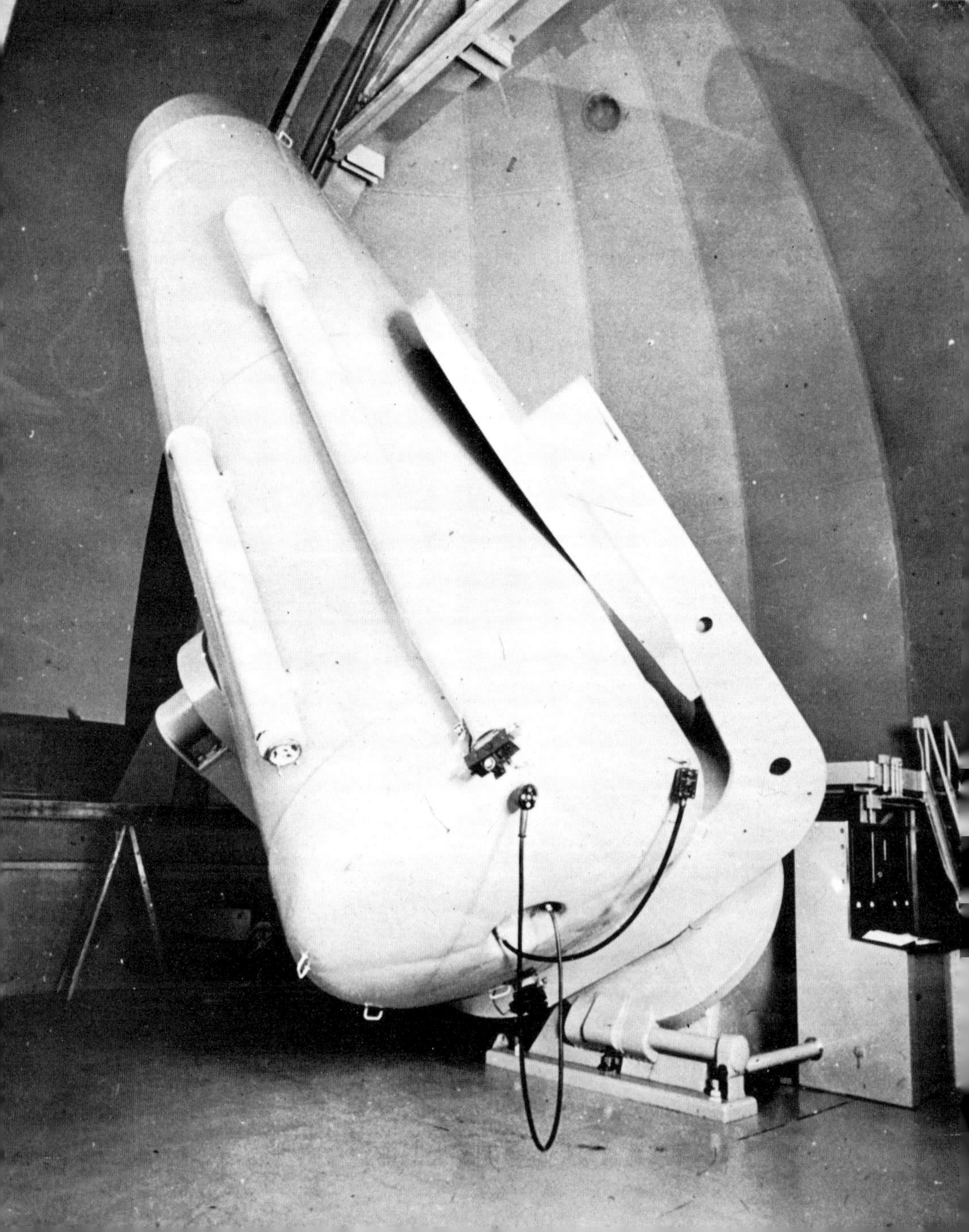

39. The 48 in. × 72 in. f/2·5 Schmidt reflecting telescope at the Palomar Observatory. (See H. C. King, *The History of the Telescope*, p. 372, London 1955.)

40. The 200 in. (508 cm.) Hale reflecting telescope. The telescope is pointing to the zenith and is seen from the south. The slot along the length of the southern side of the tube prevents obstruction of the beam from the coudé mirror in all attitudes of the telescope. The inner cylinder is a man-riding prime focus cage. (See H. C. King, *The History of the Telescope*, p. 404, London 1955.)

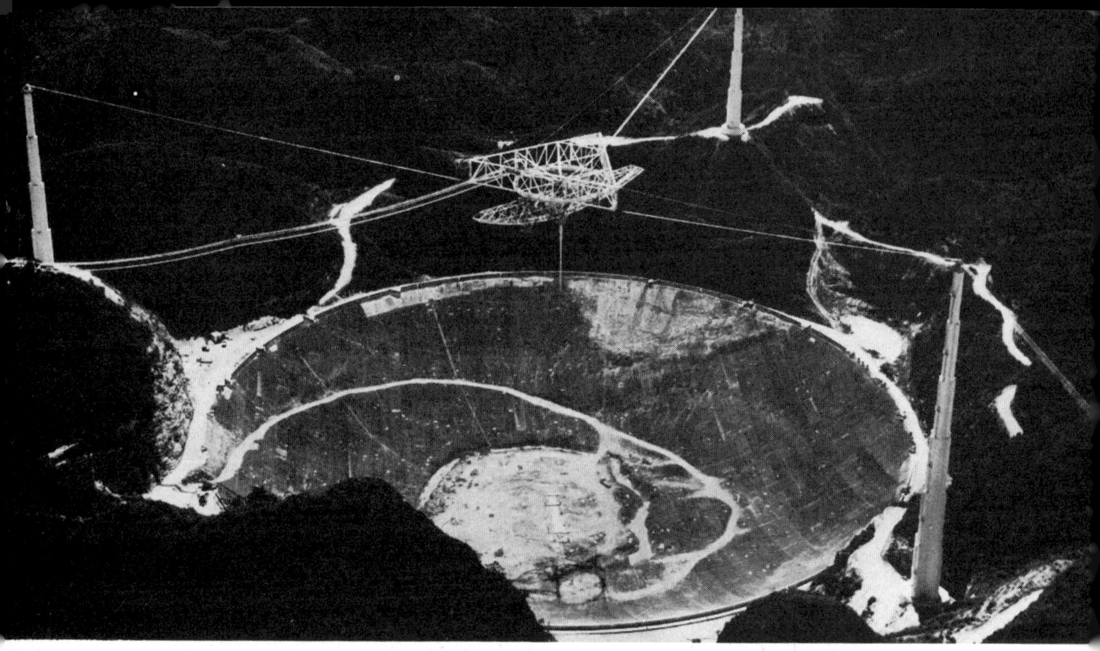

41. The dish of the Arecibo Ionospheric Observatory in Puerto Rico is the largest in the world and is used for the exploration of outer space. It measures 1000 ft. across and covers 18·5 acres. It is 158 ft. deep in the centre. A road leads down the rim to the centre of the bowl. The observatory, conceived and now operated by Cornell University, was made possible by funds totalling nearly $9 million from the Advanced Research Projects Agency of the US Department of Defence.

42. The moon at the gibbous phase.

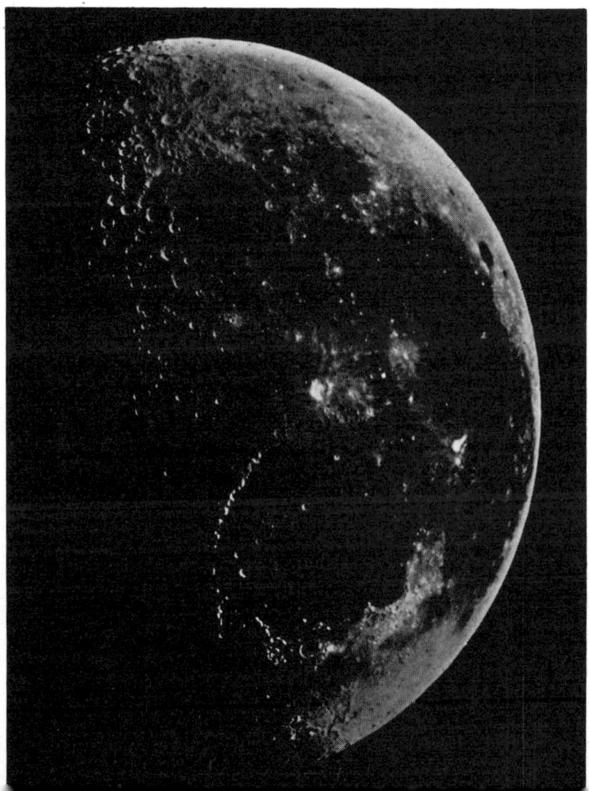

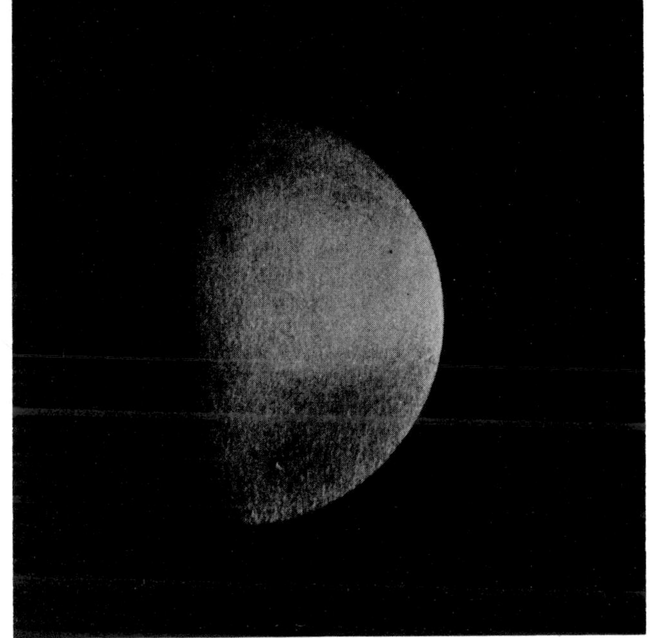

43. Mercury as a 'morning star', 1968, November 4, 13 h. 25 m. U.T., exhibiting the gibbous phase (70 per cent). Superior conjunction occured on 1968, 7 December. Drawing by J. B. Murray, 18 in. O.G. × 300.

44. Mercury: 1969, January 18, 17 h. 30 m.; 1969, April 29, 19 h. 50 m.; 1969, May 8, 20 h. 10 m.; April 29, diameter 6·6 seconds of arc, phase 0·582, distance 1·066 a.u. (18 in. O.G. × 300). Drawings by J. B. Murray.

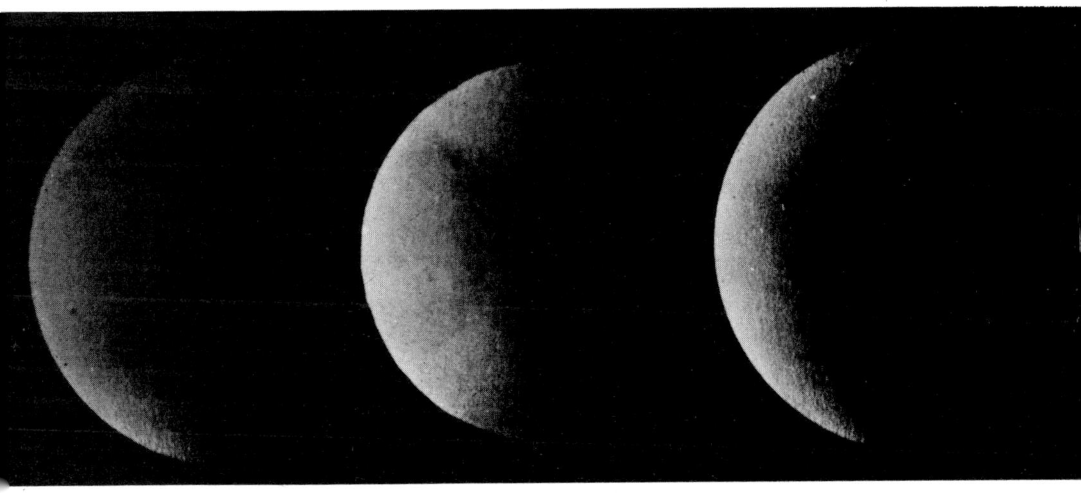

45. Drawings of Venus by J. Hedley Robinson, Devon. 1968, December 19, 17 h.; 1969, March 5, 14 h. 30 m.; 1969, March 8, 18 h. 25 m.

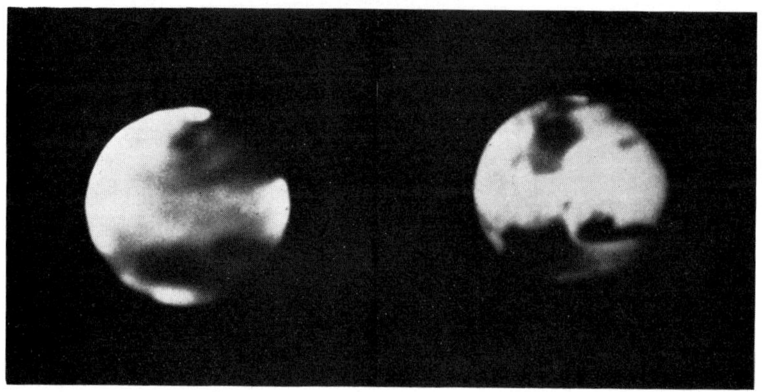

46. *Left:* Mars in blue light. *Right:* Mars in red light. The photographic determinations give a decrease in the diameter of the planet with increasing wavelength of the light used. A phenomenon discovered by Wright in 1924 and called after him the Wright Effect. The following table illustrates the phenomenon:

Wavelength (mμ)	Diameter of the planet Mars
360 u.v.	9″ 27
440 violet	9″ 13
560 yellow	9″ 17
760 Infra red	8″ 96

The photographs were taken by the 200 in. Hale reflecting telescope.

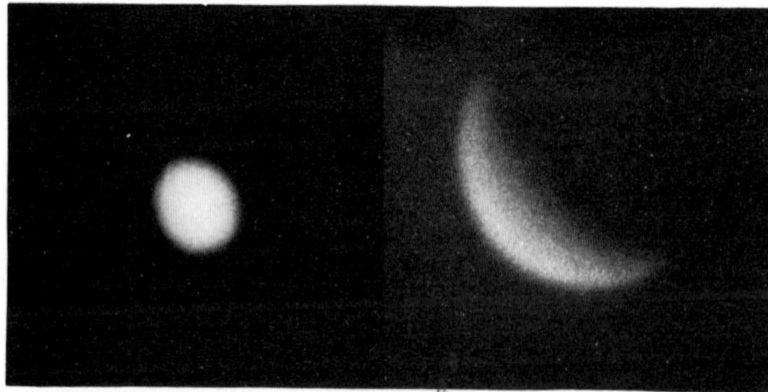

47. Venus at the gibbous and crescentic phase. *Left:* 1967, March 3 (f85 × 7) diameter 11 seconds of arc. *Right:* 1967, July 19 (f85 × 7) diameter 34·7 seconds of arc, distance 0·484 a.u. Photographs by Cdr. H. R. Hatfield, Sevenoaks, Kent with a 12 in. reflector.

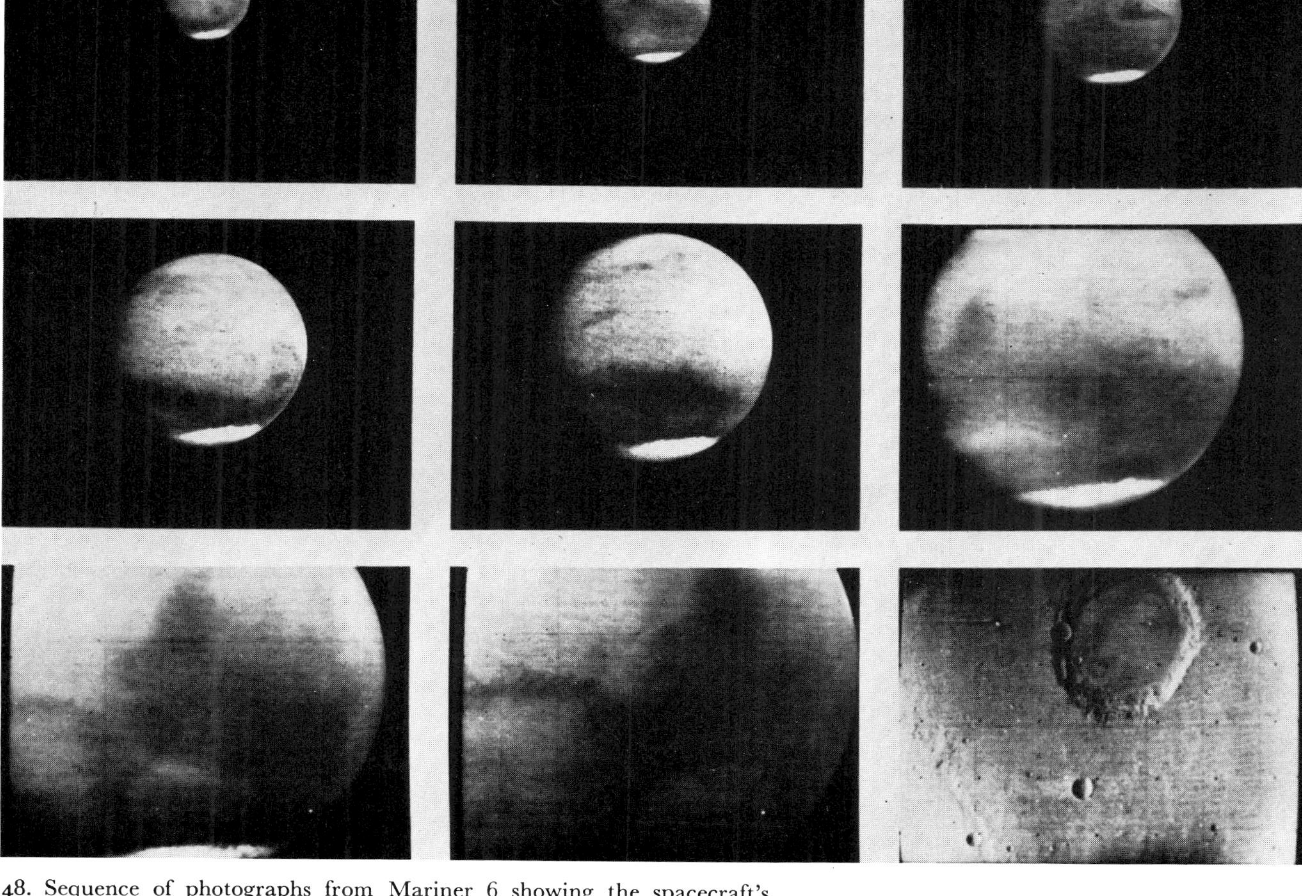

48. Sequence of photographs from Mariner 6 showing the spacecraft's approach to Mars on 1969, July 29–31. The first picture was taken from a distance of 1,241,350 km. and the last from about 3500 km. The final frame shows a crater about 38 km. in diameter. The bright area is the south polar cap.

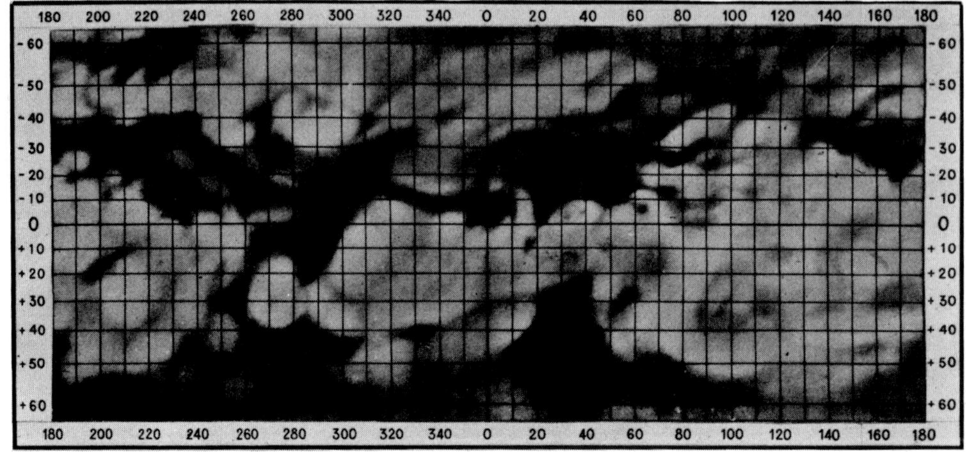

Acidalium M. (30°, + 45°)
Aeolis (215°, — 5°)
Aeria (310°, + 10°)
Aetheria (230°, + 40°)
Aethiopis (230°, + 10°)
Amazonis (140°, 0°)
Amenthes (250°, + 5°)
Aonius S. (105°, — 45°)
Arabia (330°, + 20°)
Araxes (115°, — 25°)
Arcadia (100°, + 45°)
Argyre (25°, — 45°)
Arnon (335°, + 48°)
Aurorae S. (50°, — 15°)
Ausonia (250°, — 40°)
Australe M. (40°, — 60°)
Baltia (50°, + 60°)
Boreum M. (90°, + 50°)
Boreosyrtis (290°, + 55°)
Candor (75°, + 3°)
Casius (260°, + 40°)
Cebrenia (210°, + 50°)
Cecropia (320°, + 60°)
Ceraunius (95°, + 20°)
Cerberus (205°, + 15°)
Chalce (0°, — 50°)
Chersonesus (260°, — 50°)
Chronium M. (210°, — 58°)
Chryse (30°, + 10°)
Chrysokeras (110°, — 50°)
Cimmerium M. (220°, — 20°)
Claritas (110°, — 35°)

Copais Palus (280, + 55°)
Coprates (65°, — 15°)
Cyclopia (230°, — 5°)
Cydonia (0°, + 40°)
Deltoton S. (305°, — 4°)
Deucalionis R. (340°, — 15°)
Deuteronilus (0°, + 35°)
Diacria (180°. + 50°)
Dioscuria (320°, + 50°)
Edom (345°, 0°)
Electris (190°, — 45°)
Elysium (210°, + 25°)
Eridania (220°, — 45°)
Erythraeum M. (40°, — 25°)
Eunostos (220°, + 22°)
Euphrates (335°, + 20°)
Gehon (0°, + 15°)
Hadriacum M. (270°, — 40°)
Hellas (290°, — 40°)
Hellespontica Depressio (340° — 6°)
Hellespontus (325°, — 50°)
Hesperia (240°, — 20°)
Hiddekel (345°, + 15°)
Hyperboreus L. (60°, + 75°)
Iapigia (295°, — 20°)
Icaria (130°, — 40°)
Isidis R. (275°, + 20°)
Ismenius L. (330°, + 40°)
Jamuna (40°, + 10°)
Juventae Fons (63°, — 5°)
Laestrigon (200°, 0°)
Lemuria (200°, + 70°)

Libya (270°, 0°)
Lunae Palus (65°, + 15°)
Margaritifer S. (25°, — 10°)
Memnonia (150°, — 20°)
Meroe (285°, + 35°)
Meridianii S. (0°, — 5°)
Moab (350°, + 20°)
Moeris L. (270°, + 8°)
Nectar (72°, — 28°)
Neith R. (270°, + 35°)
Nepenthes (260°, + 20°)
Nereidum Fr. (55°, — 45°)
Niliacus L. (30°, + 30°)
Nilokeras (55°, + 30°)
Nilosyrtis (290°, + 42°)
Nix Olympica (130°, + 20°)
Noachis (330°, — 45°)
Ogygis R. (65°, — 45°)
Olympia (200°, + 80°)
Ophir (65°, — 10°)
Ortygia (0°, + 60°)
Oxia Palus (18°, + 8°)
Oxus (10°, + 20°)
Panchaia (200°, + 60°)
Pandorae Fretum (340°, — 25°)
Phaethontis (155°, — 50°)
Phison (320°, + 20°)
Phlegra (190°, + 30°)
Phoenicis L. (110°, — 12°)
Phrixi R. (70°, — 40°)
Promethei S. (280°, — 65°)
Propontis (185°, + 45°)

Protei R. (50°, — 23°)
Protonilus (315°, + 42°)
Pyrrhae R. (38°, — 15°)
Sabaeus S. (340°, — 8°)
Scandia (150°, + 60°)
Serpentis M. (320°, — 30°)
Sinaï (70°, — 20°)
Sirenum M. (155°, — 30°)
Sithonius L. (245°, + 45°)
Solis L. (90°, — 28°)
Styx (200°, + 30°)
Syria (100°, — 20°)
Syrtis Major (290, + 10)
Tanais (70°, + 50°)
Tempe (70°, + 40°)
Thaumasia (85°, — 35°)
Thoth (255°, + 30°)
Thyle I (180°, — 70°)
Thyle II (230°, — 70°)
Thymiamata (10°, + 10°)
Tithonius I. (85°, — 5°)
Tractus Albus (80°, + 30°)
Trinacria (268°, — 25°)
Trivium Charontis (198, + 20)
Tyrrhenum M. (255°, — 20°)
Uchronia (260°, + 70°)
Umbra (290°, + 50°)
Utopia (250°, + 50°)
Vulcani Pelagus (15, — 35)
Xanthe (50°, + 10°)
Yaonis R. (320°, — 40°)
Zephyria (195°, 0°)

49. Named Martian markings. The main surface markings are designated by names according to the co-ordinates.

50. The close approach to Mars by the Mariner 6 spacecraft. The area of Mars displayed is 83 × 72 km. The large crater is about 24 km. in diameter. The camera distance above the surface of the planet is 3460 km.

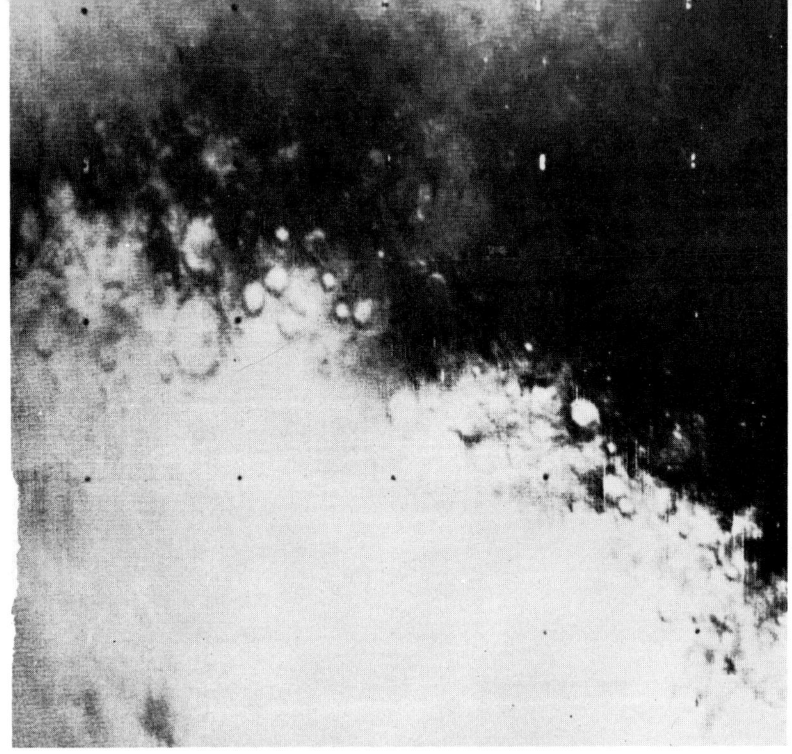

51. Mariner 7 photograph showing the edge of the south polar cap on Mars. Scientists at the Jet Propulsion Laboratory in California believe the light area to be either a layer of frozen carbon dioxide a few feet deep or a layer of water ice, less than an inch deep. Craters near the edge show covered southward-facing slopes and exposed slopes in the other direction. The spacecraft flew within 3500 km. of the planet on 1969, August 5.

52. Mariner 7 spacecraft, 1969, August 5. A television relay picture of the south polar cap of Mars. The five wide angle pictures were taken 84 seconds of time apart, left to right. The six narrow angle pictures cover at high resolution small areas within the wide angle pictures.

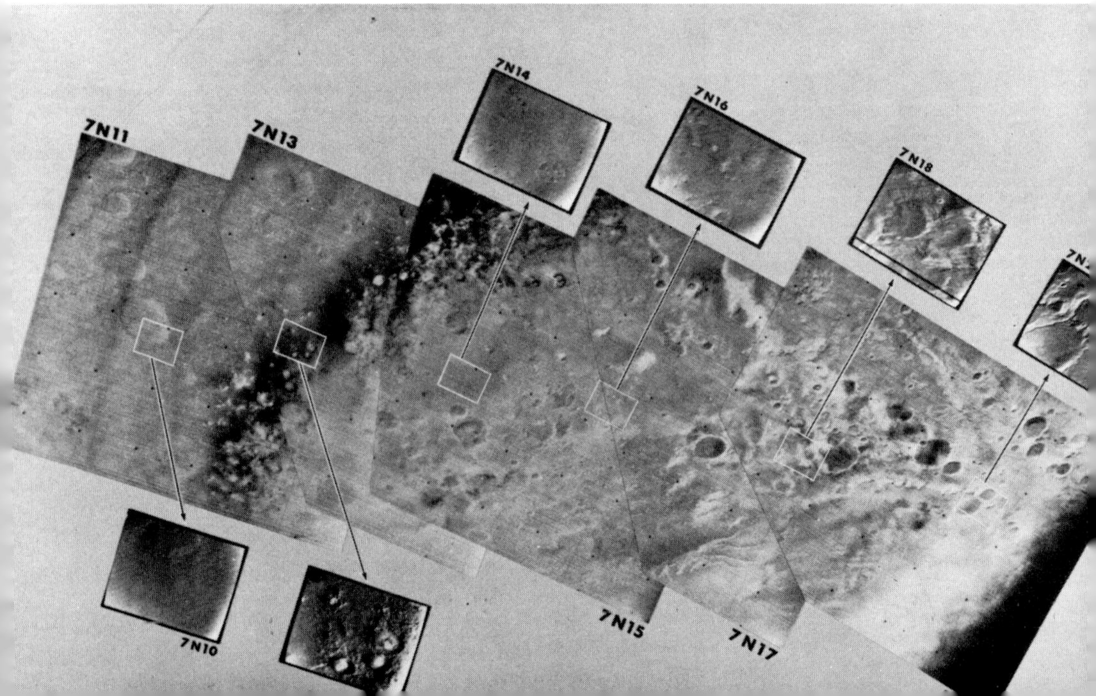

where k is Boltzmann's constant[2] in ergs per degree Kelvin, T is in degrees Kelvin, m is expressed in grams and V, the mean speed, is in cm. per sec. If now the mean speed V of the molecule is equal to the velocity of escape V_e for the planet that is host to it, then the molecule will move away into space and eventually be lost to that planet. If V is less than V_e then the molecule may, given sufficient time, diffuse away into space and the planet lose its gaseous covering, i.e. its atmosphere. From the above it is possible to calculate the absolute temperature in degrees Kelvin at which any given planet may lose specific molecules. The temperatures for escape of the molecules of hydrogen, water, nitrogen, oxygen, and carbon dioxide are shown below for the planets out to Neptune.

TEMPERATURES IN °K REQUIRED FOR ESCAPE OF MOLECULES FROM PLANETARY ATMOSPHERE

	H_2	H_2O	N_2	O_2	CO_2
Mercury	56	500	779	889	1,220
Venus	339	3,030	4,720	5,390	7,410
Earth	403	3,600	5,610	6,400	8,810
Mars	83	739	1,150	1,310	1,800
Jupiter	10,700	95,500	148,000	170,000	233,000
Saturn	3,470	31,000	48,200	55,000	75,700
Uranus	1,480	13,200	20,500	23,400	32,300
Neptune	2,000	17,900	27,900	31,800	43,800

To understand the table we need to know the temperature in those layers of the atmosphere of the planet from which the specific molecules are most likely to escape, and this is generally thought to be the exosphere that is to say, that part of the atmosphere in which the molecules suffer few collisions in their movements. We have also to invoke the much misused concept of uniformitarianism; that is to say, we assume with doubtful validity that the temperatures obtaining today in those remote bodies resemble the temperatures existing in past ages (periods in excess of 10^9 years). This is a point of some difficulty that is not to be overcome by refined instruments nor the application of mathematics. We now consider the planets individually.

MERCURY

The small planet Mercury appears to be devoid of an atmosphere. No spectroscopic bands or lines have been observed in its spectrum that differ from those observed in direct sunlight.[3] Much of the speculation concerning Mercury's atmosphere is now invalid, since it is known that the planet does *not* present at all times the same hemisphere to the Sun

[2] Boltzmann's constant $K = (1 \cdot 38041 \pm 0 \cdot 0001) \times 10^{-16}$ ergs/deg K.
[3] Kozyrev, N. A., 'The Atmosphere of Mercury', *Sky and Telescope*, p. 339 (June 1964).

and consequently its long assumed *dark side* believed to be at the intense cold of interstellar space is seen to be a fiction. The rotation and heating of Mercury, based on the clear understanding that the rotational period is coupled to its orbital period of revolution about the Sun in the ratio of 2:3, is given by Soter and Ulrichs.[4] A useful drawing (Fig. 24) taken

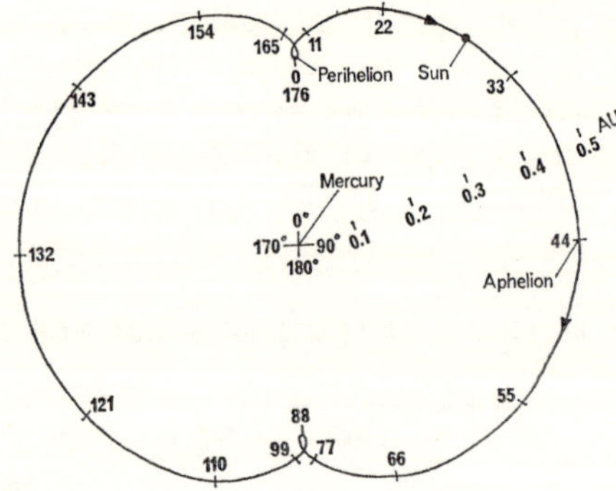

Fig. 24. Diurnal path of the Sun about Mercury, drawn to scale. The relative positions of the Sun are marked at eleven day intervals with the planet held as a fixed reference. Planetographic longitudes are indicated for Mercury. Note the loops in the apparent solar motion that occur at perihelion.

from their communication makes it clear that from a point on Mercury the Sun would be seen to move slowly retrograde through less than its own apparent diameter for 8·1 days around each perihelion passage, when the Sun's disc would hover over one or the other hot pole, so that these two regions virtually monopolize the perihelion exposure, which must be severe. How this and the now properly understood movements of the planet are to reshape the understanding of the conditions on and above the planet's surface, is still in the process of formulation. Our knowledge of the conditions on Mercury is totally inadequate and we must be grateful for the resolution of its movements after many centuries of deep misconceptions.

VENUS

The atmosphere of Venus reveals absorption bands characteristic of carbon dioxide, and these were detected as long ago as 1932.[5] It is now known that the abundance of carbon dioxide is such that it is 70,000 times that of the atmosphere on Earth. The characteristic spectroscopic indicators of nitrogen and water have also been detected from observations made from a balloon at a height above the Earth of 80,000 ft. (24,384 m.). There is weak evidence for the presence of carbon monoxide.

[4] Soter and Ulrichs, 'Rotation and Heating of the Planet Mercury', *Nature*, *214*, 1315 (1967).
[5] Adams, W. S., and Dunham, T., *Pub. A.S.P.*, *44*, 243.

114

The fly-by of the Mariner spacecraft has confirmed the existence of a hot surface at a temperature of about 700°K., which agrees well with earlier measurements made by radio techniques from Earth. The most recent report[6] compares the results of both the Mariner 5 spacecraft of the USA and the Venera IV spacecraft of the USSR, which made measurements close to the planet on successive days in October 1967. The broad conclusion reached is that the atmosphere of Venus has a temperature gradient such that the temperature increases 10 degrees centigrade for each kilometre of decrease in altitude, reaching 100° at an altitude of about 45 km. above the planet's surface. By extrapolation the surface temperature is thought to be 423°C., or some one hundred degrees above the melting point of lead. The atmospheric pressure at the surface is roughly one hundred times that of the pressure on the Earth's surface.

Measurements lead those skilled in this art to the view that Venus has less than 0·7 per cent of water vapour in its lower atmosphere, making surface water four hundred times *less* abundant on Venus than on Earth.[7] The concentration of oxygen on Venus is no more than the amount that accumulates as a result of water dissociation in a period of a year. This amount is 10^{-4} g/cm^2 and is far too small to be effective as a shielding layer against solar ultra-violet radiation and is one reason for the dryness of Venus. It is now clear that the temperature in the lower atmosphere of Venus far exceeds the temperature of a simple black body[8] at the distance of Venus from the Sun. It is now known that the Venera IV instruments did not go below an altitude of 25 km. when they recorded a pressure of 20 atmospheres and a temperature of 545°K.

Even as late as March 1969 the Mariner 5 experiment was still providing scientific dividends,[9] in that the analysis of the probe's findings

[6] Sagan, C., and Pollack, J. B., 'On the Structure of the Venus Atmosphere', *Icarus*, *10*, p. 274 (1969). *The Venus Atmosphere*. Edited by Jastrow, R., and Rasool, S., Institute for Space Studies, Goddard Space Flight Centre NASA.

[7] Since this chapter was written Rasool and de Bergh have put the chief differences between Venus and Earth to the single circumstance that Venus was formed 30 per cent closer to the Sun than was the earth; see *Nature*, *226*, 1037 (1970).

[8] *Black body*. A body which when raised to incandescence emits a continuous spectrum of light rays.

Black body radiation according to Kaye and Laby is the radiation within a completely closed cavity, the walls of which are at a uniform temperature. The same term is applied to the radiation emitted into the cavity by the enclosing wall, and such a wall, or any surface having in all respects the same emission, is called a full, black-body or Planckian radiator. The wall temperature of a closed cavity is the single parameter on which the intensity and spectral distribution of the contained radiation depends. With complete closure of the cavity there is no access to the radiation. Once a viewing aperture is introduced the radiation is modified in a way which depends on the size and shape of aperture and cavity and on the emissivity of the cavity walls. The usual experimental approximation to a closed cavity is a tube closed at one end and open or partially open at the other; the radiation from the closed end is observed through the aperture at the open end, along the tube axis (Ribaud, 'Traité de Pyrométrie Optique', *Rev. d'Optique* (1931). If the tube walls have emissivity 0·5 for all wavelengths and the tube length is ten times its radius, the observed intensity of radiation is about 3 per cent below that corresponding to a completely closed cavity. (For a discussion of the defect from full radiation of the emission from practical black-bodies of different shape and wall properties, see J. C. de Vos, *Physica*, 20, 669 (1954).)

[9] *Nature*, *221*, 1100 (1969); *J. Geol. Phys. Res.*, *74*, 1128 (1969).

showed an excess of deuterium[10] over hydrogen in the Venusian outer atmosphere. The deuterium atoms outnumber hydrogen atoms by 10 : 1 at an altitude of 6500 km. This is difficult to explain because on Earth, on the average, there is only one part of deuterium to seven thousand parts of normal hydrogen. On 16/17 May 1969 two further Russian probes,[11] designated Venera V and Venera VI, entered the Venusian atmosphere at 100 km. above the surface, but little information of the results gathered have come to the author's attention. In regard to the best information available to date from Mariner 5, we can say that the surface temperature on Venus is about 750°K. and the pressure about 90 atmospheres.

Seven experiments will be carried out on the Mariner probe that will fly by Venus and Mercury in 1974. The spacecraft will be launched in the autumn of 1973, swinging by Venus at a distance of 3300 miles and being directed by gravity alone to within 625 miles of Mercury a month later.

The seven experiments, accounting for 113 pounds of the satellite's 900 pounds weight, will take photographs of the two planets, measure the fields and particles surrounding them, and study their atmospheres and ionospheres.

Because of the two extra antennae, 210 feet in diameter, that are under construction at Tidbinbilla, Australia, and near Madrid, many of the pictures taken by the Mariner's two television cameras can be transmitted directly to Earth instead of being tape recorded and stored until the Goldstone antenna comes over the horizon. With three antennae placed round the Earth, the Mariner probe should be able to transmit many more pictures than its two predecessors that flew past Mars in 1969.

MARS

The only constituents of the Martian atmosphere that are established beyond doubt are carbon dioxide and water. The water content is very small and upper limits have been set for oxygen and ozone, N_2O, CH_4, C_2H_6, ammonia and carbon monoxide by spectroscopic observations over a protracted period from the Earth. The temperature of the atmosphere at the planet's surface is believed to be 270°K., with the temperature of the exosphere at 1100°K. In July 1965 the US spacecraft Mariner 4 passed within 6118 miles of the Martian surface in a spectacular flight and indicated that the temperature of the atmosphere at the point it had under test was about −113 degrees Celsius (160°K.). A record was made of two values for the lower atmospheric temperature viz. 160°K. and 230°K. The first relates to a winter's day on the

10 *Deuterium.* Heavy hydrogen, the hydrogen isotope of atomic mass 2.
11 'Parachutes over Venus', *Nature*, *222*, 712 (1969).

Martian surface in latitude 50° South, the second to a summer's night in latitude 60° North. Prior to the results from the Mariner 4 space probe, experiments from Earth had put the pressure of the atmosphere at the surface of Mars at 8·5 per cent of the pressure on Earth, but more recent measurements have reduced it to as small as 2·5 per cent. The space probe gives figures which suggest that the surface pressure is as low as one-hundredth that of the pressure on Earth. We have then, the interesting results that the Earth's atmospheric pressure stands midway between that of Mars and Venus. This is clearly shown in the graph below (Fig. 25).

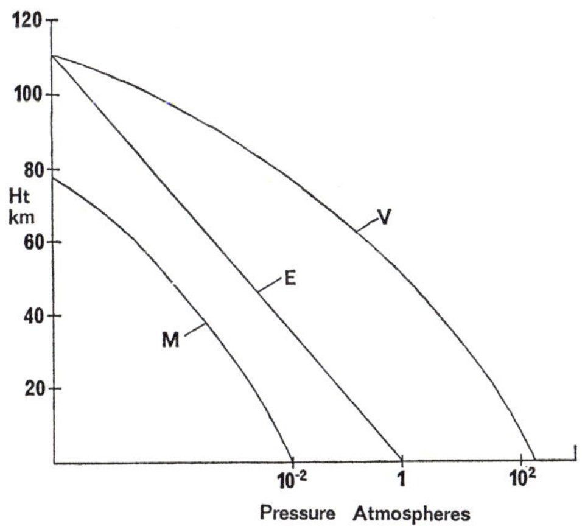

Fig. 25. Variations of pressure with height in the atmospheres of Mercury, Earth and Venus.

The latest results to be given[12] show that there is possibly a multilayer structure at the limb, from about 10 km. up to 50 km. which the infrared spectrometer experiment indicates may be solid carbon dioxide. Dr G. Pimentel has reported that there is also infrared evidence for silicate dust in the atmosphere, and a 'quite low' abundance of H_2O may have been detected. New upper limits are being placed on the extent of NO, N_2O, OCS, CH_4, O_3 and NH_3. As expected, the ultraviolet spectrometer showed a spectrum similar to the laboratory spectrum of carbon dioxide irradiated by sunlight, but what is odd is the scarcity of atomic oxygen. Carbon dioxide in the laboratory is easily broken down by sunlight. A broad absorption band at 2500 Å interpreted as ozone is detected over the polar caps, but not over the unfrosted areas. Either the polar caps are acting as a cold trap, or for some reason the band is easier to see against a light background. The ultraviolet spectrometer has also shown that aerosols of sizes less than a tenth of a micron or so are present in large quantities, scattering twice as much light as the atmosphere itself.

[12] *Nature*, *227*, 1002 (1970).

The solar system

In 1973 the US Viking Experiments are to be undertaken with an attempted soft landing on Mars, and two further Mariner space probes are planned to orbit Mars for a period of some three months in 1971 to attempt to resolve some of the present anomalies.

THE JOVIAN PLANETS

In the atmospheres of the large extra-terrestrial planets (The Jovian planets), Jupiter, Saturn, Uranus and Neptune, the presence of methane and ammonia has been detected by the spectroscope. The presence of hydrogen was discovered in 1962 and it would appear that the amount of molecular hydrogen is large, indeed, more than one hundred times the amount of methane present. The temperature is thought to be as low as 130°K.

From what little is known with certainty the three outer planets have similar atmospheres, but with lower temperatures as they increase their distances from the Sun. Ammonia bands do not show in the spectroscopic studies that have been made of the atmospheres of Uranus and Neptune. The sparse material available may be summarized in a small table:

Planet	Gases in atmosphere
Jupiter	H_2
	CH_4
	NH_3
	He
Saturn	CH_4
	NH_3
	H_2
	He
Uranus	CH_4
	H_2
	He
Neptune	CH_4
	H_2

A US 'Jupiter' deep entry probe[13] and a fly-by are planned for 1974. A Jupiter orbiting mission is planned for 1976 and a Jupiter/Saturn/Pluto mission for 1977. A Jupiter/Uranus entry probe mission for the early 1980s.

Several of these missions are to be based on what is termed the 'grand-tour' concept, that is to say, a dynamic system that makes use of the gravitational field of the planets *per se* to give the necessary impetus to direct the spacecraft into the desired path and to induce it to orbit, or pass near to, the particular planet in the recesses of the far-flung parts of

[13] See the report 'The outer Solar System, a program for Exploration', Space Science Board of the National Academy of Sciences, *c*. August 1967 (*Nature*, *223*, 661 (1969)).

the solar system. The dynamics of such a complex journey, to be successful, cannot be left to chance, and the position of the planets is conducive
to success only in the late 1970s.[14] A similar opportunity will not occur
again until the second half of the twenty-second century when a generation may have come to maturity that will frown upon such 'frivolities'
and look back upon the present unique attitude of the faustian mind[15] as
a strange aberration.[16]

If we are to learn of what obtains in the realm of the Jovian planets
we must look to the next two decades.

No atmosphere has been observed on Pluto.

[14] See 'Outer Planets Exploration 1972–85', Space Science Board, 2101 Constitution Avenue,
Washington DC 20418; See also *Nature*, *234*, 246, (1971).
[15] See Spengler, O., *Decline of the West* (1926), chapter VI, 'Makrokosmos, Apollinian
Faustian and Magian Soul.
[16] It is now seriously suggested that if man ruins his fertile home on Earth it may be possible
to keep our species going on the desert of the Moon. I am reminded of Philips' words—

> *How vain a thing is Man, whose noblest part,*
> *That soul which thro' the World doth rome,*
> *Traverses Heav'n, finds out the depth of Art,*
> *Yet is so ignorant at home?*

Chapter 7

The interiors of the planets

Peraventure . . . that her interior
iye sawe privily, and gave
to her a secrete monicion of
the great calamities.
HALL 1548

The nature of the interiors of the planets is not one that may be resolved from observation alone and it poses a particularly difficult problem to the theorist. Is it not strange to relate that we believe we know more concerning the interiors of remote stars than we do of our own Earth? Yet such is the construction of the universe as it obtrudes itself upon our particular limitations that far gaseous bodies are made familiar while the natures of the relatively close solid bodies remain singularly inaccessible to us.

The following data are of prime importance in any study of planetary interiors:

1. The mass of the planet.
2. The density of the planet.
3. The oblateness and the rotational period.
4. The surface temperature.

This chapter must be remarkably short owing to the paucity of clear knowledge and the small number of facts available to us, but let us gather what there is.

MERCURY, VENUS AND MARS

Recent determinations of the radius and the mass of Mercury and Venus have been made and the values given by Lyttleton[1] based on the figures derived from radar experiments conducted at the Lincoln Laboratory are as follows.

For Mercury the radius is 2440 ± 2 km. and the mass $0 \cdot 3297 \pm 0 \cdot 0029 \times 10^{27}$ g. The density is $5 \cdot 42$ g cm^{-3}. The mean density is thus 99 per cent of that of the Earth and the problem is how is this to be

[1] See *The Observatory*, *89*, 47 (1969).

explained for such a small planet with a mass intermediate that of the Moon and Mars. For Venus the radius is 6056 $\pm$ 1 km., the mass is $4 \cdot 866 \times 10^{27}$ g and the density is $5 \cdot 23$ g cm^{-3}.

Lyttleton postulates that Mercury consists of 60 per cent iron surrounded by a rocky exterior or other heavy elements of an unexpected kind, perhaps even with a liquid core of mercury. On any view the planet is an exceptional object.

Venus must, in Lyttleton's view, have a liquid core and be a three-zone body, if it has a constitution similar to that of the Earth. The core will contain just under 25 per cent of the total mass. Lyttleton draws up a comparison table for the four inner planets as follows:

	Mass	Mean density	Percentage mass in core	Possible explanation
Earth	1	5·52	31·4	Liquid core Fe or phase change
Mercury	0·055	5·42	60·0	Fe only
Venus	0·815	5·23	24·7	Liquid core Fe or phase change
Mars	0·108	3·92	0	No core

A model of Mars due to Jeffreys is given below.

Distance from centre km.	Density g/cm.³	Pressure 10^{12} dynes/cm.²
3,000	3·42	0·05
2,500	3·57	0·11
2,034*	3·69	0·16
2,034*	4·23	0·16
1,500	4·35	0·24
1,424*	4·37	0·25
1,424*	8·28	0·25
1,000	8·45	0·35
0	8·60	0·44

* Boundaries. For an explanation of the low density of Mars when compared with that of the Earth see *Nature*, *234*, 89 (1971); *MN*, *108*, 406 (1948); *MN*, *109*, 457 (1949).

THE INTERIORS OF THE JOVIAN PLANETS

It can be seen from the low densities of these planets that they must have a very different structure from that of the terrestrial planets[†]. The densities suggest light elements such as hydrogen and helium.

The general picture is that of a heavy core with a light outer covering, often figuratively referred to as 'a stone in a snow ball'. For models of Jupiter, Saturn, Uranus and Neptune we are indebted to the calculations of De Marcus[2] given below:

[†] Mercury, Venus, Earth and Mars.
[2] De Marcus, W. C., *Handbuch der Physik*, 52 (1959). See also the work edited by Runcorn, S. K., *The Application of Modern Physics to the Earth and Planetary Interiors*, Wiley (1969).

CALCULATED PROPERTIES OF THE INTERIOR OF JUPITER AND SATURN

Distance from centre in planetary radii R = 1	Density g/cm.	Pressure 10^{12} dynes/cm.^{2}	Percentage helium by weight
JUPITER			
1·000	0·00016	0	0
0·994	0·138	0·0068	0
0·98942	0·185	0·0200	0
0·98942	0·197	0·0200	0
0·86	0·593	1·06	0·090
0·802	0·777	1·93	0·13
0·802	1·08	1·93	0·13
0·7	1·56	5·07	0·16
0·6	2·12	9·52	0·23
0·5	2·66	15·3	0·25
0·4	3·14	21·9	0·26
0·3	3·58	29·2	0·26
0·2	4·08	37·6	0·27
0·15	4·40	43·2	0·27
0·1	19·09	63·5	1·0
0·05	27·90	96·3	1·0
0	30·84	110	1·0
SATURN			
1·00	0·00016	0·000001	0
0·990	0·092	0·00186	0
0·970	0·185	0·0200	0
0·970	0·197	0·0200	0
0·95	0·236	0·0478	0
0·90	0·293	0·137	0
0·80	0·397	0·396	0
0·70	0·498	0·775	0
0·60	0·611	1·32	0
0·5227	0·719	1·93	0
0·5227	0·999	1·93	0
0·40	1·289	3·99	0
0·30	4·155	8·74	0·976
0·20	9·445	24·0	1·0
0·10	13·92	45·2	1·0
0	15·62	55·5	1·0

DATA ON INTERIORS OF URANUS AND NEPTUNE*

Quantity	Uranus	Neptune
Possible hydrogen mass fractions	0·2–0·03	0·16–0·00
Maximum helium mass fraction	0·9	0·6
Minimum heavy element mass fraction	0	0·12

* De Marcus and Reynolds. Based on models computed on the assumption of a cold planet.

It is now thought by Hoyle and Wickramasinghe[3] that the densities of Uranus and Neptune are too high (1·70 and 2·26 respectively) for

[3] Hoyle and Wickramasinghe, *Nature*, *217*, 415 (1968).

them to contain much hydrogen or helium. The densities of these two planets are in their view commensurate, after suitable allowance for compression, with the following constituents H_2O, NH_3, N_2 and CO_2.

Another criticism comes from the fact that Jupiter is now known to emit much more energy than it absorbs and explanations of the source of this heat depends very much upon our knowledge of Jupiter's interior and of the behaviour of condensed matter at very high temperatures and pressures. Smoluchowski[4] has attempted one explanation. He objects to the De Marcus model of the interior in view of our ignorance of the equation of state of condensed matter at very high pressure and at very high temperature, in particular, the melting points at high pressures that are at best obtained by extrapolation of low pressure data by means of various semi-empirical relations such as Simon's equation,[5] he considers to be invalid. In support of his case he points to the two very dissimilar models that exist side by side, the model of De Marcus, in which essentially all hydrogen is solid, and the high-temperature model of Peebles, in which everything except a small core is assumed to be fluid to provide uniform mixing. From the energy source of the interior 10^{33} ergs/year are said to be liberated and a concentration of radioactive elements characteristic of the solar system would account only for 5×10^{-5} of the required energy. Again the highest temperatures ever suggested for the inside of Jupiter are orders of magnitude too small to sustain exothermic nuclear reactions and thus the gravitational field of the planet appears to be the most obvious source of the energy. Smoluchowski proposes a model of Jupiter that is intermediate between those of De Marcus and Peebles.[6]

[4] Smoluchowski, R., 'Internal Structure and Energy Emission of Jupiter', *Nature*, *215*, 691 (1967).
[5] Babb, S. E., junior, *Rev. Mod. Phys.*, *35*, 400 (1963).
[6] Peebles, P. J. E., 'The Structure and composition of Jupiter and Saturn', *Ap.J.*, *140*, 328 (1964).
Note: For a recent work see Dollfus, A., *Surfaces and Interiors of Planets and Satellites* (1970).

Chapter 8

Interplanetary dust and gas

O'er great, o'er small extends
his physic laws, Empalms the
empyrean or dissects a gaz.[1]
J. BARLOW

If a mote, a hair, a dust prepond
On Inclination's side, down
drops the Scale.
C. B. SOUTHEY

In considering the interplanetary dust and gas we are restricted to two major pieces of evidence: the zodiacal light and the collection of cosmic dust that falls on the Earth.

The zodiacal light is little known and only seen in the latitude of London by the most assiduous observers and then they will need to use averted vision. It appears as a diffuse cone of light rising obliquely into the sky above the fading twilight in the west or the dawn in the east. It is now known to be the light reflected from a vast assemblage of 'dust' of doubtful size. In view of the Poynting Robertson effect, which requires very small particles to spiral into the Sun so that after as short an astronomical period as 2×10^6 years all particles less than 1 cm. diameter would be removed, the 'dust' may be considered, on not very reliable evidence, to be particles of an appreciable size, with some free electrons which are possibly the outermost parts of the Sun's corona.

The zodiacal light is best seen when the angle of the ecliptic plane and the horizon are mutually perpendicular and for this reason the refined examinations of the zodiacal light have usually been conducted in tropical regions[2] and often at very high altitude. It was Cassini[3] in

[1] Gas, a word invented by the Dutch chemist J. B. Van Helmont (1577–1644).
[2] Humboldt says: 'Those who have lived for many years in the zone of palms must retain a pleasing impression of the soft beauty with which the zodiacal light, shooting pyramidally upwards, illumines a part of the uniform length of tropical nights. I have seen it shine with an intensity of light equal to the Milky Way in Sagittarius, and that not only in the rare and dry atmosphere of the summits of the Andes at an elevation of from thirteen to fifteen thousand feet, but even on the boundless grassy plains, the Llanos of Venezuela, and on the seashore, beneath the very clear sky of Cumana. This phenomenon was often rendered especially beautiful by the passage of light fleecy clouds, which stood out in picturesque and bold relief from the luminous background. In our gloomy so-called "temperate" northern zone, the

the seventeenth century who first guessed the cause of the zodiacal light. He reasoned that because the light lay along the path of the zodiac it must be a celestial rather than an atmospheric phenomenon, and he conjectured that it was caused by the scattering of sunlight by myriads of particles of dust in orbit about the Sun and in the ecliptic plane forming a lenticular disc about the Sun. In more recent years measurements of the absolute surface brightness and the polarization of the zodiacal light have been made by Blackwell and Ingham[4] at the high altitude cosmic ray station of Chacaltaya in the Bolivian Andes (17,100 ft. altitude in geomagnetic latitude 3°S). The site was chosen for good weather conditions and an atmospheric extinction that is close to that expected for a dust-free Rayleigh scattering atmosphere. Following Cassini it had been widely held, up until 1953, that the zodiacal light was caused by the scattering of dust alone in interplanetary space. But this view was modified by Behr and Siedentopf,[5] who made the first accurate measurements of the degree of polarization of the zodiacal light and put forward the hypothesis that some half of the brightness at certain parts of the zodiacal light is due to the scattering of sunlight by free electrons.

They estimated that the electron density in the ecliptic plane at 1 a.u. from the Sun is 600 electrons cm.$^{-3}$. However, more recent results suggest that the electron density at 1 a.u. is not greater than 120 cm.$^{-3}$. The spectrum of the zodiacal light was examined in $c.$ 1963 by Dr John F. James, again at the Chacaltaya station, using a large Fabry Perot interferometer and spectrometer to match. He concentrated his attention on the hydrogen line at 4861 Å in the blue-green part of the spectrum. The results showed a shift to the blue in the evening and a corresponding shift to the red in the morning (Fig. 26). This suggests that the dust cloud is moving[6] with a revolution in harmony with the main revolution of the major bodies of the solar system.

James found no clear evidence for an electron component and thus we have something of an anomaly. The Siedentopf data require an electron density of 600 electrons cm.$^{-3}$; the Mariner 2 spacecraft obtained evidence for about 500 electrons cm.$^{-3}$, and the spectroscope at Chacaltaya no more than 120 electrons cm.$^{-3}$ and probably less than 5 electrons cm.$^{-3}$. The presence of dust in the zodiacal light is certain, but the gaseous component is not firmly established. For those interested in

<hr>

zodiacal light is only distinctly visible in the beginning of spring, after the evening twilight, in the western part of the sky, and at the close of autumn, before the dawn of day, above the eastern horizon'.
[3] Cassini, J. D., *Mem. Acad. Sci. Paris*, 8, 121, 1666–99.

[4] Blackwell and Ingham, 'The zodiacal light from a very high altitude station I, II, III, IV', *Monthly Notices R.A.S.*, *122*, No. 2, 113 (1961).
[5] Behr, A., and Siedentopf, H., *Zeits. f. Astrophys.*, *32*, 19 (1953).
[6] See James, J. F., and Smeethe, M. J., *Nature*, 227, 588 (1970).

the zodiacal light it is of importance to record that Blackwell and Ingham place its plane such that it intersects the plane of the ecliptic at longitude 115° and is inclined at an angle to it of about 1·5°: indeed the plane of the zodiacal cloud is close to the invariable plane of the solar system. There is now reason to believe[7] that there is a terrestrial component in the Doppler-shifted zodiacal light, so that light scattered off particles in circum-terrestrial space may make an appreciable contribution to the zodiacal glow.

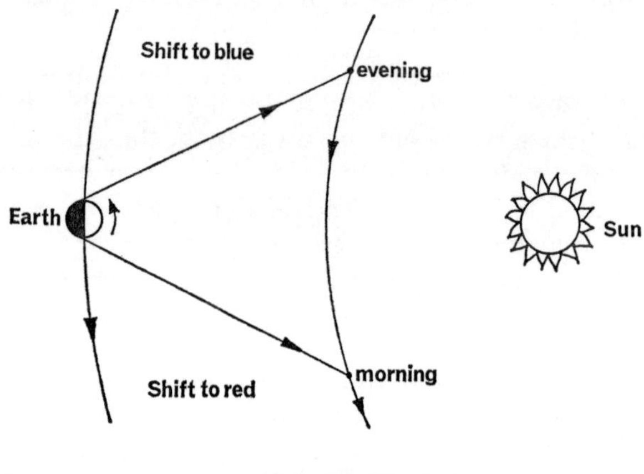

Fig. 26. Explanation of variations in the Zodiacal light.

The gaseous content of the interplanetary spaces is now more clearly to be understood from the studies of the solar wind and from the theory that there is an interplanetary plasma that is a tenuous extension of the solar corona. One further observation is of interest here, viz. the nature of the *Gegenschein*: a local brightening of the zodiacal light near the anti-solar point; whether it can be accounted for by the reflection of light from the dust of the zodiacal light is still unsettled. S. A. Arrhenius, the great Swedish physicist (*c.* 1860), has suggested that the *Gegenschein* is caused by corpuscles sent off by the Earth and repelled by the Sun (now surely by the solar wind) as with a comet tail. Soviet astronomers[8] have made a close study of the *Gegenschein*, they find it to be variable in shape, variable in intensity and in the spatial distribution of its luminescence. They have proved it to exhibit an emission spectrum and to have a parallax that places it at a distance of approximately twenty Earth radii; a result of some significance. They interpret their findings in favour of the Earth possessing a gaseous tail. In the view of V. G. Fesenkov, the tenuous outer atmosphere of the Earth has the form of a paraboloid of

[7] Harwit, M., *Nature,* *225,* 1231 (1970).
[8] *Nature, 171,* 555 (1953).

revolution, convex toward the Sun; he estimates that the density of the tail is halved every 4·7 Earth radii, but he assumes that the density is constant over any cross section. I. S. Astopovich modifies this opinion by postulating that the tail is a hollow sleeve of gas driven from the twilight rim of the Earth. Fesenkov and Divari have discovered independently a new feature of the *Gegenschein*. As it descends to the west of the meridian a broad pyramid of fainter light appears connecting it to the horizon. Fesenkov terms this phenomenon the 'false zodiacal light'. Strange to relate the false zodiacal light seen under the best conditions in Central Asia does not appear in the eastern sky when the *Gegenschein* is rising. The false light may be generated by atoms and molecules excited in sunlight but dropping to low energy levels when they pass into the great shadow cone of the Earth.

A more recent contribution by Roosen (*Nature, 229, 478* (1971)) shows that on the assumption that the dust particles in question are 10 μm in radius and have a reflectivity at opposition of 0·1, it is possible by means of a simple calculation to set an upper limit on the spatial density of particles in each case, remembering that 3 per cent of the Sun's light is blocked by the Earth's shadow for the constant density cloud at $-8°$, while 55 per cent is blocked for the cloud with density proportional to $R^{-1·5}$ at the same angle, using the phase function of 0·023 mag/deg and taking the *Gegenschein* peak brightness as 200 S_{10} (vis) units. A constant density geocentric dust cloud must have fewer than $2·3 \times 10^{-14}$ particles cm.$^{-3}$, while the maximum density just above the Earth's surface for a geocentric dust cloud for a density proportional to $R^{-1·5}$ is $4·3 \times 10^{-12}$ particles cm.$^{-3}$.

Because the actual particle sizes and reflectivities are very uncertain, the above calculation of spatial density should not be considered definitive. The important point is that the hypothetical geocentric dust cloud is undetectable in the direction opposite the Sun to a limit of a few per cent of the background sky brightness.

The dust from the interplanetary spaces need not be inferred only from the zodiacal light; it comes directly down on to the Earth and may be collected for examination. Its fall, at times of spectacular cosmic activity, has been recorded[9] over many years.

It can be shown that some meteorites, micrometeorites, are so small that they may pass through the Earth's atmosphere at high velocity without suffering ablation. These particles radiate energy away as fast

[9] Chladni, E. F., *On Fireballs and dust falls* (Vienna 1819), p. 434, ch. 6.
Arago, D. F. J., 'A list of the principal recorded showers of cosmic dust', *Astron Populaire*, 4, 208.
Nordenskiold, N. A. E., 'On the cosmic dust on the Earth', *Philos. Mag. Ser.*, 4, 48, 546 (1874a).
Yung, E., 'Falls of cosmic dust', *Comptes Rendus 97*, p. 1149.
Hodge, P. W., Wright F. W., and Hoffleit, D., 'An Annotated Bibliography on Interplanetary Dust', *Smithsonian Contributions to Astrophysics*, vol. 5, p. 85.

as they generate it. The size of such particles is calculated to be about 60μ diameter or smaller for a body entering the Earth's atmosphere at 12 km./sec. Such particles have been found in arctic ice and from this the rate of fall is estimated to be 1 per cm.^{2} per day. The size is about 3μ diameter and they generally consist of iron in a spheroidal form which strongly resembles that of meteorites. The Swedish Deep Sea Expedition of 1950 and other research workers[10] have obtained many samples of the globigerina ooze that forms the floor of the Atlantic and Pacific Oceans. The cores brought up in the sample drillings from the ocean bed show that much more nickel is present than is to be found in the Earth's crust. Also the nickel in the Pacific Ocean deposits was found to be eight times in excess of that in deposits taken from the Atlantic and this agrees with the known sedimentation rate in these ocean floors. This finding strongly reinforces the evidence from the nickel content itself which is much in harmony with nickel from interplanetary sources in the meteorites and the spherules from the virgin arctic ice.

It is now suggested from these studies that some 6000 tons per day in total of the interplanetary dust falls on the Earth and that one half of this comes from the cloud of dust in the vicinity of the Earth, viz. that shown to us by the zodiacal light and the *Gegenschein* or counterglow.

One may legitimately wonder what the solar system finds itself immersed in, or in other words, what is its 'amniotic' fluid. The short answer[11] is that the solar system appears to be immersed in interstellar hydrogen, at a very low density of about 0·06 cm.$^{-3}$.

[10] Smales, A. A., Wiseman, J. D., 'Analysis of globigerina ooze', *Nature*, *175*, 464 (1955). Pettersson, H., *et al.*, 'Exploring the bed of the ocean', *Nature*, *164*, 468 (1949); *Nature*, *166*, 308 (1950).
[11] See Fahr, H. J., *Nature*, *226*, 435 (1970).

Chapter 9

On the nature of the Sun

*We differ from the ancients only in
the richness of the language with
which we can adorn the obscurity that
surrounds us.*
PROUDHON

The Sun is at once inscrutable[1] and ubiquitous,[2] there is no man who
does not have some knowledge of him,[3] yet his intrinsic nature is and
may always remain a mystery. He is the brightest (as seen from the
Earth) of the heavenly bodies, the luminary or orb of day, the central
body of the solar system around which the Earth and other planets
revolve, being kept in their orbits by his attraction and supplied with
light and heat by his radiation. On the Ptolemaic system he was
reckoned as a planet, in modern astronomy as one of the stars.[4]

To the unaided human eye the ball of the Sun is seen to advantage
only when he is shielded by mist or smoke and it was undoubtedly in
this manner that observers with acute vision, in earlier times, were able
to detect the larger blemishes that disfigure his face. Ever since the
researches of Galileo, F. Scheiner, and Fabricus the younger, the
blemishes have been known as the Sun spots. From the earliest times the
Sun has been venerated as the life-giver to the Earth and all living things
are indebted to the munificence of his outpourings of radiant energy.
That his embrace may equally be the harbinger of death, as well as life,
is known to all who move within tropical regions of the Earth. The mainspring of his heat reservoir and his distance from the Earth have proved
to be two of the most intractable problems in physics and astronomy. So
much has been written concerning the Sun in science and in art that the
present task is singularly daunting. I seek below to give a concise picture

[1] *—the heaven's glorious Sun, That will not be deep-serch'd with saucy looks.* Shakespeare, *Love's
Labour's Lost*, Act. 1, sc. 1, 84.
[2] *His going forth is from the end of the heaven, and his circuit unto the ends of it: and there is nothing
hid from the heat thereof. Psalms*, XIX, 6.
The Sun with one eye vieweth all the world. Shakespeare, *Henry VI*, Pt. 1, Act 1, sc. 4, 84.
[3] In conformity with the gender of old English the feminine pronoun was used until the sixteenth century in referring to the Sun, since then the masculine has been commonly used,
without necessarily implying personification. The neuter is somewhat less frequent.
For yet the Sun Was not; shee in a cloudie tabernacle Sojourn'd the while. Milton (1667), *Paradise Lost*.
[4] Spectral type, G2V.

of the nature of the Sun and to give as far as I am able the recent advances in knowledge. The subject is made more tractable if it is taken under the following heads.

1. The Distance, Size, Mass, Position and Structure of the Sun
2. The Face of the Sun
3. The Solar Spectrum
4. The Composition of the Solar Photosphere and Corona
5. The Source of the Sun's Energy
6. The Solar Wind and the Van Allen Belts
7. The Sun's Rotation
8. The Radio Emission of the Sun.

1. THE DISTANCE, SIZE, MASS, POSITION AND STRUCTURE OF THE SUN

The mean distance of the Sun from the Earth is a fundamental constant, the astronomical unit (a.u.), and it is currently given as 149,600,000 km. or 92,957,209 miles, derived from the solar parallax[5] of 8·79405 seconds of arc. Owing to the ellipticity of the Earth's orbit the difference between the perihelion and aphelion distance is some three million miles (5×10^6 km.) and the Sun's disc as seen from the Earth varies in its apparent diameter of about half a degree of arc (mean 31′59″) depending upon the position of the Earth in its orbit.[6]

The mass of the Sun is $1 \cdot 990 \times 10^{33}$ g (2×10^{27} tons) and its size is immense. It would easily, for example, embrace the Earth and the complete orbit of the Moon, having an equatorial diameter of 1,392,000 km. (865,000 miles) which is 109 times the equatorial diameter of the Earth. It is often said that no departure from a perfect spherical shape can be detected but this latter-day Aristotelianism is now in doubt.[7]

The volume of the Sun is 1,300,000 times that of the Earth, but owing to the Sun's low mean density its mass is little more than 332,958 times that of the mass of the Earth. The mean density of the Sun is 1·409 against the Earth's mean density of 5·517, where the density of water is unity.

The Sun lies near the plane of the Milky Way at a distance of some 10 kiloparsecs ($3 \cdot 1 \times 10^{22}$ cm.) from the centre. It is in one of the spiral arms and revolves about the centre of the Milky Way System at a velocity of 250 km./sec. The Solar apex is 37°28′ from the Sun's axis of rotation (see *Nature*, *231*, 172 (1971).

The structure of the Sun in its various layers as presently understood

[5] See chapter 12 page 213.
[6] See *The Observatory*, *75*, 235, *re* Semi diameter of the Sun with the Airy transit circle at Greenwich 1915–37.
[7] See page 146 dealing with the Sun's rotation and solar oblateness.

is most conveniently shown from a sectorial slice taken through the ball of the Sun in which T is the temperature and p the pressure. (Fig 27).

2. THE FACE OF THE SUN

The ball of the Sun that exhibits a continuous spectrum and reveals a clear spherical outline is known as the photosphere, above it is the chromosphere and the corona. When we talk of the face of the Sun we refer to the surface of the photosphere. If we examine the photosphere with considerable care in a suitable telescope[8] it is seen to possess a granulated structure that is never quiescent. The granules when photographed from the Earth appear 35 per cent to 40 per cent brighter than the background and they cover about 35 per cent of the surface of the photosphere. They are estimated to number $2 \cdot 5 \times 10^6$. They are unstable and have an average lifetime of but a few minutes; about $8 \cdot 6 \pm 0 \cdot 2$ minutes. Their temperature difference from the background temperature is about $175°C$ when measured at sea level. The diameter of the granules is 400–1000 km. and there is some evidence for super-granules of 30,000 km. in diameter having a life of about 20 hours. In this 'rice grain' structure appear the sunspots, which are of special interest in that they have *inter alia* disclosed the rotation of the photosphere.

We have already said that keen eyed observers had detected blemishes on the face of the Sun. Schove[9] has with considerable academic skill pushed the earliest observation of a sunspot in the East back to the spring of 165 B.C. He appeals to the lost works of Wang Ch'i-ching.[10] In the West the earliest reference to sunspots is accredited to Theophrastus of Athens,[11] in the mid fourth century B.C. Russian observations of the Sunspots made in an atmosphere filled with the smoke and haze produced by the extensive forest fires of *c.* 1365–71 are recorded graphically in the Niconovsky Chronicle.[12] 'During this year (1371) there was a sign in the Sun. There were dark spots on the Sun, as if nails were driven into it.'

With the invention of the telescope the spots were seen by Galileo in the days of high solar activity between 1610 and 1611, and in 1612 he sent the following short but accurate account to Cosimo II of Florence.

Repeated observations have finally convinced me that these spots are substances on the surface of the solar body where they are continuously produced

[8] No one should turn a glass to the Sun unless it is equipped with suitable filters or a solar prism able to reduce the volume of the heat and light—otherwise the eyesight may be seriously put at risk.
[9] Schove, D. J., 'The Earliest dated Sunspot', *J.B.A.A.*, *61*, 22 (1950).
[10] See Dubs, H. H., *History of the Former Han Dynasty of Pan ku*, vol. 1 (1938); p. 258, vol. 2 (1944). Kanda. S., *Proc. Imp. Acad. Tokyo*, *9*, No. 7, pp. 293–6 (1933).
[11] Theophrastus—a successor of Aristotle in the peripatetic school, a native of Lesbos, born *c.* 372 B.C., died 287 B.C.; see Zeller, *Aristotle and the Earlier Peripatetics*, English translation by Costelloe and Muirhead, vol. II, ch. 18 (1897).
[12] See Vyssotsky, A. N., 1949, *Astronomical Records in the Russian Chronicles from A.D. 1000–1600*. Medd. fran Lunds Astro Observatorium, Scr. 2, Nr. 126; Historical Papers, Nr. 22, Lund.

and where they are also dissolved some in shorter and others in longer periods. And by the rotation of the sun, which completes its period in about a lunar month, they are carried round the sun; an occurrence important in itself and still more so for its significance.

In 1844 Schwabe, the apothecary of Dessau, disclosed in *Astronomische Nachrichten*, vol. 21 (1844), his researches into the periodicity of sunspots,

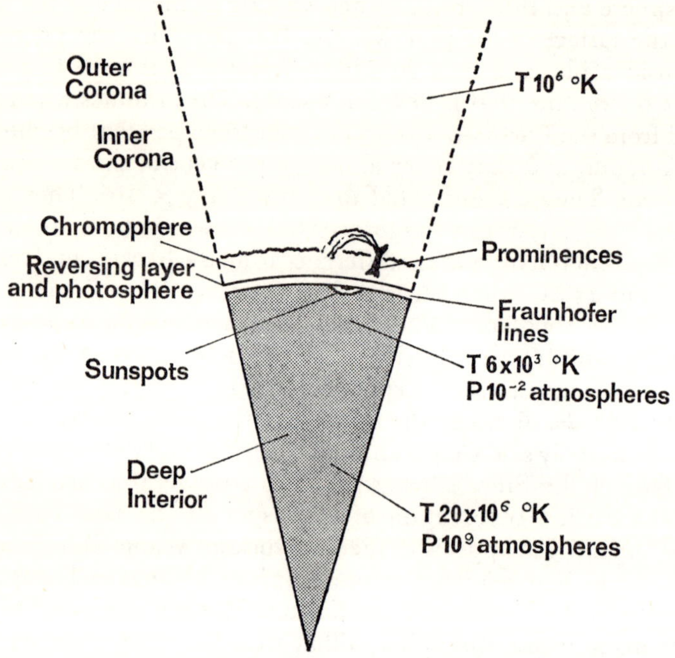

Fig. 27. Schematic view of the layers of the Sun.

but it was left to Humboldt to make it known to an astonished scientific world in the third volume of his *Cosmos* (1851) that since the days of Galileo 200 years had elapsed and yet no one had entertained the idea, let alone detected, the ten to eleven years sunspot cycle. The scientific world has no example to equal Schwabe for painstaking care coupled with humility. For twelve years he worked to satisfy himself of his discovery, six more years to satisfy and thirteen more years to convince the rest of mankind. It is said that the Sun never rose above Dessau but he was met by the imperturbable telescope[13] of Schwabe. The observations of Schwabe from 1826–51 were extended by Wolf who searched the historical records back to 1610 and arrived at a spot mean cycle

[13] Schwabe had two telescopes, both refractors, one of 3½ ft. focal length and the other of 6 ft. He had stopped them down to apertures of 1½ in. and 2½ in. respectively and viewed the Sun directly with a blue filter over the eyepiece. It may be recalled that Galileo put his eyesight at risk in observing the face of the Sun directly but Scheiner skilfully used the safe method of projecting the fierce solar image on to a white card.

period of 11·12 years. Whether the solar cycle *per se* can be set at this figure is not wholly clear since Gleissberg[14] has adduced evidence for a cycle of eighty years based on the frequency of aurora. The periods of sunspot maxima have been given by Schove[15] and his table, somewhat simplified, is reproduced below.

YEARS OF SUNSPOT MAXIMA FROM 649 B.C.

(*arranged by centuries*)

−700 B.C.					52				
−600 B.C.							78	88	99
−500 B.C.	9	19	29	39					
−400 B.C.	7				51	61			
−300 B.C.	7	17	28	39	51	64	77	86	95
−200 B.C.	8	18	28	37	51	65	75	87	96
−100 B.C.	9	18	28	38	47	58	73	84	95
0 +	8	20	31	42	53	65	76	86	96
A.D. 100 +	5	18	30	41	52	63	75	86	96
A.D. 200 +	8	19	30	40	52	65	77	90	102
A.D. 300 +	11	21	30	42	54	62	72	87	96
A.D. 400 +	10	21	30	41	52	65	79	90	101
A.D. 500 +	11	22	31	42	57	67	78	85	97
A.D. 600 +	7	18	28	42	54	65	77	89	99
A.D. 700 +	14	24	35	45	54	65	76	87	98
A.D. 800 +	9	21	29	40	50	62	72	87	98
A.D. 900 +	7	17	26	38	50	63	74	86	94
A.D. 1000 +	3	16	27	38	52	67	78	88	98
A.D. 1100 +	10	18	29	38	51	60	73	85	93
A.D. 1200 +	2	19	28	39	49	59	76	88	96
A.D. 1300 +	8	16	24	37	53	62	72	82	91
A.D. 1400 +	2	13	29	39	49	61	72	80	92
A.D. 1500 +	5	19	28	39	48	58	72	81	91
A.D. 1600 +	04·5	15·5	26·0	39·5	49·0	60·0	75·0	85·0	93·9
A.D. 1700 +	05·0	18·2	27·5	38·7	50·3	61·5	69·7	78·4	88·1
A.D. 1800 +	05·2	16·4	29·9	37·2	48·1	60·1	70·6	83·9	94·1
A.D. 1900 +	07·0	17·6	28·4	37·4	47·5	58·5	72·5	84·5	94·5
A.D. 2000 +	04·5	14·5	25·5						

Note 193 B.C. or year −192 (= −200 + 8).
  649 B.C. or year −648 (= −700 + 52).
For a more detailed tabulation see *Journal of Geophysical Research*, *60*, 127 (1965).

Carrington is the first to have noticed, *c.* 1853, the change of mean solar latitude of spots with the progress of the 11 year solar cycle, but the presentation best able to make this clear to the human eye and mind is the beautiful butterfly diagrams of Maunder *c.* 1904. The new spots appear in solar latitude 30°–35° north and south of the Solar equator and move progressively toward the equatorial region until near sunspot minimum, at about solar latitude 7° north and south, the spots disappear

[14] Gleissberg, W., 'The Eighty Year Solar Cycle in Auroral frequency number', *J.B.A.A.*, *75*, 227 (1965); *76*, 265 (1966).
[15] Schove, D. J., 'Sunspot Maxima since 649 B.C.', *J.B.A.A.*, *66*, 59 (1956). See also Waldmeier, M., *Sunspot activity in the years 1610–1960* (Zürich 1961).

and the spots of the new cycle appear in the higher solar latitudes once again. In 1908 Hale at Mt Wilson discovered the strong magnetic field centred upon a sunspot. It was subsequently found that the spots have a paired polarity. The leader-spot has an opposite polarity to the follower-spot and in any given eleven year cycle the polarities of the twin spots are reversed on opposite sides of the solar equator—what is even more remarkable, the pattern of magnetic polarity reverses with each ensuing cycle and thus the magnetic sunspot cycle may more accurately be said to be one of twenty-two years.

What of the structure of the spots themselves? Wilson[16] in 1774 discovered the phenomenon ever afterwards linked with his name, viz., that the spot area decreased toward the limb of the Sun more than would be expected in normal foreshortening due to the curvature of the ball of the Sun, and that spots were depressions in the surface of the solar photosphere.

Visually a sunspot is a dark (umbral and penumbral), relatively cool area of the Sun's disc, and it was shown by Langley[17] (*c.* 1875) using a sensitive bolometer that there is a lower temperature in spots than in the surrounding photosphere, a point not arrived at without difficulty since clearly the inner parts of the Sun must increase in temperature. Indeed as late as 1897 Evershed[18] was suggesting that the spots were hotter than the photosphere. In 1908, as previously stated, Hale[19] discovered magnetic fields in sunspots, and Evershed[20] in 1909 detected a radial[21]

[16] Wilson, A., *Phil. Trans.*, *64*, 1 (1774). See Addey, F., 'The Wilson Effect', *J.B.A.A.*, *74*, 43 (1963).
[17] Langley, S. P., *Monthly Notices of the R.A.S.*, *37*, 5 (1875).
[18] Evershed, J., *Ap. J.*, *5*, 249.
[19] Hale, G. E., *Ap. J.*, *28*, 100; *28*, 315.
[20] Evershed, J., *Kodaikanal. Obs. Bull.*, *2*, 63. See *The Observatory*, *72*, 105 (1952).
[21] Magnetic fields in sunspots and the so called Evershed effect have been the centre of a considerable controversy. The following letter (from C. E. R. Bruce) to *The Observatory*, *86*, 35 (1965) speaks for itself:
I am grateful to Bray and Loughhead[1] for resolving the one outstanding difficulty which has faced the electrical discharge theory of cosmic atmospheric phenomena since its inception in 1941. It will be evident from the view of sunspots given in the original account of the theory[2] and again more recently in an explanation of the Evershed effect,[3] that on this theory the magnetic field throughout the umbra should be transverse, and only over the penumbra should it be more or less longitudinal, or outwards from the Sun's surface. After all these years of abortive efforts to make the field come out of the spots, it was gratifying to read their note that 'The observations of Evershed and Severny thus suggest a field configuration very different from the classical picture. On the new view, a predominantly transverse field in the umbra is surrounded by a largely vertical field in the penumbra. This interpretation receives strong support from some magnetograph observations made by Severny[4] (1959: *c.f.* Fig. 2) which actually show a region of zero longitudinal field at the very centre of the umbra of a sunspot.'
Evershed's observations[5] have long been available, but this is the first occasion on which I have seen a reference to them. Ellison,[6] for example, merely describes the well-known picture originated by Hale and Nicholson.
[1] R. J. Bray and R. E. Loughhead, *Sunspots* (Chapman and Hall Ltd.), pp. 217–218, 1964.
[2] C. E. R. Bruce, *A New Approach in Astrophysics and Cosmogony* (London), 1944.
[3] C. E. R. Bruce, *Jl. Franklin Inst.*, *276*, 407 (1963).
[4] A. B. Severny, *A. Zn.*, *36*, 208 (1959).
[5] J. Evershed, *The Observatory*, *64*, 154 (1941).
[6] M. A. Ellison, *The Sun and its Influence* (Routledge and Kegan Paul Ltd.), pp. 46–48, 1955.

outflow of material so that two further facts[22] needed to be fitted into any satisfactory theory of a sunspot. There is not yet available any adequate understanding of the effect of magnetic fields on the convection currents both within a sunspot and in its boundary layers to give a complete picture. A spot shows a dark umbra and a brighter penumbral region with a clear-cut outline, and in a typical large spot the magnetic field strength increases roughly in step with the area during growth. The maximum area and field strength are reached in about twelve days, the area then diminishes steadily and the field strength remains substantially constant until the spot disappears; the decay time from maximum area is about fifty days.[23] A sunspot model by Allen,[24] based on a polar drift at the Sun's surface and a Bjerknes[25] type circulation of magnetic toroids, is able to explain the reversal of the polar magnetic fields, the sunspot polarity law, the sunspot equatorial drift and the bi-polar spot group inclinations.

Frequently, indeed every few hours, near active sunspot groups, a great outburst of energy occurs producing a solar flare. These are seldom seen in white light but the first one to be observed was seen under these conditions by Carrington in 1859. A region of the solar surface covering many thousands of square kilometres brightens in a few seconds and great clouds of gaseous matter are ejected at velocities of some 1000 km./sec. The sunspot magnetic field extends far above the solar surface that contains the spot and not infrequently this produces arched prominences as the material thrown up by the flares cools, condenses and rains down along the curved magnetic lines of force.

It is of interest to record that when the highest solar temperature was thought to be only about one million degrees, Bruce[26] deduced that the temperature in solar flares must reach, at discharge, a value of hundreds of millions of degrees. This was later confirmed in the discovery of X-rays from solar flares by Severny in 1966. These violent phenomena are associated with intense magnetic storms on the Earth and disturb its ionosphere severely.

It is uncommonly difficult to unravel the temperature of the Sun at the centre at different stages of its evolution and we are indebted to Kuroda for an analysis of the problem (see *Nature Physical Science, 230,* 40 (1971)). Kuroda has examined the relative amounts of the isotopes of xenon in available extra-terrestrial samples. He has found that they correspond to expected values for conditions of much greater temperature than 'normal' room temperature, and he concludes that they

[22] Since this note was written Mallia, E. A., *et al.*, have detected water vapour in sunspots; see *Nature, 226,* 735 (1970).
[23] See Cowling, T. G., *Monthly Notices of R.A.S., 106,* 218.
[24] Allen, C. W., 'A Sunspot cycle model', *The Observatory, 80,* 94 (1960).
[25] Bjerknes, V., *Ap. J., 64,* 93 (1926).
[26] Bruce, C. E. R., *Nature, 184,* 2004 (1959); *187,* 865 (1960); see also 'Electrical Research Association. Report 5275', (1968).

formed in the young Sun and may have been scattered through the solar system. From the many possible isotopes of xenon which can be built up in this way, he has determined the appropriate temperature of the medium in which they were mixed. It seems that the required temperature is about eight million degrees, just over half the temperature thought to exist at the Sun's centre today. Therefore it seems likely that the isotopes were processed in the Sun early in the history of the solar system, because the temperature of the Sun during its formation must have slowly risen to the value we know now. The problem of distributing these isotopes from the Sun into meteorites and the Moon is not answered by Kuroda's investigations for this relates to the formation of the planets, which is really a separate question, and one that has still not been resolved satisfactorily. It may be, however, that the solar wind carries a steady stream of these isotopes out of the relatively cool atmosphere of the Sun where they have remained for millions of years, and deposits them on the lunar and meteoritic surfaces.

3. THE SOLAR SPECTRUM

The electromagnetic spectrum emitted by the Sun extends from fractions of an Ångström (Å)[27] to many hundreds of metres. The solar spectrum is divided according to Robinson[28] into the following main wavelength regions (1 cm. $= 10^8$Å $= 10^4\mu$)

	< 10 Å	X-rays, γ-rays
10 Å	2000 Å	far ultra violet
2000 Å	3150 Å	middle ultra violet
3150 Å	3800 Å	near ultra violet
3800 Å	7200 Å	visible
7200 Å	105 μ	near infra red
105 μ	516 μ	middle infra red
516 μ	1000 μ	far infra red
	> 1000 μ	micro and radio waves.

Wavelengths greater than 30,000 Å are absorbed by the atmospheric water vapour and carbon dioxide. In the ultra violet the wavelengths smaller than 2860 Å do not reach sea level and the smaller wavelengths are absorbed by the ozone layer high above the Earth.

Above the photosphere is the reversing layer which is cooler (about 5300°K.) than the photosphere and here the dark absorption lines of the solar spectrum are produced—the famous Fraunhofer spectrum discovered by Joseph Fraunhofer[29] in 1814 and explained in part by the classic researches of Kirchhoff and Bunsen in 1859, when they showed that absorption occurs when a source of a continuous spectrum is

[27] *Ångström unit* (Å). The name of A. J. Ångström, a Swedish physicist (1814–74). A hundred-millionth of a centimetre (10^{-8} cm.), used in expressing short wavelengths of radiation.
[28] Robinson, N., *Solar Radiation* (Amsterdam 1966).
[29] See *Denkschriften der Königlichen Akademie der Wissenschaften zu München* (1817); also *Edinburgh Journal of Science* (1827–8). Wollaston discovered a few dark spaces in the spectrum of sunlight in 1802, but did not pursue the matter.

viewed through a cooler gas, and that those elements, contributing to the continuous spectrum, are absorbed at just those specific wavelengths that characterize the elements both in the said continuous spectrum and in the cooler intervening gas. Hence the dark absorption lines are pointers to the elements in the source of the continuous spectrum, i.e. in the case of the Sun, the photosphere.

4. THE COMPOSITION OF THE SOLAR PHOTOSPHERE AND CORONA

On the view that the Sun presents us with a sample for the abundances of the elements in the solar system it is of interest to give the composition of the solar atmosphere as far as it can be detected. Many research workers[30] have laboured with this difficult problem and the table overleaf is based on their contributions, but it is mainly taken from the work of Müller,[31] Urey[32] and Warner.[33]

5. THE SOURCE OF THE SUN'S ENERGY[34]

The old theories of combustion, the Kelvin–Helmholtz contraction hypothesis embracing gravitational energy as a source of the Sun's heat, are no longer satisfactory to the intellect owing to the short time scale that is derived from their postulates—that is to say, the time scale is about one hundred times shorter than the presently accepted age of the Earth's crust.

The energy is now thought to stem from nuclear reactions in which hydrogen is transmuted into helium by a process of nuclear fusion. It has been calculated that if the solar mass of hydrogen was converted into helium it would yield $0 \cdot 007 \, c^2 M$ ergs, where $0 \cdot 007$ is the mass defect in amu[35] per proton for the transmutation process and M is the total mass of hydrogen in the Sun. This process would give the Sun a life of 10^{11} years, a not inconsiderable and unattractive time scale when one reflects on the ephemeral nature of human life and the currently accepted age of the universe that stands at 3×10^{10} years.

Two processes were postulated by Bethe,[36] and these have had wide

[30] See Aller, L. H., *Abundances of the Elements* (Interscience N.Y. 1961); *Advances in Astronomy and Astrophysics* (Academic Press, N.Y. 1965).
Müller, E. A., and Mutschlechner, P., *Astrophys. J. Suppl. Ser.*, *9*, 1 (1964).
Goldberg, L., Dupree, A. K., and Kopp, R. A., *Astrophys. J.*, *69*, 140; *140*, 707 (1964).
Teplitskaya, R. B., and Vorob'era, V. A., *Soviet Astr.*, *7*, 778 (1964).
[31] Müller, E. A., *Solar Physics*, Interscience 1967 N.Y. Proceedings of NATO Advance Study, Institute on Solar Physics Lagonissi, Athens, Greece, Sept. 1965.
[32] Urey, C. H., *Quat. J. R.A.S.*, *8*, 23 (1967).
[33] Warner, B., *The Observatory*, *84*, 14 (1964).
[34] *Whence are thy beams, O Sun thy everlasting light? Thou comest forth in thy awful beauty: the stars hide themselves in the sky; the Moon cold and pale sinks in the Western wave, but thou thyself moves alone.* MACPHERSON
[35] 1 amu is $1 \cdot 65981 \times 10^{-24}$g, i.e. the mass of a proton.
[36] Bethe, H. A., *Phys. Rev.*, *55*, 103, 434.

SOLAR ABUNDANCE RESULTS

relative to Hydrogen (H) on the basis of log $(N_H) = 12$.
$N_{El} = N_{Element}$

Atomic number	Element	log N_{El} Photosphere	Corona**
1	H	–	12
3	Li	1·54	
4	Be	2·34	He–11·3
6	C	8·62	9·2
7	N	7·88	8·2
8	O	8·86	8·8
11	Na	6·30	6·5
12	Mg	7·36	8·0
13	Al	6·20	
14	Si	7·45	8·3
15	P	5·34	
16	S	7·30	7·1
19	K	4·70	5·6
20	Ca	6·04	6·8
21	Sc	2·80	
22	Ti	4·58	
23	V	4·12	
24	Cr	5·07	6·0
25	Mn	4·80	5·9
26	Fe	6·70*	7·87
27	Co	4·40	5·3
28	Ni	5·44	6·72
29	Cu	3·50	
30	Zn	3·52	
31	Ga	2·51	
32	Ge	2·49	
37	Rb	2·48	
38	Sr	2·60	
39	Y	3·20	
40	Zr	2·65	
41	Nb	2·30	
42	Mo	2·30	
44	Ru	1·82	
45	Rh	1·37	
46	Pd	1·27	
47	Ag	1·04	
48	Cd	1·66	
49	In	1·28	
50	Sn	2·05	
51	Sb	0·42	
56	Ba	2·50	
70	Yb	2·28	
82	Pb	1·63	

N.B.—Helium is not included, for its lines appear to originate in the chromosphere and its abundance cannot be derived by the usual procedures. Gold, bismuth and thorium have been detected by the lines λ 3122·97 Å   3067·7 Å   4019·140 Å respectively, but there is some doubt.

** Corona. Pottasch, S.L., *Ap. J.*, *137*, 945 (1963); *Monthly Notices R.A.S.*, *125*, 543 (1963).

* See Ross, J. E. 'Abundance of Iron in the Solar Photosphere', *Nature*, *225*, 610, 1970.

support, the carbon-nitrogen cycle of fusion and the proton-proton reaction. The former was to some extent once discredited[37] in the refined

[37] *J.B.A.A.*, *61*, 164 (1951); *Scientific American* (June 1951); *Sky and Telescope*, *10*, 138 (April 1951).

researches in 1951 by Greenstein, Richardson and Schwarzschild, who failed to find any C^{13} in the Sun,[38] and now the proton-proton reaction is thought to be the dominant one. But the matter is not so readily resolved today as it would have been before the null result in the Brookhaven[39] experiment, to which we will return. But first let us give the reactions of both cycles, after Fowler.[40]

REACTIONS OF THE CNO-CYCLE

The CNO-cycle	Energy release
$\rightarrow C^{12} + H^1 \rightarrow N^{13} + \gamma$	1·95
$N^{13} \rightarrow C^{13} + \beta^+ + \nu$	1·50
$C^{13} + H^1 \rightarrow N^{14} + \gamma$	7·54
$\rightarrow N^{14} + H^1 \rightarrow O^{15} + \gamma$	7·35
$O^{15} \rightarrow N^{15} + \beta^+ + \nu$	1·73
$N^{15} + H^1 \rightarrow C^{12} + He^4$	4·96

or (1/2200)   (6% ν-loss) 25·03 Mev

$N^{15} + H^1 \rightarrow O^{16} + \gamma$	12·11
$O^{16} + H^1 \rightarrow F^{17} + \gamma$	0·59
$F^{17} \rightarrow O^{17} + \beta^+ + \nu$	0·76
$O^{17} + H^1 \rightarrow N^{14} + He^4$	1·20

(1/2200) 15·66 Mev

$4H^1 \rightarrow He^4$ Total = 26·72

REACTIONS OF THE PROTON-PROTON OR PP CHAIN

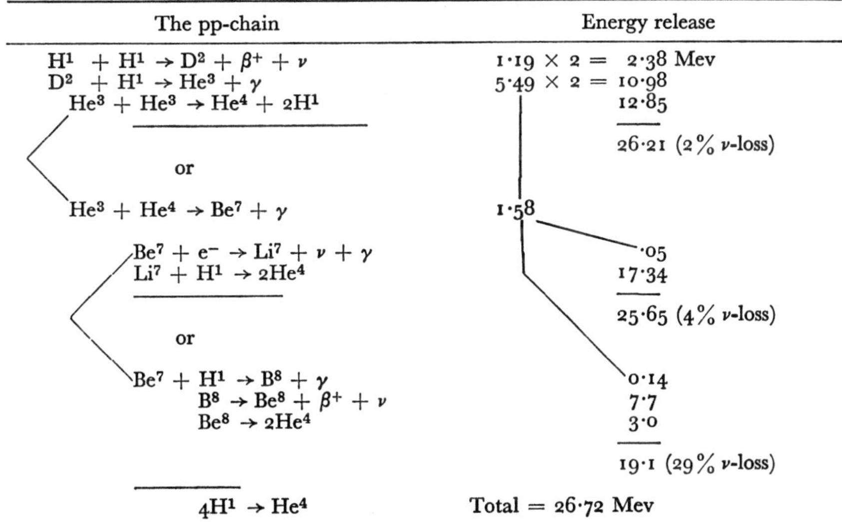

The pp-chain	Energy release
$H^1 + H^1 \rightarrow D^2 + \beta^+ + \nu$	1·19 × 2 = 2·38 Mev
$D^2 + H^1 \rightarrow He^3 + \gamma$	5·49 × 2 = 10·98
$He^3 + He^3 \rightarrow He^4 + 2H^1$	12·85

26·21 (2% ν-loss)

or

$He^3 + He^4 \rightarrow Be^7 + \gamma$ 1·58

$Be^7 + e^- \rightarrow Li^7 + \nu + \gamma$	·05
$Li^7 + H^1 \rightarrow 2He^4$	17·34

25·65 (4% ν-loss)

or

$Be^7 + H^1 \rightarrow B^8 + \gamma$	0·14
$B^8 \rightarrow Be^8 + \beta^+ + \nu$	7·7
$Be^8 \rightarrow 2He^4$	3·0

19·1 (29% ν-loss)

$4H^1 \rightarrow He^4$ Total = 26·72 Mev

[38] Note, this only eliminates a carbon-nitrogen cycle solar model in which the interior is in convective motion.
[39] Davis, R., *Phys. Rev. Lett.*, *12*, 302 (1964); Davis, R., *Phys. Today*, *21*, 73 (1968).
[40] Fowler, W. A., 'Experimental and Theoretical results on nuclear reactions in stars II', *Soc. Roy. de Sci. de Liége*, Tome 3, 207–23, (1960).

The solar system

The rate of the carbon-nitrogen cycle increases as the seventeenth power of the temperature of the Sun's core, whereas the rate of the proton-proton cycle increases at the low rate of the 3/2 power of the temperature of the core. On the basis of the neutrino flux to be associated with these two cycles it is postulated by Rouse[41] that the high energy neutrino flux will be less than expected, since in his view the Sun most probably operates on the carbon-nitrogen cycle with a core temperature of 21×10^6 degrees C. as against the presently postulated temperature of 16×10^6 degrees C. for the proton-proton cycle of operation. Rouse also postulates a solar density of 30g/cc instead of the more usually accepted 160g/cc. Clearly with the extra five million degrees of temperature the carbon nitrogen cycle predominates owing to its relation with the seventeenth power of the core temperature and that cycle would produce about 99 per cent of the solar luminosity with the proton-proton cycle relegated to but the production of about 1 per cent of the total.

The Brookhaven experiment would appear to support the solar model of Rouse. In that experiment 100,000 gallons of tetrachloroethylene were placed in a tank one mile below the surface of the Earth in a gold mine in South Dakota. This large volume of liquid was then deployed to capture neutrinos from the Sun (a neutrino can pass through 20×10^{12} miles of lead before being brought to rest). The liquid with its vast number of chlorine 37 atoms (Cl^{37}) was of special importance since on the rare occasion that a neutrino is arrested by collision with an atom of Cl^{37} the atom is transmuted into argon 37 (A^{37}). The tank was flushed with helium, which sweeps out the radioactive argon, A^{37}, and this was absorbed in charcoal and detected by means of a Geiger counter, the counting rate of which is an indication of the number of neutrinos intercepted.

In the proton-proton cycle, a branch of this cycle is completed by the production of beryllium 7 (Be^7) and boron 8 (B^8) and the flux of high-energy boron 8 neutrinos is the important component in the experiment. In the carbon-nitrogen cycle, low-energy neutrinos come from the decay of nitrogen 13 (N^{13}) and oxygen 15 (O^{15}). Now the probability of capture by the liquid increases with the square of the neutrino energy and the B^8 neutrinos are much more energetic than those produced by N^{13} and O^{15}. The maximum energy of the N^{13} neutrinos is 1·20 Mev and for the O^{15} neutrinos 1·74 Mev, whereas the energy of the B^8 neutrinos is 14 Mev. Thus it would take a much greater flux of neutrinos to produce the same Cl^{37} counting rate on the Geiger counter for the N^{13}–O^{15} neutrinos than for the B^8 neutrinos. In the result, the Brookhaven experiment does not support a proton-proton B^8 solar source but it appears to be compatible with a carbon-nitrogen source of N^{13}–O^{15} neutrinos.

[41] Rouse, C. A., *Nature*, *224*, 1009 (1969).

Since the B^8 neutrino flux is temperature dependent, one way to explain a low B^8 flux is to postulate a core temperature of the Sun of less than 16×10^6 degrees C. The problem then is that for a continuous solar model the observed solar constant would have to be less than that presently obtained or the solar model would need a different source of energy—and it is this that Rouse[42] has supplied without demanding a non-nuclear solar energy source such as gravitational contraction. But Rouse still needs to justify his assumption of gravitational sorting of elements.

In the carbon-nitrogen-oxygen (CNO) cycle the process is thought to commence with carbon in the form $_{6}^{12}C$ and a collision with a proton ($_{1}^{1}H$) forms the nitrogen isotope $_{7}^{13}N$. The mass difference between $_{6}^{12}C + _{1}^{1}H$ and $_{7}^{13}N$ is transformed into energy in harmony with the Einstein equation $E + mc^2$. The $_{7}^{13}N$ isotope is unstable and decays into the carbon isotope $_{6}^{13}C$ with the emission of a neutrino and a positron; this carbon collides with a proton and is converted into $_{7}^{14}N$ with the mass difference radiated away in gamma rays. The $_{7}^{14}N$ collides with a proton to form the oxygen isotope $_{8}^{15}O$ which is unstable and the mass difference is also radiated away in gamma rays. The oxygen isotope decays to $_{7}^{15}N$ with the emission of a positron and a neutrino. In conclusion the $_{7}^{15}N$ collides with a proton and disintegrates into $_{6}^{12}C$ and an alpha particle. The process comes full circle with the production of an alpha particle (He) with four proton collisions (H). It may be calculated from this that since the mass of four protons is $4 \times 1 \cdot 672 \times 10^{-24}$ g which equals $6 \cdot 688 \times 10^{-24}$ g and that the mass of one alpha particle is $6 \cdot 644 \times 10^{-24}$ g there is a mass loss of

$$(6 \cdot 688 - 6 \cdot 644) \times 10^{-24} \text{ g} = 0 \cdot 044 \times 10^{-24} \text{ g}$$

which is about 7/1000th of the mass of the four protons and thus only about 0·007 of the Solar mass of protons is converted into radiation.

In the proton-proton reaction the rate of successful collisions in the Sun's interior is proportional to the fourth power of the temperature. It will be noted that in the final reaction two protons appear for

42 Rouse, C. A., *Bull. Amer. Phys. Soc.*, *10*, 14 (1965); *Nature*, *224*, 1009 (1969).

further reactions of the initial type and the process is able to proceed in cascade. The mass loss calculations show that the Sun loses 4×10^6 tons/sec and this derives from the difference between

$$4\,H = 4 \cdot 03257 \text{ amu and } {}^{4}_{2}He = 4 \cdot 00389 \text{ amu.}$$

Thus in each reaction $4 \cdot 03257 - 4 \cdot 00389 = 0 \cdot 02868$ amu is lost. The power emitted by the Sun is $3 \cdot 85 \times 10^{23}$ kw or $5 \cdot 16 \times 10^{23}$ HP. The energy intercepted by the Earth is $1 \cdot 79 \times 10^{14}$ kw or $2 \cdot 40 \times 10^{14}$ HP.

6. THE SOLAR WIND AND THE VAN ALLEN BELTS

Our knowledge of the solar wind is of very recent date. It was the subject of the Halley Lecture[43] for 1969. Speculation was confirmed in 1962 by evidence from the Mariner 2 spacecraft, and it is now established that a wind[44] blows continuously from the Sun with a velocity in the range of 250 to 800 km.s^{-1} (average 525 km.s^{-1}) and that the density of electrons or positive ions near the Earth (I a.u.) lies between 3 and 10 per cc. with peak values rising up to 25/cc. The wind direction was initially thought to be radial of the Sun but the Vela series of satellites has shown that it is nearly radial of the Sun with an average direction of flow from approximately $1\frac{1}{2}°$ east of the Sun. The wind breaks away from its solar magnetic bonds at about two-thirds of the Sun's radius above the photosphere. About 5 per cent of the solar wind particles are alpha particles and there is thought to be some interaction between it and interstellar hydrogen, which according to Blum and Fahr[45] has a velocity of 10 to 40 km.s^{-1} relative to the solar system, with a shock front between 5 and 50 a.u. from the Sun.

One of the most interesting confirmations of the solar wind comes from a study of the tails of comets in which the tails are of the type known as 'straight' and constituted of ionized gas. It was generally agreed up until 1951 that the direction of these tails, away from the Sun, was due to radiation pressure, but an alternative explanation due to Biermann[46] is now thought to explain matters more correctly, in that the acceleration of the cometary particles is derived from an ionized gas streaming past the comet in the radial direction (plasma drag). It was later shown by Biermann that a *continuous* solar wind was a more reasonable explanation of cometary phenomena than appeal to the transient forces derived from magnetic storms. It now turns out that although the solar wind is a reality it is not a satisfactory explanation of the motions

[43] Cowling, T. G., 'The Solar Wind' (Halley Lecture), *The Observatory*, *89*, 217 (1969).
[44] Blum, P. F., and Fahr, H. J., *Nature*, *223*, 937 (1969); see also Chambers, W. H., *et al.*, *Nature*, *225*, 713 (1970).
[45] The wind is in reality a proton-electron gas which streams past the Earth with a mean velocity of about 500 km.s^{-1} and at a temperature of 5 ± 10^5K. See Brandt, J. C., *Introduction to the Solar Wind* (1970); also Piddington, J. H., *Cosmic Electrodynamics* (1969).
[46] Biermann, L., *Z. Astrophys.*, *29*, 274 (1951).

in the gases of comet tails, which derive their momentum transfer principally from the coulomb attraction between solar wind electrons and the cometary ions employing a coupling between the magnetic field of the interplanetary magnetic field and the cometary plasma and a further coupling between the solar wind plasma and the magnetic fields in the tail condensation. The solar wind interacts with the geomagnetic field[47] which is limited to a cavity—the magnetosphere—extending out some ten Earth radii on the sunward side of the Earth, and pulled out into a tail which extends as a regular structure out to 400,000 km. and even as far as 6×10^6 km. on the opposite side to the Sun. Wilcox[48] has shown from studies with the Interplanetary Monitoring Platform that the interplanetary magnetic field has a spiralled sectorial structure which partakes of the general solar rotation. The field pattern has been shown to exist over three solar rotations with two outward field sectors ($+$) of two-sevenths of the circle and two inward field sectors ($-$) of two-sevenths and one-seventh respectively. The solar wind velocity is the same in all the sectors, but after entry into a sector the velocity increases reaching a maximum about one third of the way through the sector, then falling off. The average magnetic field strength is about 10 per cent greater in the outwardly directed sectors. Augmenting and confirming our knowledge of the solar wind is the discovery by Van Allen[49] and his co-workers of the radiation belts about the Earth, based on researches conducted with shielded Geiger Müller counters carried by satellite 1958ε and probes Pioneer III and Explorer IV.

The atmosphere of the Earth shields us from these rays that are highly destructive of life. There are two belts widely separated and each of very high radiation intensity since they contain fast charged particles, mainly protons and electrons, with energies ranging from some tens of Kev to some tens of Mev. The belts completely enshroud the Earth and the diagram (Fig. 28) to be properly understood is to be rotated about the Earth's magnetic axis. The inner belt extends from 1000 to 4000 km. above the Earth and the outer belt from three to four Earth radii. The outer belt is the larger of the two and contains the more energetic particles; its radiation consists of electrons with energies of tens of thousands of electron volts and these produce X-rays when they impinge upon any solid object such as a space vehicle. The inner belt traps charged particles for a long time, possibly several years and a weak source of particles is sufficient to maintain a fairly high level of radiation. This source is the beta decay of neutrons from cosmic rays activity

[47] The solar wind also interacts with other planetary magnetospheres. See 'Solar Wind and Venus', *Science*, *158* (2), 1672 (1967); also Bauer, S. T., *et al.*, 'Topside Ionosphere of Venus and its interaction with the Solar Wind', *Nature*, *225*, 533 (1970); see 'Particles and Fields in the Magnetosphere', edited McCormac, B. M. (1969).
[48] Wilcox, J. M., *Science*, *152*, 165.
[49] Van Allen, J. A., 'Radiation Belts Round the Earth', *Scientific American* (March 1959).

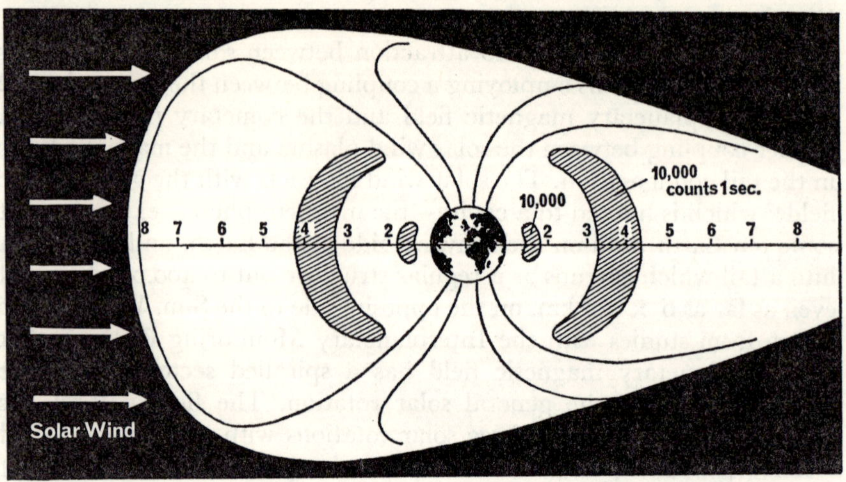

Fig. 28. Schematic diagrams of the radiation belts about the Earth.

at the outer fringes of the Earth's atmosphere; the half life of a neutron is 12 minutes and its decay leaves an electron and a proton which are trapped by the Earth's magnetic field. The understanding of these belts and their association with the solar corona and the Earth's magnetosphere are still the subject of much research and speculation.[50]

7. THE SUN'S ROTATION

The rotation of the outer parts of the Sun accessible to the viewer on Earth may be deduced from observations of the angular displacement of the sunspots, indeed this is how Galileo in 1610 deduced a rotation[51] for the Sun of about 14 degrees of arc per day, that is 360°/14 days or 25 days per rotation. Scheiner (c. 1630) had detected the faster movement of the equatorial spots, but the surprising relationship between solar latitude and the angular velocity of rotation of sunspots was reserved to R. C. Carrington[52] c. 1863 from observations made at his observatory in Redhill and his examination of the sunspot records of von Soemmerring, which went back to the year 1826. He then deduced an equatorial sidereal rotation of 24·96 days (angular velocity 14°·42) with a reduction of 1° in angular velocity at latitude 35° (25·83 days), and thereby came to the view that 'there is a different rotation period for each value of solar latitude'. In 1863 he announced that the Sun rotates according to the formula

[50] See Ness, W. N., *The Radiation Belts and Magnetosphere* (1968); see also Brandt, J. C., *The Solar Wind*, (1970).
[51] It is in the same direction as the orbital motion of the planets and the rotation of the Earth.
[52] Carrington, R. C., *Observations of the Spots on the Sun from Nov. 9, 1853 to March 24, 1861* (1863).

$$\omega = 14°\!\cdot\!42 - 2°\!\cdot\!75 \sin 7/4\,\phi$$
or
$$\omega = 14°25' - 165' \sin 7/4\,\phi$$

where ω is the rotational rate in degrees of arc per day and ϕ is the heliographic latitude.

In more recent years Newton *et al.*[53] have obtained the formula

$$\omega = 14°\!\cdot\!382 \pm 0°\!\cdot\!009 - (2°\!\cdot\!96 \pm 0°\!\cdot\!09) \sin^2 \phi.$$

This close agreement with Carrington's original work is remarkable and should not fail to excite our admiration when one considers the difficulties in this field of research for a pioneer. This type of observation is restricted to that zone of the Sun in which the spots appear and since these are seldom seen in heliographic latitudes in excess of $\pm\,30°$ it is very limited in application. We must, therefore, turn to the spectroscopic observations of Adams *et al.*[54] for results at the outer boundaries of the solar disc in the high solar latitudes. At the reversing layer they found that the rotation was greater than that detected by the observation of sunspots by $1\cdot4$ per cent and that the chromosphere was rotating 4 per cent faster than the photosphere. Many observers have worked in these difficult researches and the variation in results is said[55] to be about 10 per cent.

It is now known that the difficulty lies not only in the instruments combined with the scattering of light but also in the personal equation of the respective observers, all of which factors conspire to produce ambiguities. More recently d'Azambuja *et al.*[56] have obtained two formulae for the angular velocity based on the observation of solar filaments.

$$\omega = 14°\!\cdot\!48 - 2°\!\cdot\!16 \sin^2 \psi \text{ and}$$
$$\omega = 14°\!\cdot\!42 - 1°\!\cdot\!40 \sin^2 \phi - 1°\!\cdot\!33 \sin^4 \phi.$$

The second formula gives the smaller residuals. For the layman we may say, that the period of equatorial rotation is most easily stated as $25\cdot38$ days and that the poles rotate once in about 32 days.

There is now some doubt concerning the true meaning of these rotational periods since Ward[57] has drawn attention to the possibility of Rossby type waves in the solar atmosphere, such as are found in the Earth's upper atmosphere. Ward suggests that our analyses of motions symmetrical with the solar axis may lead to invalid results. It is to be emphasized that these researches inevitably deal with the outer parts of the Sun and what we may expect to find in the inner recesses is not something to be readily conjectured.

[53] Newton, H. W., Nunn, M. L., *Monthly Notices of R.A.S. 111*, 413 (1951); see also *The Observatory, 71*, 181 (1951).
[54] Adams, W. S., Lasby J. B., Pub. Carnegie Inst. Washington No. 138 (1911).
[55] See Kuiper, G. P., *The Sun* (p. 21) (Chicago Press 1953).
[56] d'Azambuja, L. and M. D., *Ann. Obs. Meudon*, 6, Fasc. 7 (1948).
[57] Ward, F., 'The General Circulation of the Solar Atmosphere and the Maintaining of the Equatorial Acceleration', *Astrophys. J., 141* (4), 1502 (May 1965).

The solar system

Duhem[58] says in a work of singular erudition—'the essence of heavenly things is that they merely give us an image and this is far from exact—it merely comes close—Astronomy rests with *the nearly so*.' It is clear that the Sun will not be 'deep searched', but we have not yet wholly outstripped our resources. I refer to two lines of evidence, completely independent yet both suggesting a possible high speed rotation for the innermost parts of the Sun. I speak of Professor Dicke's researches into the solar oblateness and Professor Plaskett's researches into limb darkening. Dicke *et al.*[59] have measured the oblateness of the Sun and found a difference between the equatorial and polar radii of 35 km. which amounts to 5 parts in 10^5 of the solar radius. A sun so distorted from the true sphere produces a perihelion advance of the planets, and the contribution to the perihelion advance of Mercury (see chapter 8) from an oblate sun can be determined. Indeed, Dicke deduces a contribution to the perihelion advance of Mercury of 4 seconds of arc per century and this carries with it the implication that the mean rotational period for the inside of the Sun is as small a figure as $1\cdot8$ days. Several years previously Dicke[60] in a speculative paper pointed to the possibility of solar oblateness and the distortion of the Sun's gravitational field in causing a rotation of the perihelion point of the planetary orbits, and at the same time he drew attention to the anomalous rotation of the Sun when compared with the large planets of the Sun's system,

Sun	25·4 days
Jupiter	0 d. 9 h. 50 m.
Saturn	0 d. 10 h. 14 m.
Uranus	0 d. 10 h. 49 m.

and he concluded that the rotation of the interior of the Sun may be less than 25 hours.

Plaskett's[61] researches lead him to the view that the rotation of the interior of the Sun may be about 12 hours. He comes to this surprising view from the phenomenon of limb darkening. The measures of limb darkening confirm that the poles of the Sun are hotter than the equator by some hundred degrees. Now the temperature at the surface of the Sun varies directly as the gravity and the centrifugal force. Since the gravity is the same at the equator and the poles the lower temperature of the equator must be due to the centrifugal force, and since the latter varies as the square of the angular velocity times the distance from the axis of rotation we can use the temperature difference to find the internal

[58] Duhem, P., *To Save the Phenomena*, trans. from the French of 1908 by Doland, E., and Maschler, C. (University of Chicago Press 1969).
[59] Dicke, R. H., and Goldenberg, H. M., *Phys. Rev. Lett.*, *18*, 313 (1967).
[60] Dicke, R. H., 'The Sun's Rotation and relativity', *Nature*, *202*, 432 (1964); see also *Nature*, *213*, 1077 (1967); *214*, 1294 (1967); *216*, 1283, 1206 (1967).
[61] Plaskett, H. H., *Monthly Notices R.A.S.*, *131*, 407; see also Halley Lecture, *The Observatory*, *85*, 178 (1965).

146

angular velocity. What is acceptable is the theorem that the surface brightness of the Sun varies as the effective gravity, i.e. gravitation combined with centrifugal force (see for example Eddington's *Internal Constitution of the Stars*, pp. 282–7). What is unconfirmed except for two sets of observations at Oxford is that the poles are brighter than the equator. Granted that these observations are correct a rapid internal rotation is only one way of making the poles hotter; there may well be others. Moreover there are real difficulties about such a rapid internal rotation; in particular the difference between the external and internal rotation would cause great instability.

According to Plaskett's Halley Lecture (1965), the Oxford observations give the sight-line velocities at some 900 points on the solar surface. On analysis, these show the well-known polar retardation in the rotation and the presence of meridional currents directed towards the equator. If we postulate a meridional pressure gradient to account for the latter, then a balanced motion between this pressure gradient, Coriolis force and viscosity, gives rise not only to meridional currents that increase in strength towards the equator as observed, but also in high latitudes to zonal currents moving from apparent west to east against the direction of rotation. These zonal currents in an otherwise uniformly rotating photosphere account for the apparent rotation period of 32 days at the poles as compared with 25 days at the equator. This description of the apparently non-uniform rotation of the Sun involving zonal currents is confirmed in two ways. Firstly, like well-established zonal winds in the upper atmosphere of the Earth, these photospheric zonal currents show Rossby waves. Secondly, measures of limb darkening (which have not been independently confirmed) show the poles to be hotter than the equator with a difference in temperature that quantitatively accounts for the meridional pressure gradient and the zonal currents.

It is known that polar heating can be produced by rotation. If this is the origin of the observed temperature difference between the poles and the equator it follows that the radiative envelope below the hydrogen convection zone completes a rotation in some twelve hours, thus making the angular momentum of the Sun comparable with the orbital momenta of the major planets. That the closely coupled photosphere and hydrogen convection zone rotate fifty times more slowly than the radiative envelope is a result of braking action by the zonal photospheric currents during the long life of the Sun. At the present time these currents are reducing the equatorial rotation-velocity of the photosphere by 0·001 km.s^{-1} every 50 years.

8. THE RADIO EMISSION OF THE SUN

The Sun emits radio waves and the existence of this phenomenon was appreciated by Sir O. Lodge[62] but not discovered by him, despite his investigations of the problem. The discovery of the Solar radio emission had to wait for the deployment of radar equipment in the hands of Hey[63] and Southworth[64] in 1942 with delay in publication owing to the classification of their results due to the restrictions imposed by the Second World War, and independently by Reber[65] in 1943. The 'Quiet', that is the unperturbed Sun, emits a low energy level of radio emission thought to originate in the *bremsstrahlung*[66] of electrons in collision with ions in the Solar corona and the chromosphere. Over and above the continuous spectrum of radio waves there is a discrete spectrum caused by transitions of electrons between levels of atoms and ions and the most important to be detected is the 3·04 cm. line. There is further a complex sporadic radio emission. The several distinct kinds of radio emission have been ably summarized by Denisse[67] and Zheleznyakov[68] and I give below a much simplified form of their table.

[62] See Haddock, F. T., *Proc. I.R.E.*, *46*, 3 (1958).
[63] Hey, J. S., *Nature*, *157*, 47 (1946).
[64] Southworth, G. C., *J. Franklin Inst.*, *239*, 285 (1945).
[65] Reber, G., *Ap. J.*, *100*, 279 (1944).
[66] The electromagnetic radiation produced by the sudden retardation of an electrical particle (as an electron or positron) in an intense electric field (as in the atomic nucleus).
[67] Denisse, J. F., *Paris Symposium on Radio Astronomy*, ed. R. N. Bracewell, p. 81 (Stanford Univ. Press, 1959).
[68] Zheleznyakov, V. V., *Radio Emission of the Sun and Planets*, trans. by Massey, H. S. H., p. 238 (Pergamon Press 1970).

Note: Since writing this chapter the Orbiting Solar Observatory-7 has shown Neupert *et al.* that the Sun has 'polar caps' or more specifically that there are clear boundaries between the polar and equatorial regions of the Sun's upper atmosphere, the corona. The normal temperature of the corona, except for the polar areas and a few hot spots around active regions, is about two million degrees Fahrenheit. However, the polar caps, clearly shown by the observatory readings, have temperatures of only about one million degrees. The hot spots, storm centres or active regions often associated with giant solar explosions called flares, register three to four million degrees. Flares themselves are several times hotter again, at least ten million degrees.

THE TIMES, 1971, December, 10.

For a review of advances in solar research by rockets see *Twentyfive years of Rocket and Satellite Astronomy. Nature 234*, 181, (1971).

Type of radio emission	Source of radio emission	Life	Wavelength Band	Polarization
'Quiet' Sun radio emission (B-component)	Undisturbed chromo-sphere and corona	Always present	mm., cm., dm., m.	Very weakly polarized
Slowly varying radio emission (S-component)	Regions of high density, magnetic field and temperature in lower corona above centres of activity	Few months	mm., cm., dm.	Partly polarized (with extraordinary emission predominant), degree of polarization decreases with rise in wavelength
Microwave bursts Type A (plain bursts) Type B (postburst fall) Type C (gradual rise and fall)	Generation region usually in the same place as source of S-component	Few minutes Few minutes to few hours Tens of minutes to few hours	mm., cm., dm.	Partly polarized (extraordinary emission apparently predominant)
Type I raise storms Enhanced radio emission connected with sunpots Type I bursts	Region of corona above groups of spots	From few hours to few days Fractions of a second	m.	Strongly polarized, apparently with ordinary waves predominant
Type II bursts	Region in corona moving away from chromospheric flare at a velocity of $\sim 10^4$ km/sec	Few minutes	m.	Not polarized
Type III bursts	Region in corona moving away from chromospheric flare at a velocity of $\sim 10^5$ km/sec	Few seconds	m., dm.	Sometimes partly polarized
*Type IV radio emission (follows type II burst)	Region in corona moving away from chromospheric flare at a velocity of $\sim 10^3$ km/sec	From few minutes to few hours	dm., m.	Partly polarized
Type V radio emission (follows type III bursts)	Region in corona	Minutes	m.	?

* For the discovery of *regular* pulses from the Sun in this meter-wave continuum radiation see *Nature 234*, 140, (1971).

Chapter 10

On the Earth-Moon system

Observe how system into system runs.
POPE

Art thou pale for weariness
Of climbing heaven and gazing on the earth,
Wandering companionless
Among the stars that have a different birth—
And ever changing, like a joyless eye
That finds no object worth its constancy?
SHELLEY

The Earth and the Moon behave very much as two inseparable celestial dancing partners, one of which (the Earth) makes a rapid rotation on its axis as its partner (the Moon) makes a slow revolution about it and a rotation on its axis in the same period of time as it takes to make the revolution. The two partners together make a continuous and inexorable revolution of the Sun with their orbit always concave to him. This is shown in greater detail in Fig. 29. The centre of gravity of the two partners, the barycentre[1] may be discovered by the simple law of moments and found to reside deep within the confines of the Earth, approximately 1,000 miles from the surface.

The Earth is the third planet in order of distance from the Sun and the fifth in size among the nine primary planetary bodies that orbit the Sun. It is the chief body of the four terrestrial planets.[2] The Earth is the only member of the Sun's family to have a *companion planet*; the moons of Mars, Jupiter, Saturn, Uranus and Neptune being too small when compared with their primaries to be regarded as such a companion.

The Earth travels in an orbit of eccentricity 0·0167217 and at a mean distance from the Sun of 92,957,209 miles (149,600,000 km.) so that her greatest, mean and least distances are as the numbers 1·0167217, 1., 0·9832783. The greatest and least distances from the Sun, aphelion and perihelion are readily calculated and are placed at 152,080,000 km. and

[1] See Möbius' barycentric calculus; a short reference is to be found in the Encyclopaedia Britannica, XI ed., vol. I, p. 614 a.
[2] Mercury, Venus, Earth and Mars.

150

147,090,000 km. respectively. Paradoxically in the northern hemisphere the perihelion point is reached in the depths of winter. The perihelion of the Earth's orbit is now some ten days after the winter solstice of the northern hemisphere and was (as a matter of small interest now) coincident with it in the year 1248. The perihelion point is moving away

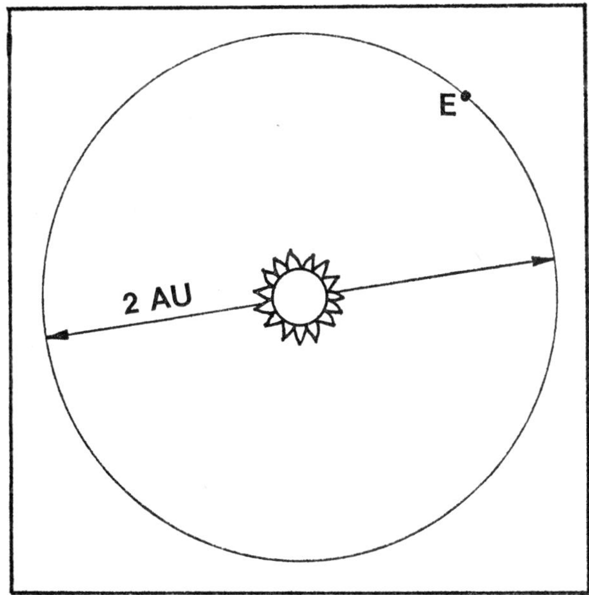

Fig. 29. Concave orbit of the Moon.

from the point of the winter solstice at the rate of $61''\cdot89^3$ per annum. It is advancing in sidereal longitude by $11''\cdot3$ and the solstice retrogrades by $50''\cdot6$ annually. The arc of separation increases one degree in 58·44 years.

It is well understood, but may be briefly stated, that the marked seasonal changes we enjoy on Earth come from the tilt of the Earth's axis to the ecliptic plane (the plane of the orbit) which is $23°27'$.[4] It is usual to consider how the Sun looks from the Earth, since it is not easy to break our anthropocentric mould, but it is more rewarding if we follow Proctor[5] and consider how the Earth looks from the Sun. This is most readily achieved in the present context by taking some well-known portion of the Earth's surface and presenting its inclination at noon in

[3] According to Newcomb it is $6189''\cdot03T + 1''\cdot63T^2 + 0''\cdot012T^3$, where T is measured in Julian centuries of 36525 ephemeris days from 1900 January 0·5 E.T.
[4] The obliquity of the ecliptic decreases by $0''\cdot47$/annum. The plane of the Earth's equator is not fixed in space, its intersections with the ecliptic have a slow retrograde motion; this phenomenon is known as precession.
[5] Proctor, R. A., *Old and New Astronomy* (London 1892).

151

midwinter and in midsummer (Fig. 30). We have in the previous chapter seen that the heat from the Sun is immense[6] and now it will be abundantly clear that the heat is proportional to the apparent area of the Earth's surface exposed to his rays. This is not the sole reason for the increase in summer warmth, however, for the days are then longer than in winter and the Sun is above the local horizon for greater periods.

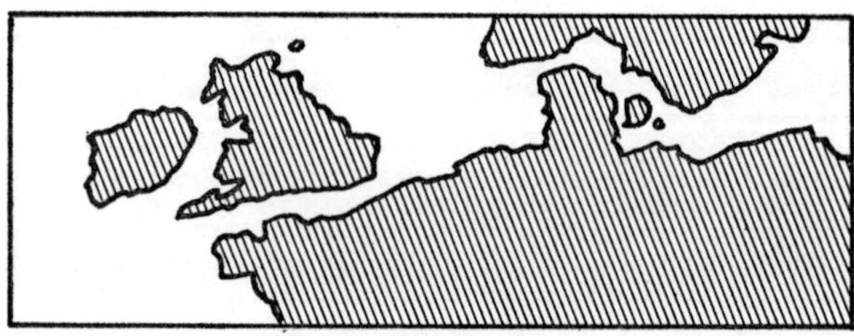

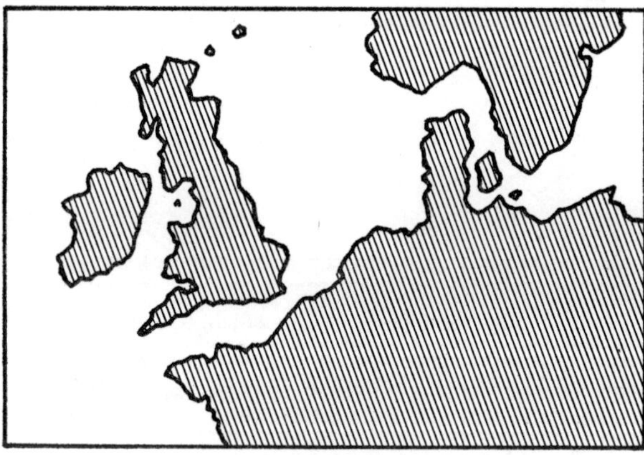

Fig. 30. Variation of area of land mass heated by the Sun with season at change.

The Moon, as the inseparable companion of the Earth in her wanderings, is an important luminary and indeed serves for the measurement of time, regulates the onset of religious festivals[7] and sways the ocean tides.[8]

[6] The radiation emitted is $3 \cdot 90 \times 10^{33}$ erg. s^{-1}.

[7] The astronomical problem is bedevilled by the fact that the two time-measuring units which made an early appeal to men, viz., the year delimited by the sun's apparent movement and the lunation delimited by the moon's phases, do not contain an exact number of days. The year (the tropical year equinox to equinox) has 365·24219 days. The lunation (new moon to new moon) has 29·53059 days. A simple system of dating can be based on either the year or the lunation and the world has long held an example of each in the wholly solar calendar of Julius Caesar and the wholly lunar calendar of Mohammed. A difficulty appears when the one is tied to the other as it is in the dating of the Passover. The Jewish months are true lunations and the 14th of Nisan is always at full moon but successive Nisans are not separated

In their combined movements the Earth and the Moon periodically produce both solar and lunar eclipses.[9] The Moon travels in practically the same orbit as the Earth and it is not unrealistic to consider her as having an orbit of her own about the Sun that is perturbed by the action of a neighbouring planet, the Earth. The disturbance to the Moon's path from the Earth's path about the Sun is seen to advantage in Fig. 29, where the barycentre of the Earth/Moon system may be considered to move in the circular orbit E. The diameter of the circle (3″) represents a true diameter of 2 a.u. and the Moon's mean orbit about the Earth is 0·005 a.u. which is about 1/130th of an inch in width on the diagram and thus less than the width of the fine line used to delineate its path in the diagram.

The literature devoted to the Earth[10] and the Moon is vast and we will not in this work deal with those many features that are to be found expressed in the general text books on Astronomy[11] but rather try to deal with some features and subtleties of the system that are neglected in standard works and widely scattered in the literature.

We should nevertheless commence our task with the vital statistics[12] of the two components of the system and these are shown below:

Tropical year, equinox to equinox $\qquad$ 31,556,925·9747s
$\qquad\qquad\qquad\qquad\qquad\qquad\qquad\qquad\qquad$ 365d·24219

Length of the day $\qquad$ 24h. in both the solar and sidereal day.

Mean solar day $\quad$ $24^h\,03^m\,56^s\!·555 = 1^d·00273791$ mean sidereal time

Mean sidereal day $\quad 23^h\,56^m\,04^s\!·091 = 0^d·99726957$ mean solar time

Sidereal rotation period of the Earth

$\qquad\qquad 23^h\,56^m\,04^s\!·099 = 0^d·99726966$ mean solar time

Length of Month

tropical (equinox to equinox) $\qquad$ 27d·32158

sidereal (fixed star to fixed star) $\qquad$ 27d·32166

synodic (new moon to new moon) $\qquad$ 29d·53059

---

by equal periods of time since the feast must fall at the full moon following the vernal equinox, a solar phenomenon. Easter has its origin in the Passover, and like it, is a luni-solar event.

A simplified approach, however, is available, owing to the genius of Gauss. Take two numbers m and n; let $m = 24$ and $n = 5$ (suitable until the year 2099). Find the remainders when the number of the year say 1964 is divided by 4, by 7 and by 19, call them a (0), b (4) and c (7). Divide $19c + m$ by 30 and call the remainder, d, (7). Divide $2a + 4b + 6d + n$ by 7 and call the remainder, e (0). So Easter Day is March $22 + d + e$. For 1964 it is March $22 + 7 + 0$, that is, March 29.

[8] The south equatorial current alone carries 6,000,000 tons of water a second northward across the equator.

[9] See Dyson and Woolley, *Eclipses of the Sun and Moon* (Oxford 1937).
[10] See Jeffreys, H., *The Earth*, 5th ed. (Cambridge University Press 1970).
[11] See for example the three very thorough works by Payne, Gaposchkin C., *Introduction to Astronomy* (1956); Jones, Spencer, H. (Barlow and Bryan), *Elementary Mathematical Astronomy*, (1946); Struve, O., *Elementary Astronomy* (1959).
[12] Mainly from *The Handbook of the British Astronomical Association* (1970).

The solar system

Earth/Moon mass ratio $\qquad$ $81\cdot302 \pm 0\cdot005$*
  Mass of the Earth $\qquad$ $5\cdot976 \times 10^{27}$g
  Mass of the Moon $\qquad$ $7\cdot351 \times 10^{25}$g
                  (i.e. $0\cdot0123$ Earth's mass)
  Mean density Earth $\qquad$ $5\cdot5$g/cm.^{3}
  Mean density Moon $\qquad$ $3\cdot3$g/cm.^{3}
Mean distance Earth to Moon $\qquad$ $384,400$ km. ($238,855$ miles) ($0\cdot0025695$ a.u.)

(Max. distance $252,700$ miles, Min. distance $221,460$ miles)

Figure of the Earth
  equatorial radius $\qquad$ $6378\cdot160$ km.
                     $3963\cdot208$ miles
  polar radius $\qquad$ $6356\cdot775$ km.
                     $3949\cdot920$ miles

the flattening is now found to be one part in $298\cdot2$; from the behaviour of close artificial satellites

Moon's radius $\qquad$ $1738\cdot0 \pm 0\cdot5$ km.
                     $1080 \pm 0\cdot3$ miles

Moon's surface not observable from Earth $0\cdot410$

The I.A.U. system of Astronomical Constants, 1964 from Wilkins, G. A., *Quat. J. Roy. Astron Soc.*, 6, 172 (1965), are:

Primary constants:

Measure of one a.u. in meters	$A =$	149600×10^6
Velocity of light in meters/sec	$c =$	$299792\cdot5 \times 10^3$
Equatorial radius for Earth in meters	$a_e =$	6378160
Dynamical form factor for Earth	$J_2 =$	$0\cdot001082\ 7$
Geocentric gravitational constant (units: m^3 s^{-2})	$GE =$	398603×10^9
Ratio of the masses of Moon and Earth	$\mu =$	$1/81\cdot30$
Sidereal mean motion of Moon in rad/sec (1900)	$n_{\mathbb{D}} =$	$2\cdot661699\ 489 \times 10^{-6}$
General precession in longitude per tropical century (1900)	$p =$	$5025''\cdot64$
Obliquity of the ecliptic (1900)	$\epsilon =$	$23°27'08''\cdot26$
Constant of nutation (1900)	$N =$	$9''\cdot210$

Derived constants:

Solar parallax	$\pi_\odot =$	$8''\cdot794$
Light time for unit distance	$\tau_A =$	$499^s\cdot012$
Constant of aberration	$\chi =$	$20''\cdot496$
Flattening factor for Earth	$f =$	$1/298\cdot25$

* From Mariner II Experiment, Anderson Null and Thornton (1963).

Heliocentric gravitational constant (units: $m^3\ s^{-2}$)	$GS = 132718 \times 10^{15}$
Ratio of masses of Sun and Earth	332958
Ratio of masses of Sun and Earth + Moon	$1/m = 328912$
Perturbed mean distance of Moon in meters	$a_{\mathbb{D}} = 384400 \times 10^3$
Constant of sine parallax for Moon	$\sin \pi_{\mathbb{D}} = 3422''\cdot451$
Constant of Lunar inequality	$L_{\mathbb{D}} = 6''\cdot440$
Constant of parallactic inequality	$P_{\mathbb{D}} = 124''\cdot986$

THE ROTATION OF THE EARTH

The rotation has, since June 1955, been studied against a caesium beam atomic clock. It is found that the Earth was decelerating until September 1957. It then accelerated until January 1962 and has since been decelerating again. The date of these changes in the future cannot be predicted. At the time of writing the period of the comparison is very small.[13]

That there has been a variation in the Earth's rotational period over the ages is clear from a number of lines of evidence. Indeed, long-term and irregular variations in the speed of rotation of the Earth have been suspected from the eighteenth century onward, but it was not until Spencer Jones's comprehensive analysis of the observed motions of the Sun, Moon and planets was published in 1939 that the necessity for a new astronomical time scale, independent of the rotation of the Earth, became evident. Ephemeris Time was recommended by the 1950 Paris conference and confirmed at the Rome meeting of the International Astronomical Union in 1952. In 1956 the Comité International des Poids et Mesures adopted the second of Ephemeris Time as defined by the I.A.U. as the unit of time interval in the Système des Unités, as 'the second is the fraction 1/31,556,925·9747 of the tropical year for 1900 January 0. at 12 h E.T.' In 1967 this definition was superseded by the definition that—'The second is the duration of 9,192,631,770 periods of the radiation corresponding to the transition between the two hyperfine levels of the ground state of the atom of Caesium 133.' This new unit is reproducible in off-the-shelf electronic apparatus with a precision of one part in 10^{11}.[14]

Munk[15] has explained that the variations of the Earth's rotation can be studied from the discrepancies of the positions of the Moon and the Sun in the observations available from *c.* 1680 to date. In this approach

[13] See Markowitz, W., 'The irregular variation in the speed of the rotation of the Earth', *The Earth Moon System* (1966); Brouwer, D., *Astron. J.*, *57*, 125 (1952).
[14] See Sadler, D. H., 'Astronomical Measures of Time', *Quat. J.R.A.S.*, *9*, No. 3, 281 (1968).
[15] Munk, W. H., *The Earth Moon System*, p. 54 (Plenum Press, N.Y. 1966).

he follows that of Munk and MacDonald in their well-known work *The Rotation of the Earth* (London 1960). The remarkable fact that the stupendous globe of the Sun, and in comparison, the minute globe of the Moon should exhibit angular diameters, as seen from the Earth, of almost the same size (about one half of a degree of arc) gives the ancient observations of eclipses of the Sun an accuracy of quite a remarkable value of precision. Dicke[16] uses this feature to discuss the secular acceleration[17] of the Earth's rotation. He considers five eclipses, used initially by Fotheringham[18] in a brilliant paper of 1920, that span a millennium viz. the eclipse of Plutarch A.D. 71, Phlegon A.D. 29, Hipparchus 129 B.C., Archilochus 648 B.C., and Babylon 1063 B.C.

Whereas Fotheringham obtained values for the secular acceleration of the Moon equal to $+ 10 \cdot 8'' \, T^2$ and the Sun $+ 1 \cdot 5'' \, T^2$, where T is expressed in centuries, Dicke obtains more refined values and proceeds to breakdown the phenomenon into eight distinct parts: viz. The transfer of angular momentum by

1. Tidal slowing of the Earth due to lunar tides, solar tides and atmospheric coupling.
2. Transfer of momentum to the solar wind.
3. Transfer of momentum to the atmosphere and oceans.
4. Electro-magnetic coupling of the Earth's core to the mantle and variations in the moment of inertia of the Earth from: (*a*) change in sea level, (*b*) continental unrest, (*c*) precipitation of the Earth's core, (*d*) Expansion or contraction of the Earth.

The effect of the solar wind is found to be too small by a factor of more than 10^5 to make any real contribution to the deceleration of the Earth's rotation; similarly the transfer of momentum to the atmosphere and oceans is also thought to be negligible, since if the atmosphere and the ocean tides were to be the 'sink' for the angular momentum lost by the Earth's mantle, then a gale of 570 km./hour at the equator would result after but a short period of 2000 years, and in less than a decade the ocean currents would double and in a few thousand years be of phenomenal proportions. Such phenomena are not observed to be occuring. The effect of continental unrest and expansion or contraction from the accumulation or loss of radiogenic heat is also shown to be negligible. Changes in sea level[19] indicated by C^{14} tests have been reviewed from all the available dated measures over the past 6000 years. It is concluded that there is a slow rise with time with an increased rate in the past. There is no general agreement on the matter and a refined analysis

---

[16] Dicke, R. H., *The Earth Moon System*, p. 98 (1966).
[17] The secular acceleration was discovered by Halley in 1693; it is ably discussed by Dyson and Woolley in their work *Eclipses of the Sun and Moon*, pp. 35–42 (Oxford 1937).
[18] Fotheringham, J., *Monthly Notices of R.A.S., 81*, 104 (1921).
[19] See Shepard, F. P., *Science, 143*, 574 (1964).

is able only to suggest the opinion that the change to the secular deceleration of the Earth's rotation is at most a few per cent, but this needs to be qualified in view of the uncertainty of the ancient eclipse data in one instance, which if differently interpreted, could put the contribution to the fluctuation up to a figure as great as 15 per cent.

Our knowledge of the Earth's interior at the present time, and in the past, is too slight and insecure to yield any meaningful figure. The core/mantle electromagnetic coupling is not yet resolved according to the consensus at the Geophysical discussion at the Royal Astronomical Society on 1970 February 20[20] from which it was clear that the behaviour at the Earth's core-mantle interface is exceedingly complex. Many important processes occur in both the core and the mantle and the interactions across the boundary may have a significant effect on the behaviour of each of the two major components. The basic cause of the variations in the Earth's rotation is thought to be due to horizontal stresses at the said boundary since they prevent the core and the mantle moving completely independently of one another. The viscous stresses are extremely weak and the electric currents from the geomagnetic field are also too weak by a factor of ten to explain the observed fluctuations. Hide[21] suggests that the stronger form of coupling, needed to offer a plausible explanation, may come from topographic irregularities or 'bumps' on the surface of the core. Calculations show that a bump of but a few kilometres in height would produce a stress of some hundred times in excess of the electromagnetic coupling. That the 'bumps' may exist has been discussed by Hide and Malin[22] and they adduce evidence in support of their broad argument. Work is in progress by others to detect the 'bumps' by an appeal to the P and PcP waves generated by nuclear explosions in the core and the mantle. Such methods are thought capable of determining bumps as small as 4 km. in height.

The tidal slowing of the Earth's rotation is difficult to resolve in view of the fact that it is not clear whether the shallow water tides are important or not. The tidal induced acceleration lies, according to Dicke, between

$$- 27 \cdot 2 \times 10^{-11} \text{ yr}^{-1} \text{ and } - 25 \cdot 1 \times 10^{-11} \text{ yr}^{-1}$$

The problem concerning shallow water tides is not serious.

Wilkins[23] explains that it is certainly correct to say that the Earth's rate of rotation is gradually slowing down; this is now clearly demonstrated by comparison of observed universal time with an atomic time scale. Between 1957 and 1967 the period of rotation increased by about

[20] See *Nature*, *225*, 793 (1970).
[21] Hide, R., *Phil. Trans. Roy. Soc.*, A 259, 615; *Science*, *157*, 55 (1967).
[22] Hide, R., 'Malin S.R.E.', *Nature*, *225*, 605 (1970).
[23] Wilkins, G. A., Private communication to the author, 20 March 1970; see also *Science*, *172*, 1022 (1971).

4×10^{-12} days each day, although most of the increase took place between 1963 and 1966. Thus a time scale, T_R say, defined by the apparent revolution of the mean sun with respect to the rotating Earth is not uniform, and so in the equations of motion of the Sun and Moon and planets we now use a (supposedly) uniform time scale which we here denote by T_E. If the rate of decrease (x, say) of the Earth's rate of rotation were constant we would be able to connect these two time-scales by a relation of the form

$$T_R = T_E - \tfrac{1}{2}x T_E^2,$$

or since x is small,

$$T_E = T_R + \tfrac{1}{2}x T_R^2.$$

Now suppose that a celestial body were moving round the Earth in a circular unperturbed orbit; then its orbital longitude would be given by a theoretical expression of the form

$$L = Lo + n T_E.$$

But if it were observed in our rotational time scale its longitude would appear to be given by

$$L = Lo + n T_R + \tfrac{1}{2}nx T_R^2.$$

It can be seen that the departure from uniform motion depends directly on n, and so the effects are more noticeable for the Moon than for the Sun. Unfortunately the actual situation is much more complicated. Firstly there are other sources of quadratic terms and secondly the Earth's rate of rotation shows irregular fluctuations superposed on the general slowing down, which probably does not take place at a uniform rate. It is therefore necessary to analyse the observations of the planets as well as those of the Sun and Moon in order to separate these effects. Spencer-Jones found that for the Sun the quadratic term amounted to $1''\cdot23$ per century[2]; this corresponds to a value of x of about $0\cdot5 \times 10^{-12}$ per day and is in agreement with the values given by Dicke. We can only say that this value is of the same order of magnitude as the value given by comparatively recent observations.

There is now evidence from palaeontological sources to show that the day in the past was not the same as the present. The Devonian month may have had $30\cdot6 \pm 0\cdot1$ days in it, though the analysis of the growth rings on corals on which this figure is based is exceedingly complex.[24] The evidence available suggests that while the period of the Earth's *revolution* about the Sun has been constant the period of *rotation* about its axis has varied considerably due in the main to the dissipation of energy by the tidal forces on the surface of the Earth giving a change of about 2 seconds of time in 10^5 years. The length of the terrestrial day has

[24] See Scrutton, C. T., *Paleontology*, 7, 552 (1965).

been inescapably increasing. Wells[25] has shown that recent corals have about 360 ridges/year while fossil corals from the middle Devonian have 400/year corresponding to a fractional lengthening of the day of $2 \cdot 8 \times 10^{-10}$ per annum. Berry and Parker[26] find from studies in Mexican *Chione* and *Idonearca vulgaris* that there was 29·65 days in the month in the Late Cretaceous (compared with 29·53 at present) which gives 370·3 days (i.e. 29·65 × 12·49) in the Late Cretaceous year. According to Munk and MacDonald the length of the day at the start of the Cambrian was 21 hours and according to Chevallier and Cailleux 21·4 hours. Kulp[27] has given a revealing graphical interpretation of the known evidence. He shows that the days per year in the various geological periods were as follows:

Cretaceous	371
Jurassic	377
Triassic	381
Permian	385
Devonian	396
Silurian	402
Cambrian	412

If we assume the year to be of a constant period we can from these data obtain some guide to the length of the day over a considerable geologic time span.

ATMOSPHERES AND INTERIORS

The Earth is massive enough to hold an atmosphere of considerable density[28] but the Moon, only one-eightieth as massive, is not able to retain an atmosphere of any gas other than the most heavy, such as stable hydrogen sulphide, krypton and xenon, but whether such exist on her surface is much to be doubted. No atmosphere of the Moon has ever been detected. Bernstein *et al*[29] have assessed the possible lunar atmospheres from the evidence available up to 1963, including the measurement taken during the flight of Lunik III. They remark that Dollfus[30] set an upper limit to any lunar atmosphere of 3×10^{10} atoms cm.$^{-3}$. From the model atmospheres computed they judge that one can expect that A^{40} is the only gas evolved from a lunar source, in this case K^{40}. All the other gases are solar wind accretions.

We should not neglect to mention briefly the observations of Kozyrev

[25] Wells, J. W., *Nature*, *197*, 948 (1963).
[26] Berry, W. B. N., and Parker, R. M., *Nature*, *217*, 938.
[27] Kulp, J. L., *Science*, *133*, 1105 (1961).
[28] The atmosphere contributes only about one-millionth of the mass of the planet.
[29] Bernstein, W., Fredericks, R. W., Vogl, J. L., 'The Lunar Atmosphere and the Solar Wind', *Icarus*, *2*, 233 (1963).
[30] Dollfus 'Recherche d'une atmosphere autour de la lune' (1956), *Ann. d'Astrophys*, *19*, 71.

who claims to have recorded an emission of gas from the well-known crater Alphonsus. More recently[31] Kozyrev has given further evidence of a true Earth/Moon system with his claim in *Novosti* that terrestrial earthquakes are synchronized with the venting of the gases, molecular nitrogen and cyanic gas, from Aristarchus. The earthquake in question was that on the Japanese coast on 1969 March 31. The venting of Aristarchus occured on 1969 April 1.

The physical model of the Earth proposed some fifty years ago is one in which there are two components, the core and the mantle, both of uniform density. The core has a radius of 3473 km.[32] More recently it has been suggested that the chemical composition of the Earth is uniform throughout. Seismic evidence suggests a rather sudden change in the state of the material (possibly olivine) at a specific depth. Hence we now consider the Earth to have a core which acts as a liquid. The surprising change in density at the 'halfway' position is shown by Bullen.

Distance from centre (km.)	Density g/cm.^{3}
6,000	3·61
5,000	4·90
4,000	5·42
3,473	5·68
Boundary of core and mantle	
3,473	9·43
3,000	10·09
2,000	11·07
1,000	16·92
0	17·20

Opinion is still divided on the nature of the core, as to whether it is made up of iron or of silicates in a metallic phase.

The pressures deep within the Earth are of a staggering magnitude— they are estimated to be about one million atmospheres. Nothing approaching this pressure can be reached in our laboratories and any picture of the behaviour of material under this pressure and at high temperature is sheer speculation. The density of the core is thought by some geophysicists to be about 10 per cent lower than that of iron and nickel at the conditions of pressure and temperature that exist there. This discrepancy is explained away by the suggestion that the core may contain quite large amounts of another lighter element such as silicon,[33] but this appears to require large scale high temperature reduction in the Earth. If the reducing agent was carbon, large quantities of carbon monoxide would have entered the atmosphere of the Earth in its early period of formation and then been lost. At the same time much of its sulphur and other volatile elements would also have been lost, but this

[31] *Nature*, *222*, 404 (1969).
[32] See Bolt., B. A., and Qamar, A., 'Upper Bound to the Density Jump at the Boundary of the Earth's Inner Core', *Nature*, *228*, 148 (1970).
[33] See Ringwood, *Geochim Cosmochim Acta*, *30*, 41 (1966).

53. Close approach to Mars by the Mariner 6 spacecraft, 1969, July 30. The large crater is 160 miles (256 km.) in diameter. The centre of the picture is at Martian longitude 15°E, latitude 16°S.

54. A close approach to Mars by the Mariner 6 spacecraft on 1969, July 30. The photograph shows a part of Deucalionis Regio under late afternoon sunlight. The large crater is about 5 miles (8 km.) in diameter.

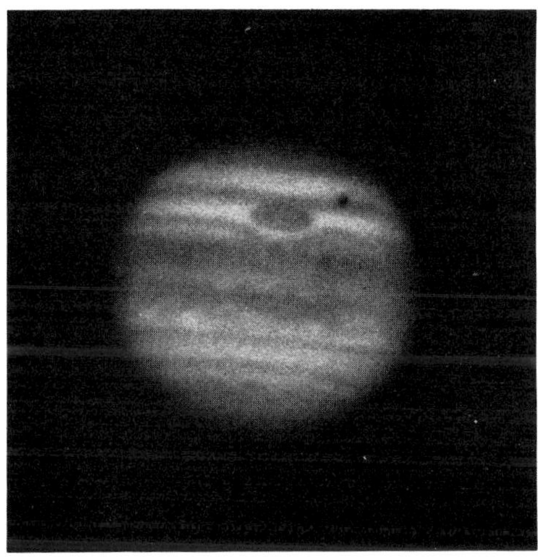

55. Jupiter, 1964, November 1, 1 h. 29 m. U.T. The red spot is near the meridian. The shadow of Satellite II, Europa appears on the disc. (15½ in. Cassegrain.)

56. Saturn, 1969, September 6. (30 cm. Cassegrain.)

57. Saturn. From a drawing by Paul Doherty, Stoke-on-Trent using a 8½ in. refractor (magnification × 274), 1968, August 4, 2 h. 20 m. U.T.

58. Comet Ikeya-Seki, 1965. Perihelion passage October, 21·1831 U.T. Photograph by Alan McClure, Mount Pinos, California, taken with a 10 in. f/3·8 lens, panchromatic plate, 1965, November 1, 12 h. 43 m.–12 h. 54 m.

59. Meteor trail in Pleiades, 1932, January 11.

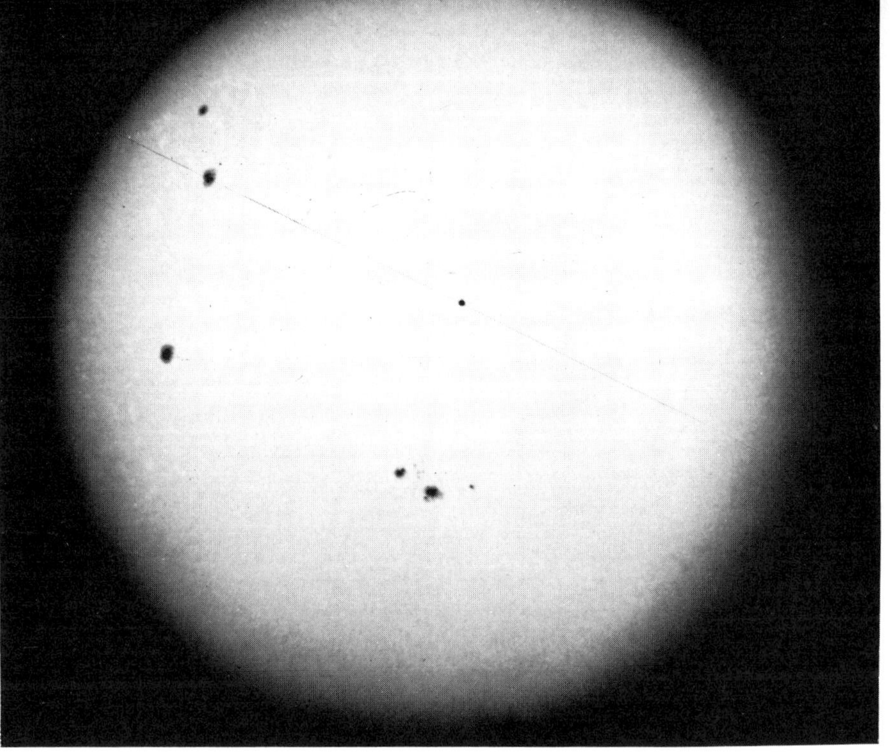

60. Transit of Mercury across the Sun's disc, 1970, May 9, 8 h. 22 m. U.T. On the photograph the diameter of Mercury's disc is approximately 14·5 seconds of arc. Taken with a 10·6 cm. refractor, 50 mm. f/1·8 camera.

61. Mars taken with the 200 in. Hale reflecting telescope at Palomar.

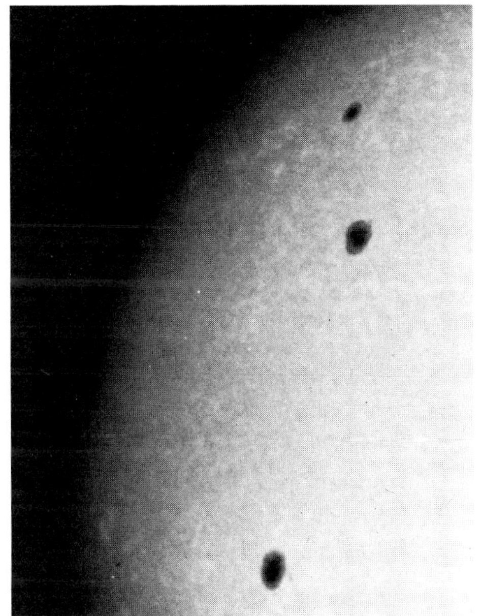

62. (a) Mercury at the third contact, 1970, May 9, 12 h. 9·50 m. U.T.

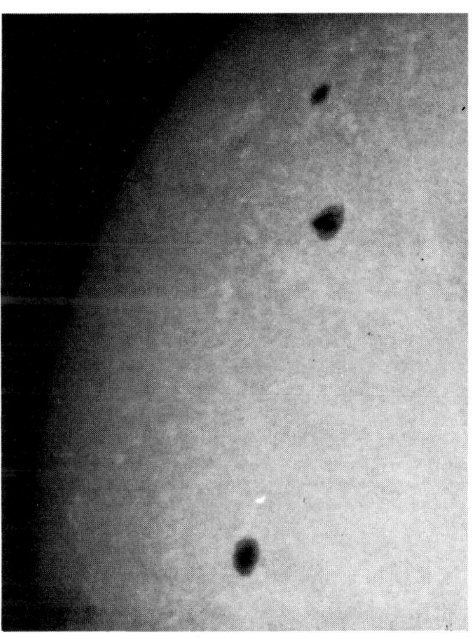

(b) Mercury superimposed on the Sun's disc, 1970, May 9, 11·11 h. U.T.

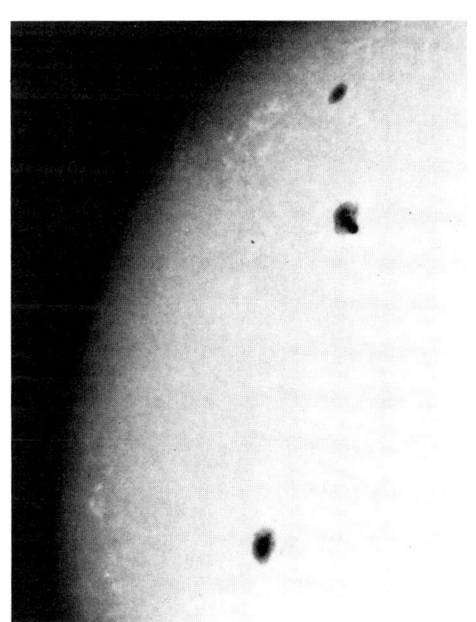

(c) Mercury superimposed on the Sun's disc, 1970, May 9, 11·04 h. U.T.

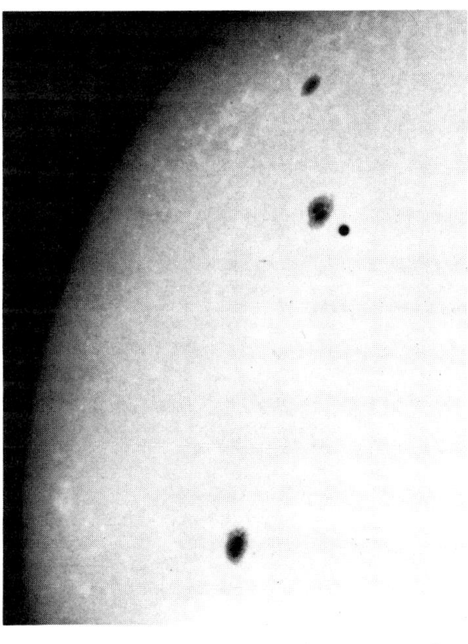

(d) Transit of Mercury, 1970, May 9, 10·57 h. U.T. Mercury's disc is about 10·9 seconds of arc on the photograph.

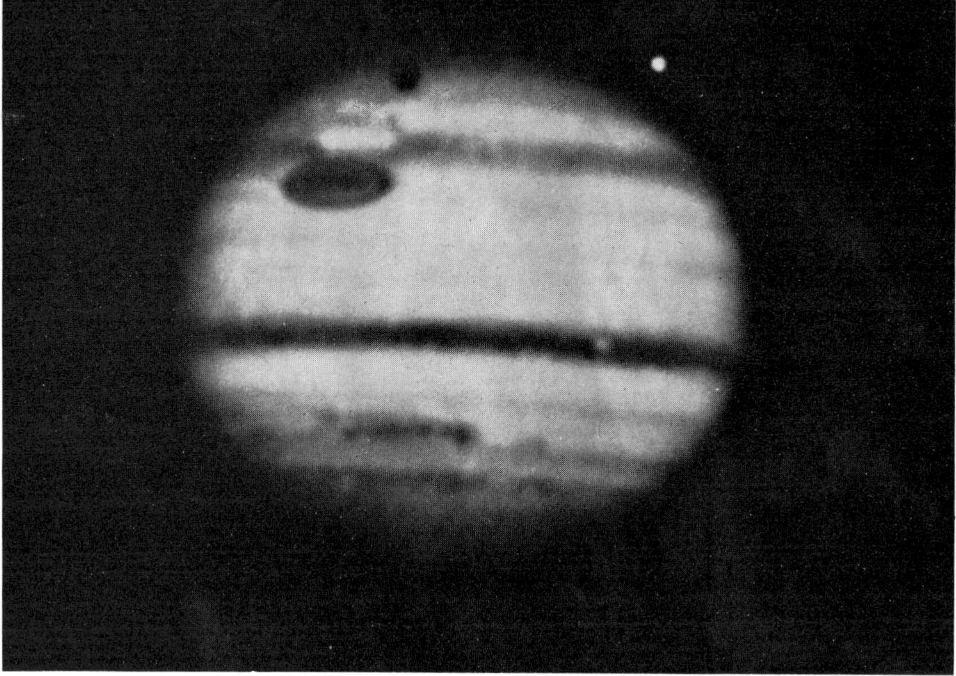

63. Jupiter. A photograph taken with the 200 in. Hale reflecting telescope. The great red spot is clearly seen, Ganymede appears above the planet and its shadow falls on the planet's face above the red spot.

64. Saturn taken with the 200 in. Hale reflecting telescope.

65. The sun-god at Sippar, Babylonia, on a stone tablet by Nabu-Apal-Iddina (ninth century BC).

66. A solar prominence, 1969, April 30, 10 h. 40 m. U.T., 30,000 miles (48,270 km.) high.

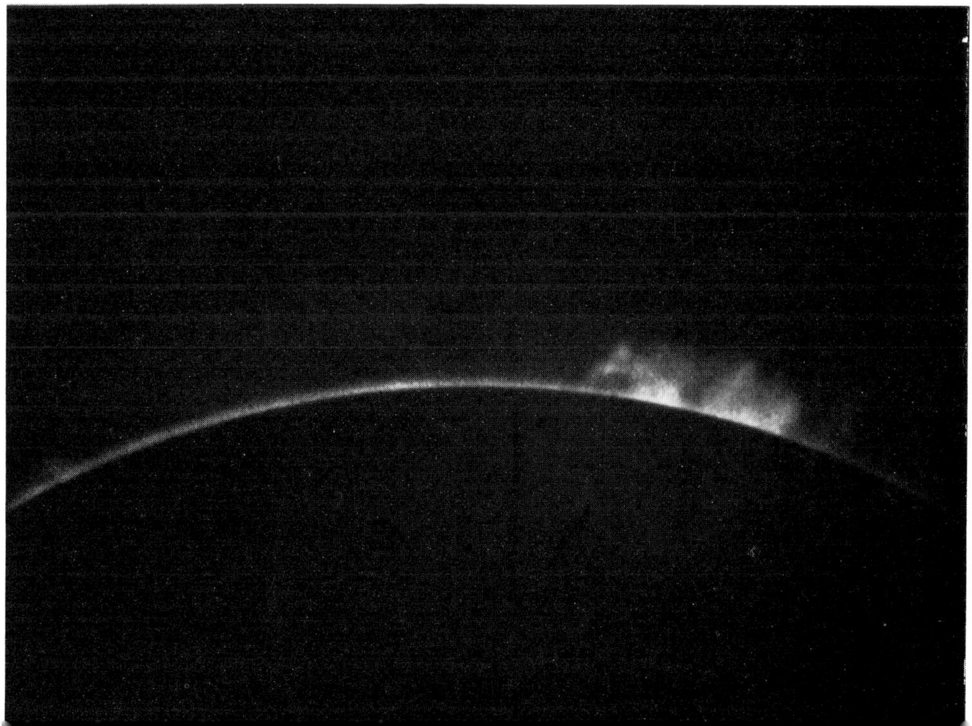

67. One of the twenty-four great wheels to the Sun-god at Konarak, Temple of the Sun (diameter 10 ft., 3050 km.). (See Max-Pol Fouchet, *The Erotic Sculpture of India*, Allen & Unwin, London, 1959.)

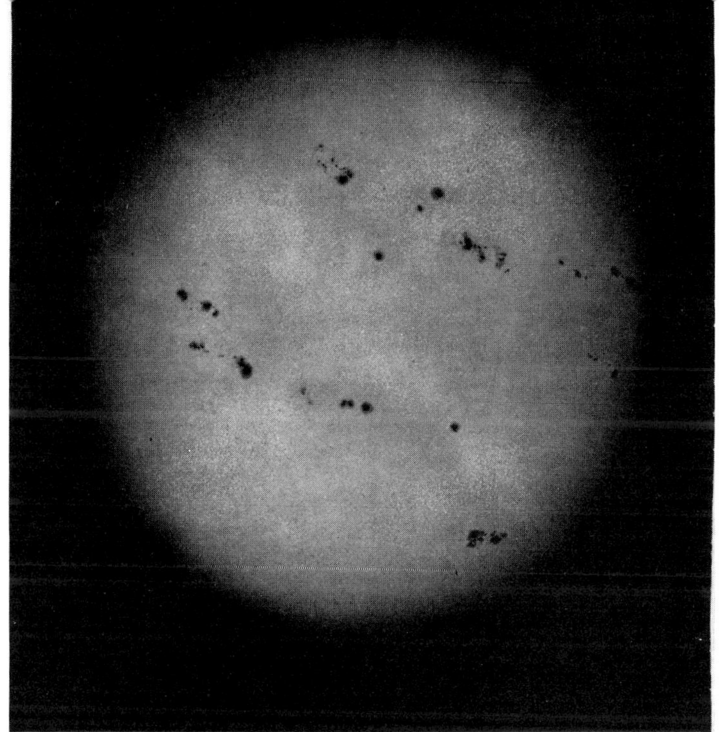

68. The sun's disc seen from Greenwich at Sun spot maximum 1947, May 24.
The smallest spot visible covers several millions of square miles and is much
larger than the planet Earth would appear if placed in such a position.

69. The face of the Sun, 1967, May 29.

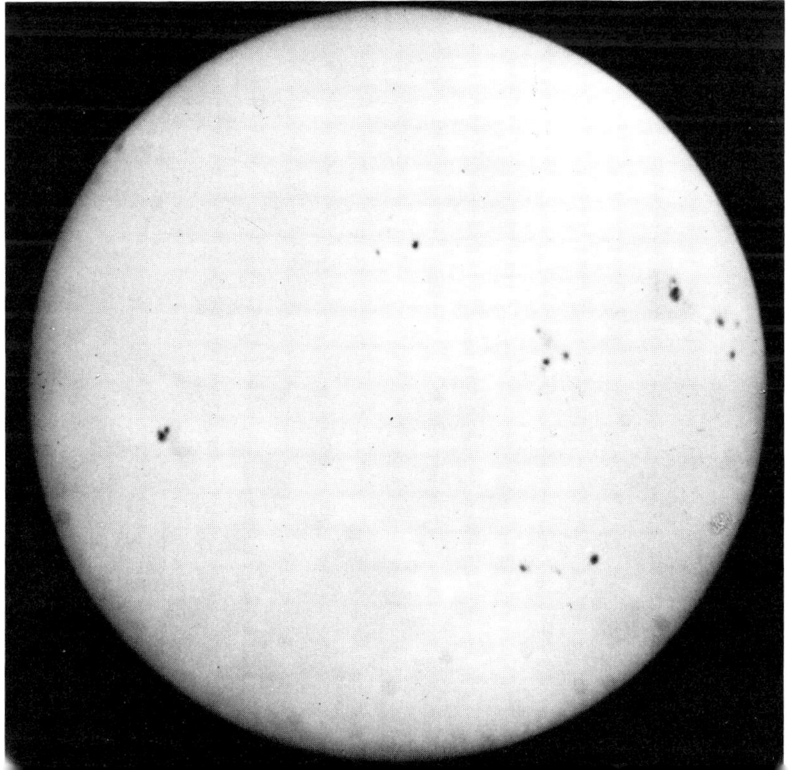

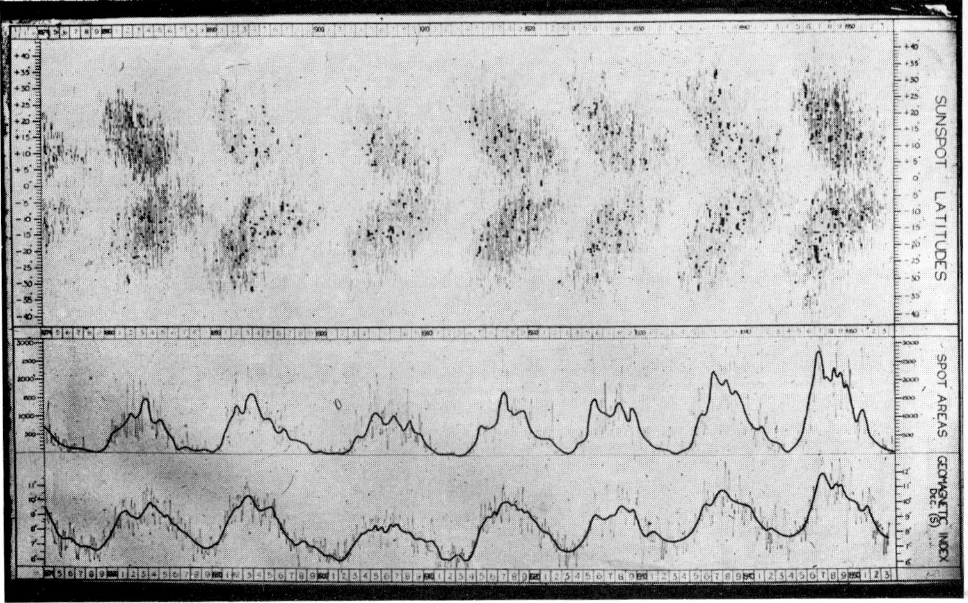

70. 'Butterfly' diagram showing changes in spot latitude during the solar cycle.

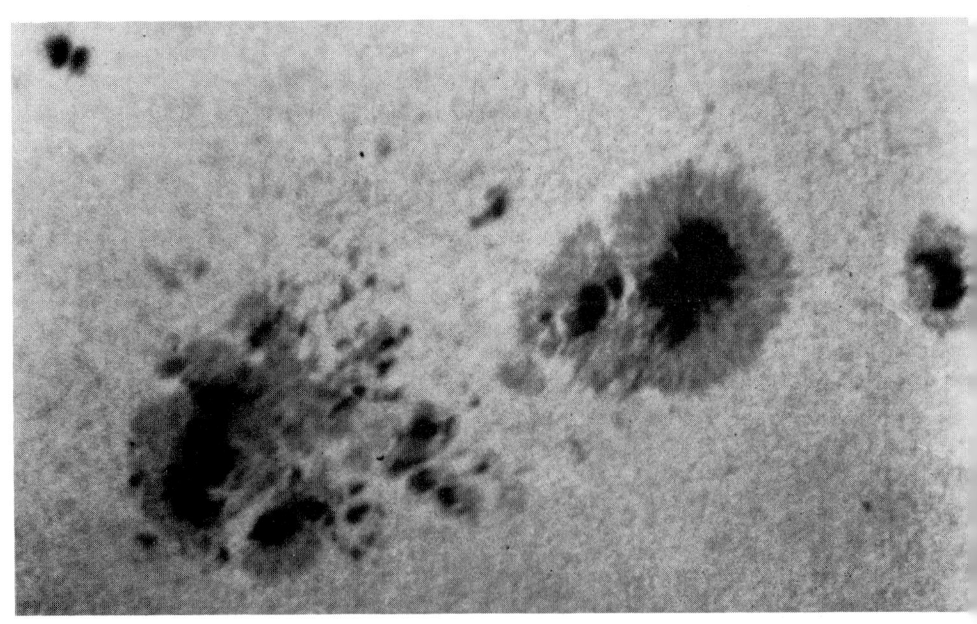

71. A sun spot group, 1959, September 10. Scale: Sun 48 in. (1220 mm.) diameter, Earth 0·44 in. (11·2 mm.) diameter.

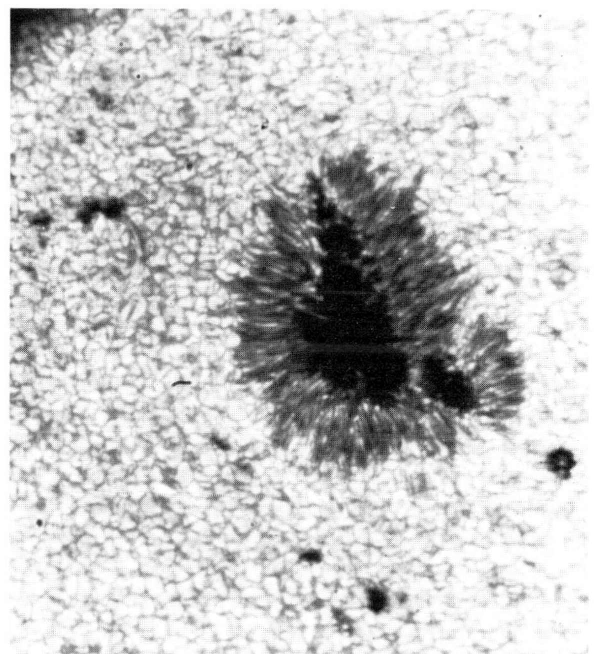

72. Solar granulations and sunspot, 1959, August 17, taken with the stratoscope 12 in. reflecting telescope carried by balloon to 80,000 ft. above the Earth.

73. An etching by Charpentier 1775, showing the burning device with a great liquid lens built in 1774 to the order of the Academie Royale des Sciences, Paris, for the investigations of Lavoisier *et al*. The lens consisted of two spherically curved cast-glass plates which fitted closely together, the space between them (4 ft. in diameter, more than 6 in. thick in the middle) being filled with alcohol and later with turpentine. (From *Oeuvres de Lavoisier*, vol. III, Paris 1865.)

74. Laboratoire de'energie solaire, Citadelle de Mont Louis. View of the oven floor and the large parabloidal mirror.

75. The largest radio telescope at Effelsberg, West Germany, for the Max-Planck Institute for Radio Astronomy. The parabloidal bowl is 100 m. (328 ft.) in diameter and constructed from 2400 rhomboidal petals which are flexibly mounted to give it an accurate figure. The average parabloidal deflection tolerance for each petal is 0·3 mm. (0·012 in.). The range of the instrument is said to be 12 billion light years.

76. Laboratoire de l'Energie Solaire Citadelle de Mont Louis, France.

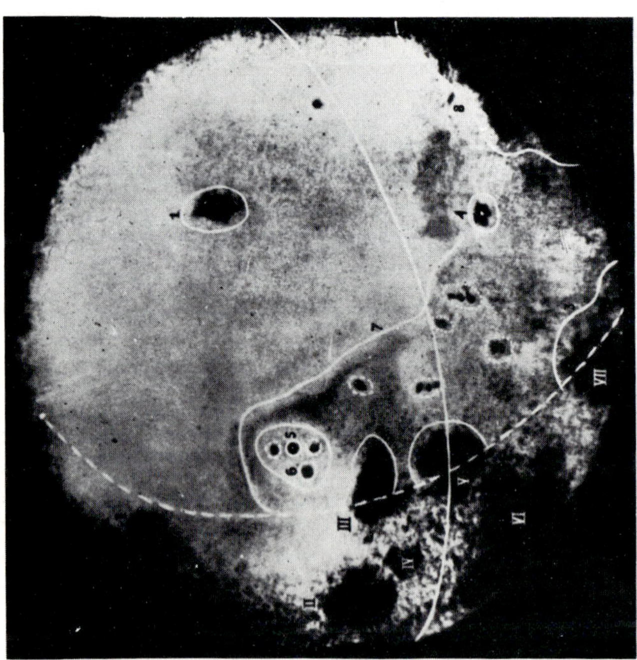

77. The first photograph of the far side of the Moon, forever hidden from the Earth, was obtained by the Russian spacecraft Lunik III on 1959, October 7 (north is at the top). The main features are identified by Roman and Arabic numerals. I Mare Humboltianum, II Mare Crisium, III Mare Marginis, IV Mare Undarum, V Mare Smithii, VI Mare Foecunditatis, VII Mare Australe, 1 Mare Moscovianum, 2 Sinus Astronauticae, 3 Part of Mare Australe, 4 Ziolkovsky, 5 Lomonosov, 6 Joliot-Curie, 7 Sovietsky Mountains, 8 Mare Somnii.

is not what we find from other studies. An alternative view due to Rama *et al.*[34] is that the other element in the core is sulphur and on this view one does not require such a catastrophic event as that for silicon. The balance of possibilies points to the view that much of the sulphur originally present in the crust and the mantle of the Earth has segregated to the core. Laboratory experiments with iron-rich melts up to a pressure of 60 kbar show that of all the iron-rich melts, including iron itself, the Fe-FeS eutectic has the lowest melting point. This means that in certain condition of temperature variation, assuming the experimental findings hold good, at the higher pressures postulated for the core, a sulphur-rich iron melt will be the first liquid formed and this, owing to its low viscosity and high density, would then sink into the core.

Seismic waves give some guide to the constitution of the Earth, and Lyttleton[35] records seven layers or regions designated for convenience Regions A B C D E F and G.

Region A, consisting of the crustal layers and extending from the outer surface to a depth of 33 km.

Region B, forming (with A) the outer shell extending down to 413 km. where the so-called 20°-discontinuity first occurs. (The pressure at this level is $0 \cdot 141 \times 10^{12}$ dyn cm.$^{-2}$.)

Region C, a transition region representing the outer part of the mantle down to about 984 km.

Region D, the main part of the mantle extending down to depth 2898 km., where a sharp change to the liquid state occurs. (The pressure at this core-mantle boundary is $1 \cdot 37 \times 10^{12}$ dyn cm.$^{-2}$.)

Region E, the outer liquid core extending to depth 4982 km.

Region F, a shallow transition region in liquid form from 4982 km. to 5121 km.

Region G, the innermost core of radius 1250 km. (which may be in solid form).

The question of the Moon's interior is equal in difficulty to that of the Earth's, but it is argued that it cannot have a large liquid core like that of the Earth. The minimum pressure in the moon, from calculation in view of its radius of about 1738 km. and assuming uniform density, is $0 \cdot 047 \times 10^{12}$ dyn cm.$^{-2}$, a pressure that is reached on Earth at a depth of about 150 km. The evidence suggests that the Moon is a one-zone body. A resolution of the problem is now possibly a little nearer with the installation of instruments on the Moon's surface. Especially important in this regard is the landing of a seismometer for continuous operation,

[34] Rama *et al.*, *Phys. Earth Planet Interiors*, 2, 276 (1970).
[35] Lyttleton, R. A., *Mysteries of the Solar System* (Oxford 1968).

able to send signals back to earth. This was achieved at both the Apollo XI and Apollo XII lunar landings.

The Earth's atmosphere[36] is complex in structure and some five distinct regions are now recognized

1. The troposphere.

2. The stratosphere.

3. The mesosphere.

4. The thermosphere.

5. The exosphere.

THE TROPOSPHERE extends to some 7–18 km. above the Earth's surface depending upon latitude. It is a region in which the temperature falls (from 300°K. to 200°K.) with increase in height. THE STRATOSPHERE is an ill-defined region from 20 to 40 km. above the Earth, often considered to be an isothermal region with a temperature of about 230°K. THE MESOPHERE extends upward to 90 km. above the Earth and the temperature fluctuates between 280°K. and 200°K. THE THERMOSPHERE is a region of increasing temperature, from 200°K. to some 500°K., extending from 90 km. to 200 km. above the Earth. Outside of this region there extends THE EXOSPHERE in which the temperature rises to as high as 2000°K.

Below about 90 km. the atmosphere consists essentially of molecular nitrogen, oxygen, argon and carbon dioxide with traces of Ne, He, CH_4, H_2O, Kr, N_2O, H_2, CO, O_3, Xe and NO_2. Above 90 km. ozone (O_2) formation occurs and nearly all oxygen is in the atomic form. Above 200 km. oxygen is ionized and the temperature increases as stated above.

The ionosphere extends from 70 to 80 km. above the Earth to an indefinite height. It is divided into two classic layers, the D and the E layers. The D layer is the lower of the two with a maximum electron concentration of $1 \cdot 5 \times 10^4$ per cm.^{3} at noon, and this disappears at night. The E layer is above about 100 km. with an electron concentration of $1 \cdot 5 \times 10^5$ per cm.^{3} with a night-time reduction to below 1×10^4 per cm.^{3}. The constituents of the layer are ionized by X-rays in the wave length 10 → 100 Å. A hypothetical region in the upper atmosphere known as the Chapman layer is presently recognized in which the distribution of electron density with height can theoretically be described by an equation. Some of the basic assumptions used to develop the equation are, that the ionizing radiations from the Sun are essentially monochromatic, that the ionizing constituents are distributed exponentially (with a constant scale height) and that there is an equili-

[36] For an informed note on the nature of the terrestrial atmosphere of the Early Earth see *Nature*, *226*, 927 (1970).

brium condition between the creation of free electrons and their loss by recombinations.

The heating of the troposphere occurs largely by convection from the ground but the stratosphere receives heat by absorption of infra-red radiation by CO_2 H_2O and O_3; the radiation coming from both above and below. In the mesosphere the temperature rises from heat supplied from the ultra-violet absorption by ozone. The thermosphere receives its heating from the interplanetary medium and the solar ultra-violet radiation. The exosphere receives its heating from the Lyman $\propto$ radiation at 1216 Å first observed in 1955 from rocket flights above the Earth.

Recent analyses of satellite orbits[37] have indicated that the upper atmosphere at heights between 200 and 400 km. is on average rotating faster than the Earth. The rotational speed of the Earth's surface is 400 m s^{-1} at latitude 30°. It is now shown[38] that the atmosphere is rotating faster than the Earth down to heights as low as 170 km. For a discussion of the primordial rare gases in the atmosphere of the Earth see Wasson, J. J., *Nature, 223.* 163. (1969); Earth Planet Sci. Lett., *6.* 213, (1969), *8,* 77, (1970).

THE GEOMAGNETIC FIELD AND PALAEOMAGNETISM

The Earth has a core composed of material of an appreciable conductivity and a magnetic field of considerable intensity, subject to broad variations that may be attributed at least in part to the nature of the core. The field is found today to be predominantly a dipole of about 0·6 Gauss at an angle of 11·5° to the axis of the Earth's rotation. This field has over a period of 135 years from the present been found to be decreasing in its intensity and moving westward at a rate of 0·07° per year. By the stable remanent magnetization of the Earth's rocks it is thought to be possible to record changes in the Earth's magnetic field through the geologic past. On the basis of a plausible assumption that the Earth's magnetic field has been very close to that of a geocentric dipole during geologic time, the variations in the record can be explained in terms of relative movement of the various land masses—a view expressed by Wegener[39] and often referred to as the Wegenarian concept of continental drift.[40]

The first estimates of the geomagnetic field intensity, with the aid of

[37] King-Hele, D. G., and Allen, R. R., *Space Sci. Rev.,* 6, 248 (1966); see also *Nature, 231,* 79, 109 (1971); Proc. Roy, Soc. A., *253,* 529 (1959); Planet Space Sci., *18,* 1433 (1970).
[38] King-Hele, D. G., *Nature, 226,* 439 (1970).
[39] Wegener, A., *Die Entstehung der Kontinente und Ozeane* (Vieweg Brunswick 1941); *Origins of Continents and Oceans* (Methuen, London 1924).
[40] See Athavale, R. N., Verma, R. K., Bhalla, M. S., Pullaiah, G., 'Drift of the Indian Sub Continent since Pre-Cambrian times', *Palaeogeophysics,* Runcorn, S. K., p. 291 (London 1970). Van der Voo, R., 'New Palaeomagnetic Evidence for the rotation of the Iberian Peninsula', ibid., p. 319.

the thermo-remanent magnetization of baked earths, was initiated by Thellier and Thellier[41] in 1956. This work was extended for estimating the geomagnetic field intensity during pre-historic and geologic time not only by use of baked earths but also by use of igneous rocks by Smith[42] and others.

The Earth's field in the past is reported according to Kaula[43] to have varied in the following manner:

1. Over the past 2500 years, the dipole intensity has decreased at an average of about 7 γ/yr (γ represents 10^{-5} gauss) so that the field was 50 per cent more intense in 500 B.C. However, about 5000 years ago the field had about the same intensity as now. About 10,000 years ago, the time scales associated with the secular variation were about the same as it is now.

2. In the last million years, the dipole field has maintained the same polarity it has now, wobbling with periods up to 10^5 years and amplitudes greater than 15 degrees about an average orientation displaced 5 degrees toward the Pacific from the present rotation axis. Throughout this time, the nondipole field in the Pacific has been relatively mild, as it is at present.

3. In the last 20 million years, there have been at least 60 reversals in polarity of the dipole field. The duration of the reversal process—i.e., the time during which there is no pronounced dipole—appears to be less than 10,000 years. During this period the average orientation of the pole was still within 5 degrees of the present rotation axis, and the wobbles were within 20 degrees of this average.

4. In the last $1\cdot0$ Æ (1 Æ represents 10^9 years) there have been many reversals of polarity; however, for two periods of about 50 million years each (Permian and Jurassic), the number of determinations of the same polarity is exceptionally high. During this time the apparent magnetic pole has wandered far from the present rotation axis, with inconsistencies between continents.

The science of palaeomagnetism as it is called now provides an impressive body of evidence for continental drift and sea floor spreading. Indeed the idea of global mobility has become a central dogma of Earth science (see the footnote on page 182).

Palaeomagnetic data are not the only evidence for global mobility, but they have the advantage of being quantitative. Although palaeontological, geological and palaeoclimatological data are mostly consistent with continental movements defined by palaeomagnetic poles, they usually have the status of supporting evidence simply because they are more difficult to quantify. Thus the doctrine of continental drift is consistent with what is known of the Earth's climatic zone pattern, but the broadness of the climatic zones means that it has been difficult to make

41 Thellier, E., Thellier, T. O., *Ann. Geophys.*, *15*, 285 (1959).
42 Smith, P. J., *Geophys.*, *12*, 321; *13*, 417; *13* 483.
43 Kaula, W. M., 'An Introduction to Planetary Physics', *The Terrestrial Planets*, Wiley, p. 133 (1968).

164

precise use of climatological criteria. It has therefore always been an open question as to what would happen if anyone were to compare the results of palaeomagnetism more quantitatively with evidence from other sources.

The assumption that the Earth's magnetic field has been axially dipolar, like the assumed constant radio flux[44] which is the shaky base for radio-carbon dating, over many millenia is now suspect. Clearly without such proof palaeomagnetism falls to the ground. Stehli[45] now claims to have some highly disturbing evidence. His analysis is based on the observation that certain brachiopods are temperature dependent. The lower the temperature at any given place, the fewer brachiopod families there will be, which means that the diversity of brachiopod families will be a maximum around the equator but will fall off as the latitude increases. That this is the case is shown by the behaviour of Recent clams, whose family diversity falls off quadratically with increasing latitude. The problem with Permian brachiopods from the northern hemisphere is that family diversity does not vary quadratically with palaeomagnetic latitude but does so when plotted on the present latitude grid. When plotted on the Permian palaeomagnetic latitude grid the Permian diversity bears no particular relationship to latitude, just as when the Recent clam diversity is plotted on the Permian palaeo-latitude grid there is, as would be expected, no simple relationship. In short, the Permian axial field is not consistent with the Permian brachiopod diversity whereas the present latitude grid is. And if this is true for the Permian it could be true for other geological periods. Stehli's paper which is bound to have many repercussions is summarized to use his own words as follows:

Quantitative paleontologic data sensitive to the planetary temperature gradient are used with similar data for living organisms to test two possible Permian latitude models for the Northern Hemisphere. The Permian paleontologic data applied to a present-earth model yield a pattern strikingly similar to that of living organisms on the same model. Permian paleontologic data applied to a Permian paleomagnetic[46] earth model show a pattern which is again strikingly similar to that exhibited by living organisms when plotted on this (for them patently incorrect) framework. The results indicate that the present-earth latitude model could be correct for the Permian but do not prove that this is so because data are sparse, the noise level is relatively high, and neither North America nor Eurasia contains the equator for this model of the earth. The data show that the Permian paleomagnetic models tested are as inappropriate for the Permian as for the Recent data. The fact that

---

[44] This is not the place for a discussion of the difficulties of radio-carbon dating but that 'science' as with palaeomagnetism makes a total appeal to the specious idea of uniformitarianism.

[45] Stehli, F. G., *J. Geophys. Res.*, 75, 3325 (1970).

[46] There are two spellings, palaeomagnetic and paleomagnetic, the former English, the latter American. The former is much to be preferred on etymological grounds. The word is not known to either the *O.E.D* or *Webster's Third International Dictionary*.

the paleomagnetic models are found inappropriate in this test may be interpreted to indicate that (1) the paleontological data are inadequate, (2) the paleomagnetic model used is not correct because data needed to define it are lacking in large areas, or (3) the Permian magnetic field was not axial.

THE ORIGIN OF THE EARTH/MOON SYSTEM

The Moon is a unique body in the solar system. The Earth alone possesses a satellite with an orbital angular momentum about its primary exceeding the rotational angular momentum of the primary *per se*. The origin of the Moon still excites much discussion especially when we consider that it is well established that angular momentum is transferred from the Earth's rotation into the Moon's orbital motion and that the result is the Moon's recession from the Earth. Clearly if these conditions have existed over a considerable period of time then a simple extrapolation of the present values discloses an Earth/Moon system in the past very different from what we see today—with the Moon very much closer to the Earth $1 \cdot 6 \times 10^9$ years ago. The discussion of this intriguing subject has been revitalized in recent years from calculations and analyses made of the past history of the Earth/Moon system by *inter alia* Gerstenkorn (1955),[47] Slichter[48] (1963), and Alfvén (1963).[49] On the basis of Gerstenkorn's results Alfvén has enlarged on Gerstenkorn's hypothesis of capture, in which the Earth captures the Moon in a highly eccentric retrograde orbit that is gradually transformed into a close circular *direct* orbit and then transformed to that which we observe today.

Since then further calculations by MacDonald[50] 1964 and Sorokin[51] (1965) have been evaluated critically by Ruskol[52] (1966). The conclusion reached is that there is now more evidence in favour of the origin of the Moon in the close vicinity of the Earth than in its capture in a *ready made form*. We will look briefly at the evidence presently advanced for capture; for the formation of the Moon close to the Earth and for its formation by fission. The task is made more easy for us since the publication of Wade's[53] review article, which surprisingly, however, does not refer to the powerful criticisms raised by Ruskol.

The hypothesis of capture has assumed an air of scientific respectability mainly from a paper of singular erudition by Gerstenkorn. It is pointed out that the Moon is now at a distance of 60 Earth radii from the Earth and that owing to the retardation of the Earth's rotation from, *inter alia*, the frictional drag of the tides and the attendant compensation in the angular momentum of the Earth/Moon System, the Moon is

[47] Gerstenkorn, H., *Z. Astrophys.*, *36*, 245.
[48] Slichter, L. B., *J. Geophys. Res.*, *68*, 4281.
[49] Alfvén, H., *Icarus*, *1*, 357 (1963).
[50] MacDonald, *Revs. Geophys.*, *2*, 467.
[51] Sorokin, N. A., *Astron. Zh.*, *42*, 1070.
[52] Ruskol, E. L., *Icarus*, *5*, 221 (1966).
[53] Wade, N., 'Three origins of the Moon' *Nature*, *223*, 243 (1969).

receding from the Earth at a rate calculated to be about 300 cm. a century. This transfer of momentum, the increase in the length of the day and the increase in the lunar distance were considered by Darwin[54] in 1879. He, however, came to the idea of fission for the origin of the Moon, but that conclusion in its simple form is now discredited.[55] Gerstenkorn's hypothesis is anything but simple, from a dynamical standpoint, since he proposes the capture of the Moon in a retrograde orbit, that is to say, opposite to the rotation of the Earth, so that the Moon is then *drawn-in* by the secular deceleration of the Earth and eventually its orbit is tilted to such a degree that the Moon is tilted over the poles of the Earth and placed in a direct, or what is often now referred to as a prograde orbit in harmony with the rotation of the Earth. This elegant yet catastrophic manoeuvre is shown from the vector analysis given below (Fig. 31.) Once the Moon is in a prograde

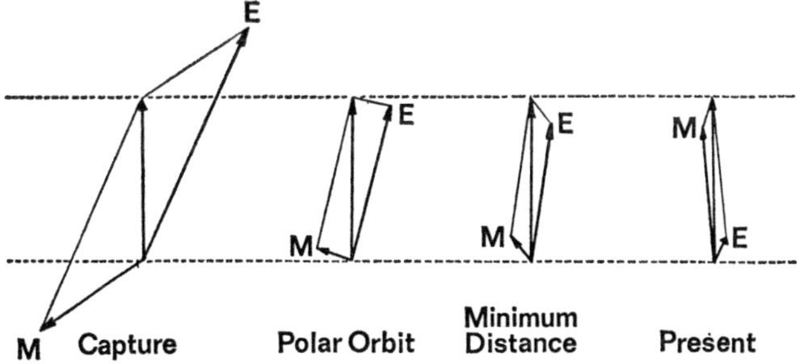

Fig. 31. Angular momentum of the Moon's orbital motion and the Earth's rotation. The vector sum is constant and is represented by the vertical arrow. Four stages in tidal evolution are shown.

orbit about the Earth the secular deceleration of the Earth drives the Moon outward away from the vicinity of the Earth. Alfvén[56] deals with this catastrophic event. He shows graphically the common density of the Earth's crust and that of the Moon and Mars (Fig. 32). He provides

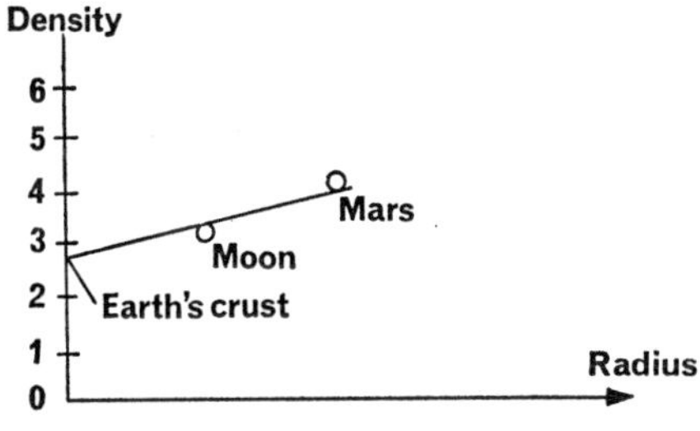

Fig. 32. Density of Earth, Moon and Mars.

[54] Darwin, G. H., *Phil. Trans. Roy. Soc.*, P II, 170, pp. 447–530.
[55] See Lyttleton, *Science Journal*, p. 53 (May 1969).
[56] Alfvén, H., *Icarus*, *1*, 357 (1963).

in a concise table the lunar orbital elements since the capture of the Moon and these are shown below:

LUNAR ORBIT AND ROTATION OF EARTH SINCE THE
CAPTURE OF MOON

	Capture in retrograde orbit	Polar orbit	Minimum distance	Present orbit
Lunar orbit				
'Distance' $a (1 - e)$	26	4·7	2·89	60
Eccentricity	1·0	0·001	0·00007	0·05
Inclination towards invariant vector η	124°	76°	38°	4°
Orbital momentum of Moon, D_M/D	0·78	0·23	0·18	0·84
Sidereal month		13^h	6·9^h	27·3^d
$\omega_M = 2\pi/T_M$	0	0·48	0·91	0·01
Inclination of axis towards invariant vector $\epsilon - \eta$	25″	14″	8″	20″
Rotation of Earth				
Inclination of axis towards lunar orbit normal ϵ	149°	90°	46°	24°
Rotational momentum of Earth D_E/D	1·58	0·97	0·86	0·17
Sidereal day	2·6^h	4·3^h	4·8^h	23·9^h
$\omega_E = 2\pi/T_E$	2·4	1·4	1·3	0·26

He then considers what will happen when the Moon approaches the Earth and comes eventually within the Roche limit given by the formula

$$^R\text{Roche}/_{R\oplus} = 2·44 \sqrt[3]{P\oplus/P_{\mathbb{C}}}$$

where $P\oplus$ for the Earth is 5·52
and $P_{\mathbb{C}}$ for the Moon is 3·34.
This point of criticality according to Gerstenkorn is ^{R}Roche $= 2·89$ $^{R\oplus}$ and took place at a distance in time of some $2·5 \times 10^9$ years from the present.

This entering of the Moon within a zone of disintegration must entail, according to Alfvén, a loss of material from the Moon. The loss may have been but a small part of the Moon's total mass or a substantial fraction of it. If the latter, then this loss could form the outer crust of the Earth, which is shown to have a density similar to that of the Moon and to be widely different from the density below the Mohorovicic discontinuity which separates the terrestrial crust from the mantle. The fragmentation debris, some of which may have been of a not inconsiderable size, may then have bombarded the Moon's surface and this would offer one attractive explanation for the gigantic Imbrium collision, the results of which are still clearly to be seen on the surface of the Moon. Some evidence from the Earth is available to support a cataclysmic event at a date of around 1·3 aeons ago (1·3 × 10⁹ years). This comes from the outcrops of the anorthosite massifs,[57] which when plotted on the continental masses of the Earth in their pre-drift positions, form two broad belts across the northern and southern hemispheres of the Earth. A further line of supporting evidence comes from the work of Ulrych[58] who shows from lead isotope measurements, derived from volcanic products from the young mantle in an uncontaminated form, that the mantle of the Earth is essentially a two-stage system as far as U/Pb ratios are concerned. The formation of the second stage appears to coincide in time with the global Anorthosite Event dated at 1·3 × 10⁹ years from the present, as stated above.

That the Moon may have had its genesis in a nucleation process as a separate body from the Earth is considered by Orowan[59] in a communication that is the more attractive for its elegance and simplicity. Orowan explains the low density of the Moon by appeal to the plastic ductility of iron at low temperatures. If it is assumed, he argues, that the planets have agglomerated from solid particles in a cloud condensed from a gas atmosphere around the Sun, then metallic particles would be expected to coalesce when they collide, possibly by either cold or hot welding, whereas silicates, which are brittle, would in contradistinction disintegrate. Thus the build-up of the planets may not unreasonably be expected to commence with the agglomeration of metallic particles that would then more readily accept non-metallic particles in an outer shell. In this manner the cloud would be depleted of metallic particles and thus a body forming from it at a later stage would of necessity be an agglomeration of non-metallic particles. It follows, from this line of reasoning, that the planets of high density are older than the Moon, and that, if iron and nickel and other heavy metals are some kind of nucleating agents for the inner planets, the position of the agglomeration and the birth of a planet within the cloud may be due primarily to the local concentration of metallic particles, which would in turn be closely related to the local vapour pressure of *inter alia* iron and nickel in the initial gas cloud.

An alternative approach to the capture hypotheses of Gerstenkorn and Alfvén is that advanced by Bailey[60] who shows that the orbital

[57] See Herz, N., 'Anorthosite Belts. Continental drift and the Anorthosite Event', *Science, 164,* 944 (1969).
[58] Ulrych, T. J., *Nature, 224,* 766 (1969).
[59] Orowan, E., *Nature, 222,* 867 (1969); see also *J.B.A.A., 80,* 299 (1970), which came to my attention long after this note was written.
[60] Bailey, J. M., *Nature, 223,* 251 (1969).

The solar system

velocities of the planets lie on a smooth curve (Fig. 33) with the exception of that of Mercury, which lies high above the curve (position A). If Mercury is placed (position B) to give the best fit to the curve at orbit

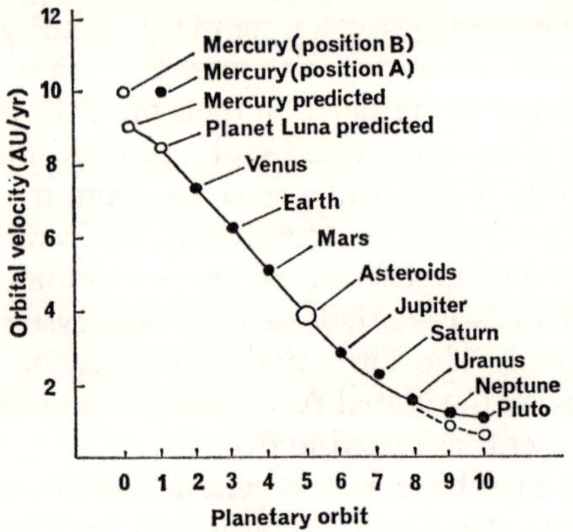

Fig. 33. Orbital velocity relationship for the planets of the solar system. Orbital velocity ($2\pi R/P$), where R is the radius of the planetary orbit in a.u., and P is period in years, plotted against the number n, of the planetary orbit, from Mercury ($n = 1$) to Pluto ($n = 10$) (solid circles). The open circles for Mercury, Luna, Neptune and Pluto are the values predicted by the Bode-Titius law, $R = 0\cdot4 + 0\cdot3(2)^{n-2}$.

O on the abscissa then there is no planet to fill the position at orbit 1. Bailey postulates that this unfilled position was filled by an hypothetical planet, LUNA. Bailey gives the following interesting table:

PERIODS AND ORBITAL RADII FOR THE INNERMOST PLANETS

	Period P (yr)		Radius R (a.u.)	
	Predicted	Found	Predicted	Found
Mercury	0·327	0·244	0·475	0·39
Planet Luna	0·408	—	0·55	—
Venus	0·617	0·617	0·72	0·72
Earth	1·0	1·0	1·0	1·0

and proceeds to show that there is an instability in the orbit of the missing planet Luna in that 0·408 yr is 5/4th of Mercury's period (0·327 × 5/4 = 0·408 approx) and that it is also 2/3 of Venus' period (0·617 × 2/3 = 0·411) and that these commensurabilities would induce eccentricities into the orbit of Luna. In appreciation of the relative closeness of the orbits of Luna and Mercury (7,000,000 miles) Bailey considers that close encounters must have taken place and he notes that Mercury's present orbit has a large eccentricity and does not satisfy the Bode Titus[61] relation. If then the present eccentricity of Mercury's

[61] See chapter 12, page 220.

170

orbit resulted from a disturbance of a previously circular orbit then the present aphelion point probably represents the radius of the earlier orbit, namely 0·467 a.u., which is close indeed to the value predicted by the application of the Bode-Titius relation which requires a value of 0·475 a.u. to satisfy it. Bailey then considers the possibility of the capture of Luna by the Earth, using Hill's curves of zero velocity.[62] He concludes that capture is feasible and would explain the closeness of the Moon and the Earth and the relative large mass of the Moon, which, as previously stated, is unlike any other secondary body in the system of the Sun and often most satisfactorily referred to with its larger companion as a double planet. This hypothesis removes the anomaly, noticed by some, that if the rates of tidal friction are extrapolated into past ages they point to a maximum age for the Earth/Moon system that is less than half of that postulated for the Earth herself. It is concluded that the Moon may be the former planet Luna that once occupied an orbit between that of Venus and that of Mercury at a mean distance from the Sun of 51,000,000 miles (82×10^6 km.).

Ruskol[63] favours a genesis of the Moon by accretion or similar growth in a close terrestrial orbit but outside the Roche Limit, at about 5 to 10 Earth radii. The Moon is presently situated at 60 Earth radii. Ruskol is highly critical of the capture hypotheses in view of new calculations which appear to show that the assumptions made by the previous investigators are open to objection, especially with respect to the secular variations of the lunar orbit owing to tidal friction. Ruskol concludes from his examination of the data that the closeness of the Moon to the Earth in past ages would produce a more circular and less inclined orbit with respect to the Earth's equator than is seen at present. These differences favour, according to Ruskol, a formation of the Moon in the vicinity of the Earth rather than its capture in a ready-made form. Ruskol performs a very real service since he highlights the weaknesses[64]

---

[62] Hill, G. W., *Amer. J. Math.*, *1*, 5 (1878).
[63] Ruskol, E. L., *Icarus*, *5*, 221 (1966).
[64] Ruskol's criticisms are as follows:

The Earth/Moon system is generally considered as an isolated two-body system, with an invariant vector angular momentum. The Sun is treated only as a body which produces tides on the Earth of secondary importance with respect to tides of lunar origin.

However, at present the modulus of the vector angular momentum of the Earth/Moon system and its direction in space suffer periodic changes due to solar perturbations of the lunar orbit. In calculations, an averaged vector for the present lunar orbit is taken as invariant for the whole time interval. However, in the past, the Moon was a closer satellite of the Earth with another position of the proper plane of its orbit. Whether that vector remained invariant also for close distances of the Moon needs further investigation.

The disturbing tidal potential is expressed by a single term P_2 of the development of potential function in spherical polynomials. This is valid as long as the Moon is a distant satellite. For the state of the system with the Moon as a close satellite, higher-order terms must be included.

The averaged orbital elements of the Moon, i.e. semi-major axis, a; eccentricity, e; and inclination, ϵ with respect to the Earth's equatorial plane, are used.

This assumption is justified as long as the time intervals in which tidal variations of a, e, ϵ,

of the capture hypotheses of the earlier investigators, showing how various are the calculations derived and how discordant the chosen parameters.

The Fission hypothesis is widely thought to be the *one* attractive explanation of lunar genesis, at least if one's reading is confined to popular texts. The broad idea 'originated' in 1898 with the views expressed by George Darwin who pointed out that if the Earth/Moon system were combined into one single body then the angular momentum would be such that the rotational period would be exceedingly short, indeed, so short as to produce a sidereal day of about four hours. He then showed that the resonant frequency would be about half of this period and that other factors would conspire to produce a tide of such proportions as to free a piece of the primary body to form a satellite. Other investigators believe this satellite to have taken its parturition from the Pacific basin, which remains as a post-natal scar still clearly delineated on mother Earth. The objections to this simple view of the Moon's birth have come from many dynamicists and these are cogently expressed by Mac-Donald,[65] who finds it difficult to accept any theory for the history of the Earth/Moon system that requires an initial rotational period for the Earth of less than ten hours, without at the same time explaining the anomalously high initial density of rotational angular momentum. Wise[66] makes what he calls a 'quiet plea for caution' in accepting these criticisms *in toto* and shows that the hypothesis of fission has many attractive features and in this his case is augmented by the views of Cameron.[67]

The hypothesis would appear to have been given new vitality from the investigations of Lyttleton[68] and McCrea.[69] Wise appeals to what may be called a modified Darwinian view of fission in which the rapidly spinning Earth making a rotation in 2·65 hours, frees part of the

---

become noticeable are much greater than periods of the changes of lunar orbit resulting from perturbations by the Sun, planets, and the Earth's oblateness. At present this condition is amply fulfilled. It is not obvious, however, whether it was fulfilled at the time when the Moon was very close to the Earth.

In all calculations it was assumed that tidal changes of a, e, and ϵ, during one revolution of the Moon are negligible. At present $\triangle a/a \approx 10^{-11}$ in a month, and even for the closest approach of the Moon $\triangle a/a \lesssim 10^{-3}$ for one revolution. Therefore this assumption is admissible.

All results on tidal changes of orbital parameters of the Moon are obtained by numerical methods, the mathematical form of equations being different in all papers quoted above.

The masses of the Earth (M) and Moon (m) were taken as constant during the entire time interval considered, that is during $2 \div 3 \times 10^9$ years. This seems quite admissible because the change of their masses owing to infall of meteorites was small and could not lead to appreciable changes of the Earth-to-Moon distance.

[65] MacDonald, G. F. J., *Science*, *145*, 881 (1964).
[66] Wise, D. U., in *The Earth/Moon System*, by Marsden *et al.*, p. 213 (New York 1966).
[67] Cameron, A. G. W., *Icarus*, *2*, 249 (1963).
[68] Lyttleton, R. A., *Mysteries of the Solar System*, p. 42 (Oxford 1968).
[69] McCrea, W. H., 'The Moon and Mars', *Nature*, *223*, 253 (1969).

equatorial region from its gravitational attraction and then, with time, produces a heterogeneous body having density stratification. The stratification decreases the moment of inertia of the Earth and reduces the rotational period, so that a change in angular momentum is available to engender a rotational instability and finally the breakaway of a satellite body. The dynamical progression is vividly expressed in the figures shown below (Fig. 34) in which the Earth is postulated to assume

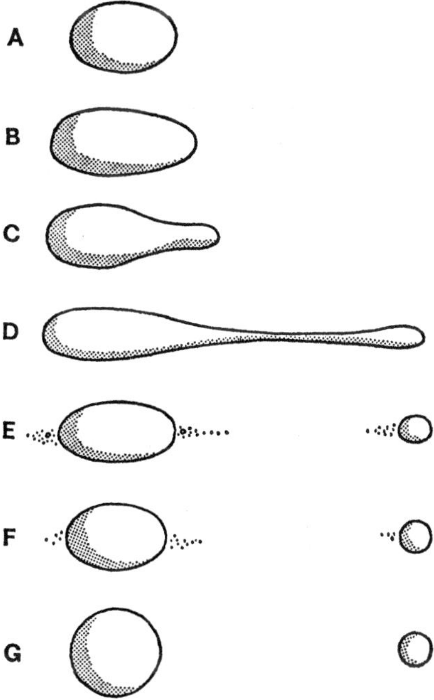

Fig. 34. Sequence of forms in lunar origin by fission from the Earth during formation of the Earth's core.

the shape of various MacLaurin oblate spheroids with ever reducing minor axes until the limiting value is reached viz. 7 : 12 for the relation between the polar and equatorial diameters. At this critical point the MacLaurin spheroid changes into a triaxial Jacobian ellipsoidal body in which the point of criticality is reached at a ratio of 8 : 10 : 23 when the body develops an attenuation and this leads to an unstable series of bodies of the Poincaré series of pear-shaped ellipsoids leading to fission and the break-away of a satellite body.

The Poincaré[70, 71] figure is shown in the diagram (Fig. 34) from which

[70] Note the MacLaurin, Jacobi and Poincaré ellipsoids are mathematical conceptions of considerable refinement and complexity. For an elegant discussion concerning their nature see

it will be seen that the figure has to be capable of attenuation beyond the Roche limit (2·7 Earth radii) to achieve success; thereby permitting the nascent Moon to form and reshape itself hydrostatically from the extremity of the attenuation of the ellipsoid and not immediately be disrupted. The Moon newly formed is then, with time, driven away from the vicinity of its birthplace by the tidal deceleration of the Mother body and its inexorable effect upon the distance of separation and the angular momentum of the pair. The subsidence of the remaining portion of the attenuation onto the Earth is an important part of the hypothesis, put forward by Wise, since this portion accelerates the Earth with respect to the Moon and ensures that the Moon's rotation is less than that of the Earth and therefore the Moon is not drawn back on to the Earth by the combination of forces associated with the event. Wise also makes an appeal to the specific gravity of the Moon at 3·34 and the value of that for the Earth's mantle at 3·32 as support for his fission hypothesis; but we have already seen that this is not a conclusive argument for fission in view of the ideas put forward by Orowan. Wise is on more satisfactory ground in support of fission when he points to the great difference in character between the two faces of the Moon, that facing the Earth and that away from the Earth. The visible face is claimed to be the healed-over heavier roots of the mantle of the Earth material from the attenuation of the Poincaré figure; whereas the 'far side' would contain the residue of any primordial crust on the outermost part of the mantle of the Poincaré figure at the time of separation. The Moon's early history under this mode of genesis begins with the falling-in of fragments to make the cratered ancient highlands and then the later capture of the orbiting fragments which produce first the maria basins punched through the ancient surface, and finally the general meteoric impact sites.

McCrea and Lyttleton envisage the formation of the Earth/Moon and Mars by fission processes from a single rotationally unstable planet. We have already seen that Alfvén groups the Earth's crust with that of the Moon on the similarity of their densities, and now Lyttleton would include the planet Mars for the identical reason. The respective specific gravities of the three bodies are, Earth's crust 3·32, Moon 3·342, Mars 3·94. If we now follow Orowan's ideas, previously expressed, we can envisage a body with a metallic (iron) core and a silicate (olivine)

Jardetsky, W. S., *Theories of Figures of Celestial Bodies*, (Interscience Inc, N.Y. 1958), and Chandrasekhar, S., *Ellipsoid Figures of Equilibrium* (Yale 1970).

[71] As a note of confusion I should point out that Jardetzky, G. V., the last pupil of the great mathematician Liapounov, points out that the Poincaré pear-shaped figure is according to the investigations of Liapounov *unstable* and one should *not* use it to explain the separation of the Moon and the Earth. This appears to stem from the fact that the instability implies rapid complete ejection to infinity. In this sense Wise's hypothesis would appear to be untenable. I should say that having discovered Jardetzky's strictures I am indebted to Prof. R. A. Lyttleton for a short discussion on this point.

mantle. Lyttleton[72] pairs the Earth and Mars as components resulting from a single rotationally unstable planet since the mass ratio is as high as 9:1. He regards the Moon as a droplet formed between the Earth and Mars after the larger planet is spun-off. Thus Mars and the Moon would have not dissimilar compositions since both are formed from the outer part of the mantle of the proto-Earth. The Earth would retain the heavy core and thus we can at once explain two bodies in neighbouring orbits of widely different composition; a feature difficult to reconcile with the view that the planets are formed directly out of the gas and dust in orbit about the Sun.

We leave to the last our discussion of the lunar surface and the interior. Conjecture has been rife and seldom unabated for many years, indeed, the arguments could without exaggeration be said to have continued for over two hundred years. It would be foolish to reiterate them, even in a concise form, now that actual exploration of the lunar surface by the crew members of Apollo XI, XII and XV has taken place.[73] The results of the examination of the lunar samples takes time to disseminate and a detailed account of all the material is not presently available[74] (March 1970). Apollo XI landed its lunar module in the northern part of Mare Tranquillitatis and the principal rocks recovered by the crew were both fine and coarse grained igneous rocks, breccias and fines. Apollo XII landed its module in Oceanus Procellarum and the rock samples collected contained far less dust than those of Apollo XI. The Apollo XII 'haul' was something close to 34 kg., half as much again as that obtained by the crew of Apollo XI.

The present consensus[75] concerning the Apollo XI samples is as follows. The lavas congealed not less than $3 \cdot 5 \times 10^9$ years ago at temperatures between 1200°C. and 1100°C. The rocks collected are the crystallized late residual fraction of that liquid and very different from the original lava in composition. The crystallization observed in the volcanic rocks is said to have occurred in strongly reducing conditions not encountered amongst terrestrial rocks. Ferric iron is absent, metallic iron is present and the Cr^{2+} ion is reported in olivine crystals. Two new minerals came back to Earth in the Apollo XI samples, a magnesian-ferro pseudobrookite, $MgFeTi_4O_{10}$ and a calcium-iron metasilicate with the pyroxmangite structure. The solid rocks at Tranquillity Base are found to be covered with a 4–6 m. thick regolith or dust layer, composed

[72] Lyttleton, R. A., *Mysteries of the Solar System*, p. 42 (Oxford 1968); *Monthly Notices of R.A.S.*, *121*, 551 (1960).
[73] The Russian Luna 16 has brought back a sample of moon rock. The spacecraft was unmanned (1970 September 24). The Lunokhod vehicle was landed on the moon by Luna 17 (1970 November 17).
[74] *Science*, vol. 167, 30 January 1970 (No. 3918), devoted its entire 335 pages to a detailed report of the Apollo XI samples.
[75] See *Nature*, *225*, 321 (1970); *226*, 925 (1970).
Since writing these notes Mason, B., and Melson, W. G., have produced their book *The Lunar Rocks* (Wiley 1970).

of local rock fragments and a high proportion of shock-altered rock chips and spheres of glass. The glasses [76] are thought to be formed by remelting of crystalline rocks by the intense heat generated when the rocks suffered impact with hyper-velocity meteorites. One feature of considerable interest lies in the fact that the regolith[77] contains anorthosite thought to have come from the lunar highlands. The Surveyor[78] analysis showed that the highland material is composed predominantly of gabbroic anorthosite.[79] Mica, amphibole[80] and aragonite[81] have been found but there is no evidence of diamonds or fossil organisms to date.

The rocks from Mare Tranquillitatis ($4000-4500 \times 10^6$y) appear to have a greater age than those from Oceanus Procellarum (3000×10^6y).[82] Isotopic analysis of the rocks recovered confirm the age of the lavas. The rocks are thought to be of local origin and of an age of $3 \cdot 8 \times 10^9$ years. Analysis of lunar lead shows that compared with terrestrial samples of that element it is rich in isotopes formed from uranium and this poses difficult questions about the origin of the Moon. The mineralogy of the Apollo XI samples is in the broadest sense basaltic; the peculiar characteristics being controlled by their unique chemistry, that is to say from the fact that they have high iron and titanium contents and low sodium and potassium contents when compared with terrestrial basalts. The solid particles of a size less than 1 cm. (the fines) have a mean density of 3 gcm.$^{-3}$ and the chief constituents are rock fragments of clinopyroxene, feldspar, ilmentite or cristobalite with broken crystal fragments, breccia fragments, dark glass containing an abundance of nickel-iron spheres and translucent coloured glass in the form of spheres or dumbbells. The particles exhibit micro-pits believed to be caused by impact with micrometeoroids. One of the most important discoveries is the interparticle adhesion which is significantly different from the behaviour of systems of similar sized particles on Earth. The general amounts of the elements in the Apollo XI lunar samples are given below:

[76] See Mueller, G., Hinsch, G. W., 'Glassy Particles in Lunar Fines', *Nature*, *228*, 254 (1970).
[77] *Regolith*. (Greek *rhēgos*, blanket, and *lith*, stone). Mantle rock.
[78] Surveyor I, first US soft landing on the Moon, 1966 May 30.
[79] *Anorthosite*. A coarse-grained rock, derived from gabbroic magma, consisting almost exclusively of plagioclase, near labradorite in composition. Rare in Britain; but important in certain areas of Pre-Cambrian rocks, e.g. the Canadian Shield.
[80] *Amphibole*. A mineral, hornblende. So named by Haüy 1801, in allusion to the protean variety in composition and appearance, assumed by the mineral genus to which he gave the name, and which Dana takes as the type of his first group of Bisilicates, including under it many species and varieties, as actinolite, asbestos, hornblende, tremolite, etc.
[81] *Aragonite* (named by Haüy, 1800, from *Aragon* or *Arragon*, a province of Spain, where first found). A carbonate of lime, crystallizing in orthorhombic prisms and many derived forms, whence several varieties are distinguished.
[82] The age of the solar system is currently put at 4700×10^6y. One of the anomalies from the Moon samples is an age of $4 \cdot 65 \times 10^9$ years for the soils with an age for the rocks in the range $3 \cdot 59$ to $3 \cdot 93 \times 10^9$ years; NATO Advanced Study Institute on the Moon and Planets (University of Newcastle upon Tyne, April 1970); see *Nature*, *226*, 212 (1970). The reader should be reminded that rock datings are notoriously inaccurate.

Amount	Element
10–100 per cent.	O, Si, Ca, Fe
1–10 per cent.	Mg, Al, Ti
0·1–1 per cent.	S, Na, K., Cr, Mn
100–1000 p.p.m.	C, N, P, Cl, Sr, Y, Zr, Ba
10–100 p.p.m.	F, Sc, V, Co, Ni, Zn, Nb, La, Ce, Pr, Nd, Sm, Gd, Dy, Er, Yb, Hf
1–10 p.p.m.	Li, Be, B, Cu, Ga, Ge, Rb, Eu, Tb, Ho, Tm, Lu, Ta, Pb, Th
0·1–1 p.p.m.	Se, Br, Mo, Cd, Sn, I, Cs, W, Os, U
< 0·1 p.p.m.	As, Pd, Ag, In, Sb, Re, Ir, Au, Hg, Bi

It is surprising to see the high level of titanium, zirconium and the rare earth elements.

The discovery that some parts of the Moon are 'paved' with pieces of glass supports the view that the Moon has suffered impacts of a very energetic nature. The report of these findings in *The Times* of September 2, 1969, drew the following letter from Mr D. O'Brien of Gonville and Caius College Cambridge:

Sir: It is satisfying to read your report today that the Moon is made largely of glass. This is just what Empedocles in the fifth century before Christ said it was made of.

The theories of crater formation that imply for the Moon a violent birth, either by impact or by volcanic processes,[83] are often difficult to reconcile with observations pointing to a comparatively gentle mode of origin. The student should not neglect to consider the hypothesis of Mills[84] in which the craters are shown to be producible by fluidization phenomena such as those observed in beds of fine particles of crushed diorite when levitated by rising gas currents. Mills points with particular emphasis to twin lunar craters, such as Azophi and Abeneera, where violent mechanisms would be expected to cause destruction of the dividing wall, yet such structures are readily produced by gentle, long-continued endogenous processes such as fluidization. There is further a generation of considerable electrostatic potentials in fluidized beds and these could account for the evanescent localized glows seen on the Moon, especially at Aristarchus, Alphonsus and Schroeter's Valley.[85]

McCall[86] offers clear evidence for a volcano-tectonic origin of Mare

[83] The impact or ballistic theory finds its most erudite expression in the hands of Baldwin, R. B., *The Face of the Moon* (Chicago 1949). A modern review of the *Lunar Controversy* is to be found under that title by McCall, G. J. H., 'A competent geologist', in *J.B.A.A.*, *80*, 19, 100, 190.
[84] Mills, A. A., 'Fluidization Phenomena and possible implications for the origin of Lunar Craters', *Nature*, *224*, 863 (1969).
[85] Mills, A. A., 'Transient Lunar Phenomena and Electrostatic Glow Discharges', *Nature*, *225*, 929 (1970).
[86] McCall, G. J. H., *Nature*, *223*, 275 (1969).

'terrain' on the Moon. He points to one of the Apollo 8 series of photographs, covering the Cauchy Fault, Crater and Scarp area of the lunar surface (plate 88) showing a pattern of faulting that is virtually identical with the uppermost Pleistocene-Holocene 'grid' or 'touche-de-piano' (keyboard) faulting in the Baringo sector of the Rift Valley in Kenya (plate 89), faulting that has been compared with the platforms at Clapham Junction station. It is not just the step effect that is similar— the persistence of a master trace with branching traces fading out, the passage into an echelon conformation at one end and into monoclinal flexuring near the other, the crescentic steps with the convex curve towards the downthrow side, the cross-over effects, the close-spacing of the faults forming narrow ledges—all these are characteristics shared with the rift valley fault structures on the south-eastern slopes of the terrestrial Silali Volcano. The Cauchy Scarp is actually the more perfect fault zone structure of the two, and it clearly involves gravity faulting on very steep fault surfaces, just like the rift valley faulting (the surfaces are close to vertical in most cases).

Some of the Silali scarps show alignments of regularly spaced blow-hole craterlets, clearly visible in the plate. McCall believes that the Apollo 8 photograph can only be interpreted as evidence of a lack of material emission associated with the endogenous vulcanicity. The lack of obscuration of the Cauchy Scarp right up to the rim of the small crater that straddles it reveals that nothing material ever came out of the crater except gas. Some sort of blowhole effect, of a quiet nature, must be responsible. The photograph suggests to McCall that the hypothesis of a largely gaseous volcanic emission, accompanied by sublimation, may prove to be the genesis of these features.

THE LUNAR MASCONS

The discovery of the mascons (lunar mass concentrations) is due to Muller and Sjogren.[87] Their evidence for these unusual areas comes from gravity anomalies associated with the orbit of Lunar Orbiter Spacecraft. They are able to specify the following twelve sites:

Mare Imbrium	Aestium Medii
Mare Serenitatis	Grimaldi
Mare Crisuim	Mare Humboldtianum
Mare Nectaris	Mare Smythii
Mare Humorum	Unnamed area 27°E 5°S
Orientale	Unnamed area 70°E 15°S

What the mascons are is an open question. Stipe[88] has calculated that the depth of penetration x by a projectile of diameter d and mass m

[87] Muller, P. M., and Sjogren, W. L., *Science*, *161*, 680 (1968); see *Nature*, *222*, 414 (1969).
[88] Stipe, J. G., *Science*, *162*, 1402 (1968).

striking a target of compressive strengths and density p with a velocity v is

$$\left(\frac{x}{d}\right) = 0 \cdot 0615 \left(\frac{\frac{1}{2}mv^2}{sd^3}\right) + 1 \cdot 10 \left(\frac{m}{pd^3}\right)^{1/3}$$

where the symbols have such meanings that each term in parentheses is dimensionless. He calculated further that meteorites embedded in the Moon may give the changes in gravity recorded by the spacecraft in orbit. His table is reproduced below:

Mare	Meteorite		
	Diam. (km)	Depth (km)	Mass ($\times\ 10^{15}$ kg)
Imbrium	61·2	450–670	930
Serenitatis	36·7	270–400	200
Crisium	27·2	200–300	82
Humorum	21·3	155–233	39
Nectaris	16·0	116–175	17
Ptolemaeus	10·0	73–109	4

Muller and Sjogren give 20×10^{-6} lunar masses for the largest mascon, whereas the largest meteorite listed above has a mass of about two-thirds of that value; corresponding to a sphere of diameter only 14 per cent smaller than that of an iron sphere of mass 20×10^{-6} lunar masses. The object supposed to lie under Mare Imbrium, however, is equivalent to a nickel iron sphere 100 km. across and centred at a depth of 50 km.; yet a high velocity asteroid would surely have blown itself to pieces at, if not prior to, impact. One of the most attractive alternative theories comes from Gilvarry,[89] who in three papers shows that the hypothesis that there was once water on the Moon can be made to account for the lunar mass concentrations without appeal to violence. In effect he argues that the maria are sediment plains and that the Moon had a primordial atmosphere and hydrosphere. This view leads to the conclusion that the mascons are isostatically in equilibrium with the supporting rocks beneath. Gilvarry then proceeds to derive the internal temperature of the Moon from the inferred creep rates of the said mascons. In contradistinction Goudas[90] points with elegant simplicity to the view that the gravity anomalies depend on the local density of the attractive mass as well as on its *elevation* and thus the maria may be at a higher altitude than previously entertained by theorists. Conel and Holstrom[91] also dislike the meteorite hypothesis and would prefer a near surface interpretation of the mascons. They appeal to the following geologic

[89] Gilvarry, J. J., *Nature*, *221*, 732 (1969); *Nature*, *223*, 255 (1969); *Nature*, *224* 968 (1969).
[90] Goudas, C. L., *Nature*, *220*, 1111 (1968).
[91] Conel, J. E., Holstrom, G. B., *Science*, *162*, 1403; see also Howard, K. A., *Nature*, *226*, 924 (1970).

pattern of mascon formation. The submare and adjacent rim material are of low density because they have been brecciated and pulverized by impact events. The mare-fill is denser material, perhaps solid basalt and this high-density-fill may have its origin from flows of melted local material or that produced from meteoritic impact and subsequent solidification to form a high density plate-like element which could be equated gravitationally with a high density sphere.

THE RADIO EMISSION AND MAGNETIC FIELD OF THE MOON

The emission of radio waves from the Moon has been detected over a range of wavelengths from 1·5 mm. to 1·5 m. The discovery of a radio emission in the wavelength 1·25 cm. was made by Dicke and Beringer[92] as long ago as 1946. Coates[93] has shown that the radio 'brightness' over the lunar disc is determined by the illumination of the Sun's rays and the visible features. The infra red emission is thought to come from a very thin layer on the Moon's surface and the longer radio wavelength from a thicker layer of lunar surface rock.

Nature, 230, 146 (1971), reports that there is no evidence to suggest that there is a significant lunar magnetic field at present. A magnetic intensity of 38 ± 3 gammas was recorded at the Apollo XII landing site, for example; and rock samples brought back by both Apollos XI and XII have been found to contain magnetization. Both of these pieces of evidence show that some parts of the Moon are permanently magnetic —a phenomenon which leads one to expect the existence of a magnetic field at the time the particular rocks were formed. But there is also another likely consequence of permanent magnetism on the Moon, as Mihalov *et al.* point out (*Science, 171,* 892 (1971)). The magnetic fields set up by the permanently magnetic rocks should, if large enough, deflect the charged particles of the solar wind which impinge on the Moon.

Because there is no overall lunar magnetic field the Moon does not, of course, have a magnetosphere like the Earth; but, as the Explorer 35 results showed, there is a diamagnetic solar wind cavity behind the Moon which is bounded by a rarefaction wave. Mihalov *et al.* have found that adjacent to this wave there are sporadic but persistent maxima in the interplanetary magnetic field. If these maxima are produced by sources on the Moon itself (this is, of course, an assumption) then it is likely that the sources lie around the boundary of the Moon's hemisphere struck by the undeflected solar plasma. This is merely the result of the fact that the perturbations along the edges of the cavity

[92] Dicke, R. H., and Beringer, R., *Ap. J., 103,* 375 (1946).
[93] Coates, R. J., *Ap. J., 133,* 723 (1961).

formed by the Moon are most likely to be caused by effects at the edge of the lunar disk as 'seen' by the solar wind moving outwards from the Sun. And given that the sources lie around the lunar boundary, it is possible to extrapolate the field maxima backwards to determine just where on the lunar surface the interactions with the solar wind take place.

Mihalov and his colleagues have shown that most of the lunar magnetic sources lie on the far side of the Moon and therefore by implication in highland regions rather than in the maria. The two largest, for example, each about 10^5 square kilometres in area, lie at about $0°-15°$ S, $165°-180°$ E and $15°-30°$ S, $135°-150°$ E, respectively. They do not therefore seem to be related directly to the mascons which tend to lie in the centres of ringed maria on the near side of the Moon.

NAMES OF LUNAR FEATURES

No attempt is made in this book to give the named lunar features on that side of the Moon directed continuously toward the Earth. The reader interested in this branch of selenography should consult a good Lunar Atlas.[94] The best known to me that is available at a small cost, is *Lunar Atlas*, Alter D, 154 plates, Dover Publications Inc. (Dover Edition 1968.) The features on the so-called far side are excellently portrayed on the map of *The Earth's Moon*, compiled by the Cartographic Division of the National Geographic Society for *The National Geographic Magazine* (Washington 1969). The student interested in the names *per se* of the lunar features should not fail to consult 'Who's Who in the Moon'. Comprising notes on the names of all lunar formations adopted in 1935 by the International Astronomical Union, which appear in the *Memoirs of the British Astronomical Association*, Vol. 34, Part 1, 1938. The naming of the 'new' features on the far side came before the International Astronomical Union's meeting in Prague in 1967 and is charged to Commission 17 of the I.A.U. meeting in Brighton in 1970 August 18–27.[95]

---

[94] The best Lunar Atlases are that of Kuiper, G. P., *Photographic Lunar Atlas*, 230 sheets, The University of Chicago Press, Chicago, 37 (1960), and *Orthographic Atlas of the Moon*, ed. by Kuiper, G. P., and compiled by Arthur, D. W. G., and Whitaker, E. A.

[95] *The Times* reported on 20 August 1970 as follows:

After Thursday, the craters on the far side of the moon will no longer have to go nameless. A list of 513 names has been agreed by lunar astronomers at the International Astronomical Union meeting in Brighton, and is expected to be ratified at a final meeting.

Among the names are 49 British people, mostly scientists but including H. G. Wells, the novelist, and Chaucer. The list sets a precedent by including a handful of names of people who are still alive.

Three craters are to be named after the American astronauts who were the first to fly behind the moon, and three are named after the moon-landing team, Neil Armstrong, Edwin Aldrin, and Michael Collins. Six Russian cosmonauts are included in the list.

A group of four lunar scientists were responsible for drawing up the names under the chairmanship of Professor Donald H. Menzel, of Harvard College Observatory.

The group began with 1,200 names put forward by national academies of science. In six meetings the names were pared down to produce the most distinguished, representative, and

The solar system

truly international list. Among British scientists who will have craters named after them are Professor Sydney Chapman who died earlier this year and William Harvey, the 17th century physician who discovered the circulation of the blood.

Yuri Gagarin, the first man to orbit the earth in 1961 and who was killed in an aircraft training flight, is commemorated. In the same way part of the moon has been named Apollo after the American moon-landing programme. But Ernest Rutherford, the New Zealand physicist, who was on the original list, had to be omitted because there is a Crater Rutherfurd on the near side. The committee that selected the names wanted to eliminate any possibility of confusion. In the same way, E. O. Lawrence, the American physicist, had to be omitted because of possible confusion with the Crater Lorentz.

To assign the names in the list to craters that have been identified on the far side of the moon the names were put into five categories of distinction. Then the craters were divided into five categories of relative size. The committee then drew lots to match the names to the craters.

Note: Since writing this chapter the Harold Jeffreys Lecture for 1971 has been delivered and published. (See Press, F., 'The Earth and the Moon' *Quat. J.R.A.S.*, *12*, 232 (1971). Further the validity of continental drift has been explored by Wesson, P. S. in a remarkable paper 'The Position against Continental Drift', *Quat. J.R.A.S.*, *11*, 312 (1970). 'The Mineralogy of Apollo 15415 "Genesis Rock" Source of Anorthosite on the Moon', is given in *Nature 234*, 138, (1971); see also Mason, B., and Melson, W. G., *The Lunar Rocks*, Wiley, (1970).

Chapter 11

On comets,* minor planets, meteors, meteorites and tektites

We are merely the stars' tennis-balls,
Struck and banded
Which way please them.
JOHN WEBSTER

Comets and meteors have been seen by mankind from early times but the first of the minor planets had to await the arrival of the telescope and was not revealed to human eyes until Piazzi on new year's day of 1801 discovered Ceres in the clear skies over Palermo. We will deal with these bodies under separate headings, but it may ease our task if we begin by defining them, so that their distinctive characteristics are left in no doubt.

A comet (from the Greek komētēs, long haired) is a celestial body moving about the Sun under the influence of his attraction in a greatly elongated elliptical, or parabolic orbit, and consisting, when near to the Sun, of a bright star-like nucleus surrounded with a misty light and having a train of light or 'tail', sometimes of enormous length and usually directed away from the Sun. A comet remains visible from Earth only for a short time, i.e. while it is in a part of its orbit near to the Sun.

A minor planet, or asteroid,[1] is one of a large number of very small planetary bodies revolving about the Sun between, in the main, the orbits of Mars and Jupiter.

A meteor is an ephemeral luminous body seen temporarily in the sky and supposed to belong to a lower region than that of the heavenly bodies; a fireball or shooting-star. A meteorite is a meteor, a mass of stone or iron, that has fallen from the sky upon the Earth; a meteoric stone or, very loosely, a meteor or meteoid.

Tektites (Greek tektos-melted) are bodies of silicate glass believed to

[1] The name 'asteroid' was coined by W. Herschel in 1802, *Phil. Trans.*, xcii, 228.
* See Appendix C for a list of comets within the confines of the Solar System.

be of extra-terrestrial origin and found in several widely separated areas of the Earth's surface. They are essentially extra-terrestrial hyaline rocks with unique shapes and sculptures of various complex forms associated with rapid melting and some aerodynamic shaping, such as spheroids, dumb-bells and apioids.

COMETS

Aristotle believed that comets were terrestrial phenomena, and it was not until the observations of Tycho Brahe[2] demonstrated that the comet of 1577 penetrated the interplanetary spaces that this fiction was laid aside. We are indebted to Gibbon[3] for a dissertation on comets in the most felicitous of language, as is to be expected from one of the greatest masters of the English tongue. He tells us that in the fifth year of Justinian's reign and in the month of September—

a comet[1] was seen during twenty days in the western quarter of the heavens, and which shot its rays into the north. Eight years afterwards, while the sun was in Capricorn, another comet appeared to follow in the Sagittary: the size was gradually increasing; the head was in the east, the tail in the west, and it remained visible above forty days. The nations, who gazed with astonishment, expected wars and calamities from their baleful influence; and these expectations were abundantly fulfilled. The astronomers dissembled their ignorance of the nature of these blazing stars, which they affected to represent as the floating meteors of the air; and few among them embraced the simple notion of Seneca and the Chaldæans, that they are only planets of a longer period and more eccentric motion. Time and science have justified the conjectures and predictions of the Roman sage: the telescope has opened new worlds to the eyes of astronomers;[ii] and, in the narrow space of history and fable, one and the same comet is already found to have revisited the earth in *seven* equal revolutions of five hundred and seventy-five years. The *first*, which ascends beyond the Christian era one thousand seven hundred and sixty-seven years, is coëval with Ogyges, the father of Grecian antiquity. And this appearance

[2] See Dreyer on the eighth chapter of Tycho Brahe's *Dani De Mundi* (1587).
    'The "æthereal world," is of wonderfully large extent; the greatest distance of the farthest planet, Saturn, is two hundred and thirty-five times as great as the semi-diameter of the "elementary world" as bordered by the orbit of the moon. The moon's distance he assumes equal to fifty-two times the semi-diameter of the earth, which latter he takes to be 860 German miles. The distance of the sun he believes to be about twenty times that of the moon. In this vast space the comet has moved, and it therefore becomes necessary to explain shortly the system of the world, which he had worked out "four years ago".'
[3] Gibbon, E., *Decline and Fall of the Roman Empire*, vol. 4, p. 366 (Everyman Edition 1910).
[i] See John Malala, tome ii, p. 190; 219, p. 454, 477, ed. Bonn. Also Theophanes, p. 154, tome i, p. 278, ed. Bonn.
[ii] Astronomers may study Newton and Halley. I draw my humble science from the article *Comète*, in the French Encyclopédie, by M. d'Alembert.
    [The identity of the comet of A.D. 1680, with the comets of A.D. 1106, A.D. 531, 44 B.C., etc., was an ingenious speculation of Halley. The observations made upon the eccentricity of a comet's orbit, whether it was a parabola or an ellipse with great eccentricity, have recently been made with such accuracy as to warrant almost exact conclusions being obtained. Cf. John Williams, *Observations of Comets from Chinese Annals.*—O. S.]

explains the tradition which Varro has preserved, that under his reign the planet Venus changed her colour, size, figure, and course; a prodigy without example either in past or succeeding ages.[iii] The *second* visit, in the year eleven hundred and ninety-three, is darkly implied in the fable of Electra, the seventh of the Pleiads, now have been reduced to six since the time of the Trojan war. That nymph, the wife of Dardanus, was unable to support the ruin of her country: she abandoned the dances of her sister orbs, fled from the zodiac to the north pole, and obtained, from her dishevelled locks, the name of the *comet*. The *third* period expires in the year six hundred and eighteen, a date that exactly agrees with the tremendous comet of the Sibyl, and perhaps of Pliny, which arose in the West two generations before the reign of Cyrus. The *fourth* apparition, forty-four years before the birth of Christ, is of all others the most splendid and important. After the death of Cæsar, a long-haired star was conspicuous to Rome and to the nations during the games which were exhibited by young Octavian in honour of Venus and his uncle. The vulgar opinion, that it conveyed to heaven the divine soul of the dictator, was cherished and consecrated by the piety of a statesman; while his secret superstition referred the comet to the glory of his own times. The *fifth* visit has been already ascribed to the fifth year of Justinian, which coincides with the five hundred and thirty-first of the Christian era. And it may deserve notice, that in this, as in the preceding instance, the comet was followed, though at a longer interval, by a remarkable paleness of the sun. The *sixth* return, in the year eleven hundred and six, is recorded by the chronicles of Europe and China: and in the first fervour of the Crusades, the Christians and the Mahometans might surmise, with equal reason, that it portended the destruction of the Infidels. The *seventh* phenomenon, of one thousand six hundred and eighty, was presented to the eyes of an enlightened age.[iv] The philosophy of Bayle dispelled a prejudice which Milton's muse had so recently adorned, that the comet, 'from its horrid hair shakes pestilence and war'.[v] Its road in the heavens was observed with exquisite skill by Flamsteed and Cassini: and the mathematical science of Bernoulli, Newton, and Halley investigated the laws of its revolutions. At the *eighth* period, in the year two thousand three hundred and fifty-five, their calculations may perhaps be verified by the astronomers of some future capital in the Siberian or American wilderness.

Gibbon appears to rely on d'Alembert for his dates and these are not now readily related to the dates of the returns of Halley's comet. For those who find this subject of more than passing interest and in view of

[iii] A Dissertation of Fréret (*Memoires de l'Académie des Inscriptions*, tome x, pp. 357–377) affords a happy union of philosophy and erudition. The phenomenon in the time of Ogyges was preserved by Varro (*apud Augustin. de Civitate Dei*, xxi, 8), who quotes Castor, Dion of Naples, and Adrastus of Cyzicus—*nobiles mathematici*. The two subsequent periods are preserved by the Greek mythologists and the spurious books of Sibylline verses.

[iv] This last comet was visible in the month of December 1680. Bayle, who began his Pensées sur la Comète in January, 1681 (*Œuvres*, tome iii,), was forced to argue that a *supernatural* comet would have confirmed the ancients in their idolatry. Bernoulli (see his *Eloge*, in Fontenelle, tome v, p. 99) was forced to allow that the tail, though not the head, was a *sign* of the wrath of God.

[v] *Paradise Lost* was published in the year 1667; and the famous lines (l. ii, 708, etc.), which startled the licenser, may allude to the recent comet of 1664, observed by Cassini at Rome in the presence of Queen Christina (Fontenelle, in his *Eloge*, tome v, p. 338).

the expected return of the comet in *c.* 1986, I give below[4] the dates of the last twenty-nine returns of this celebrated celestial visitor.

The comet of December 1680 was not comet Halley but comet Kirch, which made its perihelion passage on December 18, and mankind had to wait another two years for Halley's comet to return in the September of 1682. The return of 1682 must be considered one of the most fruitful of all astronomical events and we are reminded of this by the contemporaneous account from Frederic Kaiser, the director of the observatory of Leyden, who writes 'These two comets[5] are the most remarkable among all that have appeared in earlier and later times, in as much as the first caused Newton to apply the principle of gravitation to the theory of comets while the other enabled Halley by the discovery of its period of revolution to put the most beautiful seal on the theory of comets founded by Newton.'

Halley[6] remarks

At length, came that prodigious Comet of the Year 1680, which descending (as it were from an infinite Distance) perpendicularly towards the Sun, arose from him again with as great a Velocity . . .

Not long after, that great Geometrician, the Illustrious *Newton*, writing his *Mathematical Principles of Natural Philosophy*, demonstrated not only 'that what *Kepler* had found, did necessarily obtain in the Planetary System; but also, that all the Phaenomena of Comets wou'd naturally follow from the same Principles; which he abundantly illustrated by the Example of the aforesaid Comet of the Year 1680, shewing, at the same Time, a Method of delineating the Orbits of Comets Geometrically; therein solving (not without meriting the highest admiration of all Men) a Problem, whose Intricacy, render'd it scarce accessible to any but himself. This Comet he prov'd to move round the Sun in a Parabolic Orb, and to describe Areas (taken at the Center of the Sun) proportional to the Times.

Wherefore (following the Steps of so great a Man) I have attempted to bring the same Method to arithmetical Calculation; and that with all the Success I cou'd wish.' He then gives a table with orbital elements of 24 comets, 'the Result of a prodigious deal of Calculation. . . . in the making of which, I spar'd no Labour, that it might come forth perfect, as a Thing consecrated to Posterity, and to last as long as Astronomy it self.'

[4] See Cowell, P. H., and Crommelin, A. C. D., *Monthly Notices of R.A.S.*, *68* who give the period as 76·04 years (the period lengthens by 4·1 days at each return see *The Observatory, 91*, 174, 1971).

240·4 B.C.	A.D. 374·1	A.D. 912·5	A.D. 1456·43
163·4 B.C.	451·5	989·7	1531·65
87·6 B.C.	530·9	1066·23	1607·82
12·8 B.C.	607·2	1145·30	1682·70
A.D. 66·1	684·8	1222·69	1759·19
A.D. 141·2	760·4	1301·81	1835·87
A.D. 218·3	837·1	1378·85	1910·30
A.D. 295·3			

[5] The comets of 1680 and 1682. The former was discovered by Godefroi Kirch of Coburg Saxony on 1680 November 14 and the latter by Jesuit priests in Orleans in the Summer of 1682—now always given Halley's name.

[6] Halley, E., *A synopsis of the Astronomy of Comets* (1715).

He then explains that

no hyperbolic velocities have been found and that, consequently, it is probable that the comets would really move in very excentric ellipses, and thus, he says, 'make their Returns after long Periods of Time: For so their Number will be determinate, and perhaps, not so very great. Besides, the Space between the Sun and the Fix'd Stars is so immense, that there is Room enough for a Comet to revolve, tho' the Period of its Revolution be vastly long. . . . And, indeed, there are many Things which make me believe, that the Comet which *Apian* observ'd in the Year 1531, was the same with that which *Kepler* and *Longomontanus* more accurately describ'd in the Year 1607; and which I my self have seen return, and observ'd in the Year 1682.'

He continued to explain that also the comets of 1456 and 1301 were likely to have been returns of the same comet, and he predicted that it would return again in 1759, a prediction which was beautifully fulfilled.

The nature of an average comet is still not fully resolved. The best authorities[7] visualize it as an assemblage of pieces of matter of different sizes, comprising a central region of greater concentration containing one or more large components of as much as a few miles in diameter interspersed and surrounded by a cloud of small stones, dust particles and gas molecules. We observe when we examine an average comet in a good glass a head or coma, a nucleus and, when it is bright, a tail. The coma is usually of considerable size and of a nebulous nature; it may exceed 100,000 miles (160,935 km.) in diameter. The coma of Halley's comet has exhibited a diameter of 248,000 miles (400,000 km.) or 30 Earth diameters. The tails may extend to 1 a.u. and while usually single they are seen in some instances to be of a multiple nature; indeed Borrelly's[8] comet of 1903 was seen to possess nine separate tails. It would seem from many observations that comets may well be the largest bodies in the solar system with the single exception of the Sun himself; but their masses are found to be extremely small and no gravitational effect on the Earth/Moon system or other celestial bodies has ever been recorded. Wurm and Balazs[9] have examined the dimensions of cometary heads, using several photographs of Comet Brooks 1911 V obtained with the Crossley reflector of Lick Observatory, and are able to show that the emission head extends to about 5×10^5 km. from nucleus. Using plates of comet Halley 1910 II, taken with the reflector at Helwan Observatory, they are able to show that the head emission reaches out to distances of $1 \cdot 1 \times 10^6$ km. from the nucleus. Comet Tago-Sato-Kosaka (1969g) and comet Bennett (1969i) have been found[10] sur-

[7] Oort, J. H., *The Observatory*, *71*, 129 (1951); Merton, G., *J.B.A.A.*, *62*, 6 (1952). Since writing these words Lyttleton, A., has given the Halley Lecture on the subject of comets; see *The Observatory*, *90*, 178 (1970).
[8] Borrelly's comet is designated 1903 IV. Its perihelion passage occurred on August 28 of that year.
[9] Wurm, K., and Balazs, B., *Icarus*, *2*, 334 (1963).
[10] See *Nature*, *225*, 413 (1970); *226*, 313 (1970); *230*, 156 (1971).

prisingly to possess an enveloping cloud of hydrogen gas of a size similar of that of the Sun. These results come from orbiting geophysical observatories using instruments able to detect the short wavelength (Lyman $\propto$ 1216Å) generated by the hydrogen clouds. These waves do not penetrate the Earth's atmosphere. The tail and coma are usually transparent and stars may be seen through them without any appreciable change to their brightness. The number of gas molecules in the coma is estimated to be about 10^6 per cm.^{3} which is 10^{13} times less than in a centimetre cube of air at sea level. The spectroscope discloses, to the skilled observer, that the compounds C_2, CN, CH, CO, C_3, NH, NH_2, OH are often present and Urey has suggested that acetylene, C_2H_2, may be present and would explain some of the pyrotechnic type of cometary outbursts. The evidence available suggests clearly that the head of a comet is composed of transient particles, that is to say, particles that are continually being lost to interplanetary space. Oort has calculated that the upper limit of mass of Halley's comet (from the 1910 return) is 2×10^{19} g or one three hundred millionth of that of the Earth. Clearly such a body cannot keep a coma of gas owing to the small velocity of escape which is but 18m/sec., and the ordinary temperature motions of the molecules would exceed this by a factor of 30. We are led to see then that all molecules that are detached from the solid nucleus will inevitably escape freely into space. Few measurements and estimates appear to have been made, but Encke's[11] comet is said to have a mass of 8×10^{16} g and a diameter of 3 km.; and comet Pons Winnecke[12] a diameter as small as 0·4 km.

Whether a comet nucleus is single or particulate is not known with any certainty. In 1843 the comet known as the Brilliant comet passed so close to the Sun at its perihelion passage (1843 Feb. 27) as to enter the corona of the solar furnace to within a depth of one-tenth of the Sun's diameter and yet escaped dissolution. Minnaert has calculated that to survive such an experience the comet must possess solid masses with a dimension of at least a kilometre. Some comets are not as fortunate as the Brilliant Comet and the most celebrated of the suicidal group is Biela's comet (period 6·62 years) discovered in 1772 that made four returns, but on the return in 1846 February, at its perihelion passage on the 11th, it broke into two parts and on its return in 1852 presented to the viewers on Earth, two parts separated by $1·5 \times 10^6$ miles ($2·4 \times 10^6$ km.). According to Merton, Ensor's comet (1926 III) reached naked-eye brilliance before perihelion passage at a distance of 30×10^6

---

[11] The mass of Encke's comet was measured by Cunningham. This celebrated comet was making its forty-ninth return in 1970 after its discovery in 1786. It is known in the USSR as Comet Encke Backlund. Its period is 3·3 years.
[12] Comet Pons Winnecke at its return in 1927 was measured by Baldet. The comet was discovered in 1819 and has made seventeen returns. It was not seen in 1957, but was recovered in 1964 February and in 1970. Its period is 6·344 years.

miles (48×10^6 km.) from the Sun and disintegrated completely except for a faint trail seen as a stain on the photographic plate used to record that part of the sky that should have contained it. The probability of the Earth colliding with a large comet is thought to be vanishingly small. Urey[13] has shown that assuming completely random orbits the chance of a comet hitting the Earth is $\dfrac{\pi a^2}{4\pi R^2} \times 2$ where a is the Earth's radius and R the distance of the Earth from the Sun. The factor 2 takes account of the chance of collision while the comet approaches the Sun and when it leaves it once again. Approximately the factor is a^2/R^2 and this equals $1 \cdot 8 \times 10^{-9}$. With ten comets arriving a year in the neighbourhood of the Sun only one collision in 50 million years could be expected. But clearly if the comet disintegrates into a large swarm of particles then the chance of a collision with some part of the swarm is much increased. We have already referred to the break-up of Biela's comet and the Earth is thought to have encountered its debris on 1827 November 27 with an attendant meteor shower of unusual magnificence.

This subject is one of peculiar fascination and has been studied closely by Porter.[14] It was Schiaparelli in 1866 who announced that the August Perseids (the meteors known as the 'Tears of St Lawrence'), were moving in the same orbit as the comet[15] of 1862 and Leverrier and Peters who identified the orbits of the Leonid meteor shower of November with that of Tempel's[16] comet of 1866. The idea that comets cause meteor streams is not established but it is clear that meteor streams and certain comet orbits are in some kind of relationship, possibly each has a common origin. Porter gives the following table and states that there is a high probability that the Taurid meteor showers are associated with the famous periodic comet of Encke and the May Aquarid shower with Halley's comet.

THE MAJOR SHOWERS

Shower	Comet	Date	Radiant
Lyrids	1861 I	Apr. 21	$271° + 34°$
June Draconids	Pons-Winnecke	June 30	$208° + 54°$
Perseids	1862 III	Aug. 11	$45° + 58°$
October Draconids	Giacobini-Zinner	Oct. 10	$262° + 54°$
Leonids	1866 I	Nov. 15	$151° + 23°$
Andromedids	Biela*	Nov. 30	$23° + 44°$

It has recently been suggested[17] that a new meteor stream may be associated with comet Mellish 1917 (period 145·3 years).

[13] Urey, H. C., *Nature*, *179*, 556 (1957).
[14] Porter, J. G., *Comets and Meteor Streams* (London 1952).
[15] Comet, Swift/Tuttle, 1862 III period, 119·6 years.
[16] Comet, Tempel-Tuttle, 1866 I period, 33 years.
[17] *Nature*, *225*, 1232 (1970).
* See footnote on page 212.

The solar system

Whipple[18] has proposed a model for a comet in which ices and meteoric dust are formed into a conglomerate and these are found to provide an adequate supply of gas. As the nucleus evaporates the meteoric material forms an insulating layer that retards further evaporation. The Whipple model is in 'reality' a dirty glacier flying around in the interplanetary spaces. The tail grows brighter as the comet approaches the Sun and when it comes to within about $\frac{3}{4}$ a.u. the intensity of the solar radiations upon it produces the emission lines of metals in the spectrum of the comet head, the yellow line of sodium being prominent and sometimes those of magnesium, nickel and iron. The tail is directed away from the Sun and this phenomena, which was incorrectly ascribed for many years to the effect of light pressure, is now believed to result from the directing effect of the ubiquitous solar wind.

The origin of comets has occupied a number of distinguished minds and we are indebted to Laplace, Svedstrup, Oppenheimer, Hurnik and van Woerkom for ideas on this most difficult subject, but rather than trace their ideas over the centuries it is most convenient today to take up the subject with Oort's[19] celebrated paper of 1950. Oort reminds us that we have fairly reliable orbits of over 500 comets (i.e. in 1950), whereas Halley had but 24. As a brief summary we now know some 40 comets that have short periods of between 5 and $7\frac{1}{2}$ years, while their orbits all extend fairly close to the orbit of the massive planet Jupiter. For this reason they are most frequently called the Jupiter family of comets. They all have direct orbits with small inclinations to the plane of the ecliptic. About 80 per cent of all comets have periods over a century and their orbits extend to enormous distances. Two-thirds of the long period comets have orbits that extend to more than 2000 a.u., and one third more have orbits that extend to more than 40,000 a.u. that is to say, more than one thousand times the distance of the furthest planet. The orbital planes of these comets are oriented in planes at random to the ecliptic. One may be led from this to give the comets an interstellar home since the nearest star is at a distance of 300,000 a.u.; yet it is readily shown that they are members of the solar system. This is achieved by an appeal to the velocity of the body in question, since if it came into the solar system from outside it would have an hyperbolic velocity and none is found to exist when all the orbital refinements are properly applied. This fact alone forces us to look within the solar system for the immediate origin of the comets we observe. In short, Oort maintains that the combined effects of the stars and of Jupiter appear to determine the main statistical features of the orbits of comets.

From a score of well-observed original orbits he shows that the 'new'

<hr>

[18] Whipple, F. L., *Astrophys. J.*, *111*, 375 (1950); *Astrophys. J.*, *113*, 464 (1951).
[19] Oort, J. H., 'The Structure of the Cloud of Comets Surrounding the solar system and an hypothesis concerning its origin', *B.A.N.*, *11*, 91 (1950 January 13).

190

long-period comets generally come from regions between about 50,000 and 150,000 a.u. distance. The sun must be surrounded by a general cloud of comets with a radius of this order, containing about 10^{11} comets of observable size; the total mass of the cloud is estimated to be of the order of 1/10 to 1/100 of that of the earth. Through the action of the stars fresh comets are continually being carried from this cloud into the vicinity of the sun.

Oort indicates how three facts concerning the long-period comets, which hitherto were not well understood, namely the random distribution of orbital planes and of perihelia, and the preponderance of nearly-parabolic orbits, may be considered as necessary consequences of the perturbations acting on the comets.

The theoretical distribution curve of $1/a$ following from the conception of the large cloud of comets is shown to agree with the observed distribution, except for an excess of observed 'new' comets. The latter is taken to indicate that comets coming for the first time near the sun develop more extensive luminous envelopes than older comets. The average probability of disintegration during a perihelion passage must be about 0·014.

The existence of the huge cloud of comets finds a natural explanation if comets (and meteorites) are considered as minor planets escaped, at an early stage of the planetary system, from the ring of asteroids, and brought into large, stable orbits through the perturbing actions of Jupiter and the stars.

The important quantity 1/a is the reciprocal value of the semimajor axis of the comet orbit and is expressed in terms of the mean distance of the Earth to the Sun (1 a.u.) and is proportional to the total energy of a comet in its orbit. The overall action is such that a swarm-trap is provided and this operates through the combination of two effects, the perturbations caused by Jupiter and the perturbations caused by the fixed stars (Fig. 35). Oort's picture of the life history of a comet is as follows:

It was probably formed in the same period and in the same region where the planetary system was born, as a small fragment moving in an orbit that was less regular than that of the large planets. Relatively soon after its formation it was thrown out through perturbations and caught in the big swarm. Since that time stellar perturbations have steadily been directing some members of this swarm back to our vicinity, where under the influence of solar radiation they develop comas and tails, and become true comets. They end up by being dissolved into gas and meteorites or by being definitely thrown out of the solar system.

Oort's hypothesis is not without its critics and Lyttleton[20] calls attention to what he considers to be three 'fatal' objections.

[20] Lyttleton, R. A., *Monthly Notices of R.A.S.*, *139*, 225 (1968).

The solar system

(1) For the long period comets $1/a$ can undergo changes of the order of $\pm\ 500 \times 10^{-6}$ a.u.$^{-1}$ at each return to the Sun through planetary action. (2) The time taken to come in close to the Sun from a cloud of the radius of 50,000 a.u. would be only about one-thousandth of the age of the solar system and for every one such comet coming in for the first time there may be in sum ten or even one hundred coming in for the second or third time and it would be necessary to explain what becomes of these cometary bodies. (3) The $1/a$ plot of Oort is thought to contain an epistemological error seen from the work of Galibina.[21]

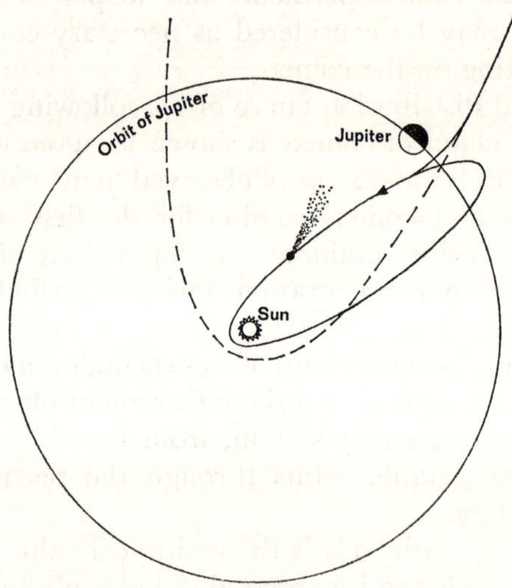

Fig. 35. Orbit of a short-period comet and a schematic illustration of the capture process (actually many successive perturbations of Jupiter are required).

Lyttleton's criticisms are disputed by the Leyden group Le Poole and Katgert[22] who champion Oort. They point out that all one needs are the direct observational facts which are in support of a large number of comets whose orbits, when computed back, extend to distances of more than 20,000 a.u., and that from this evidence alone the existence of a reservoir of comets at large distances from the Sun follows without hypotheses or assumptions. They argue strongly for Lyttleton's criticisms not to be entertained as serious objections to Oort's theory, which has up to now had a clear run for nearly twenty years. It seems to me, if I may be so rash as to express an opinion, that, if Lyttletons third objection can be supported, then it is such as to strike at the roots of Oort's cloud and despite an astronomical life of two decades it may have to disappear.

[21] Galibina, M., *Bull. Inst. theor. Astr.*, *9*, 496 (1963).
[22] *The Observatory*, *88*, 164 (1968).

Lyttleton[23] has advanced his own theory for the origin of comets, which bears some resemblance to that of Bobrovnikoff and Nölke. Lyttleton consid erscomets to be formed from accretion within an interstellar dust cloud, of a density in the range 10^{-25} to 10^{-23} gcm.$^{-3}$, which is entered by the Sun at a slow speed in its great orbit about the centre of the Galaxy which occurs at a velocity of 250 km. s^{-1}. The dust particles of the cloud would, it is suggested, be pulled into hyperbolic orbits relative to the Sun and streams of them from opposite sides of the Sun would converge in the direction of the axis of the Sun's movement. In one numerical example Lyttleton shows that the birth of 5×10^7 comets is not unreasonable but all of these would not remain as permanent members of the solar system, since, to begin with, when the periods are short the rate of expulsion by planetary action will be rapid and some will be swept into the Sun, but others will achieve increasingly long periods until even they are ejected from the system. This theory would appear to stand or fall on the ability to explain the accretion of cometary bodies, and if one is prepared to accept Whipple's icy agglomerate model it is difficult to see how the condensations will occur especially if large amounts of kinetic energy are present. Moreover, it is difficult to understand the structure of meteorites by a process of accretion and if this is not accepted then we have to postulate different origins for meteorites and comets.

It would seem attractive with present knowledge to believe that meteors are the debris of comets and at the same time to accept the probability that comets are born among the planets but before we offer any conclusion on this difficult matter we should consider the minor planets since they also may be linked with the comets.

THE MINOR PLANETS

Before the discovery of the trans-saturnian planets Kepler had remarked on the gap between the orbits of Mars and Jupiter and speculated on this lack of harmony in the system of the Sun. In 1772 demands of the newly published Bode-Titius law were a more respectable catalyst to thought than today, and with the discovery of Uranus in 1781 (which exactly fitted the 'series') it was difficult not to express surprise at the unfilled orbit between the paths of Jove and the God of War. Such was the speculation that an unseen planet may be discovered by assiduous observation that a band of twenty-four dedicated German observers was gathered together by De Zach to track down the fugitive. It was with the application of that unconcern with which nature favours the well equipped that the first sight of the minor planets was reserved to the

<hr>

[23] Lyttleton, R. A., *Mysteries of the Solar System*, p. 145 (Oxford 1968); see also *Monthly Notices of R.A.S.*, *108*, 465 (1948); see reference 7, p. 187.

sharp eyes of Piazzi scanning the stars of Taurus, in the clear skies above Palermo, on New Year's Day, 1801. Piazzi then became seriously ill and the planet was only recovered by De Zach on the last day of that year, after Gauss had, by the application of his mathematical genius, devised a wholly new method for calculating the orbit of the planet from but three observations, an achievement that has not lost its lustre with the passage of time.

Piazzi conferred on the new planet the name of Ceres Ferdinandea, an allusion to the titular goddess of Sicily. The search that had been made for Ceres in its period of uncertainty had familiarized Olbers with the faint stars in the ambit of its geocentric path and on the March 28, 1802, he was rewarded with the discovery of the minor planet Pallas, this time against the stars of Virgo. It was with some audacity that Olbers now put forward the view that these two bodies were parts of a disrupted planet which had once filled the now unfilled orbit between Mars and Jupiter, as required by the Bode-Titius law. Few then shared his opinion but with the steady and continuous discovery of yet more and more of these small planetary bodies the idea takes upon itself a note of realism. By 1889 ten planetary 'fragments' had been discovered and named.

3. Juno	4. Vesta
5. Astraea	6. Hebe
7. Iris	8. Flora
9. Metis	10. Hygeia

It may surprise those who seldom ponder these matters that, to date, above 2000 of the minor planets are known and have orbits assigned to them. This difficult task is carried on by the Observatory of Cincinatti.

It is estimated that the Hale telescope could photograph 100,000 asteroids, indeed, one plate of the 48" Schmidt telescope on Palomar Mountain has disclosed 90 of these small bodies. Hubble estimated that 30,000 were within reach of his instruments, and Baade was of the opinion that 44,000 were available for observation on statistical evidence. In view of the ninety-first fragment of *Heraclitus*[24] I would simply point out that with the ever-increasing resolving power of telescopes and the new techniques now to hand it will be remarkable if the numbers do not multiply to the consternation of future astronomers and the complete unconcern of the rest of mankind. Not all of these bodies move within the zone between the orbits of Mars and Jupiter but a great number do find themselves confined to it. Without doubt their movements are dominated by Jupiter and this is seen from their remarkable grouping

---

[24] Heraclitus says 'You never go down to the same stream twice'; in this case not only is the universe never the same but the instruments and the astronomers employed have such varying propensities. Now a new survey has been carried out with the Schmidt telescope at Mount Palomar; see Van Houton, *et al.*, *Astron. Astrophys. Suppl.* **2**, 339 (1970).

together. This feature was noticed by Kirkwood in 1870, though at that time the number of asteroids known was insufficient to illustrate the matter in its true nature[25] (Fig. 36). There is some confusion about the grouping even today and it is well to recall the exposition of this phenomenon given by the great astronomer Simon Newcomb:

the seeming fact pointed out by Kirkwood was that, when these bodies are arranged in the order of their mean motions, there are found to be gaps in the series at those points where the mean motion is commensurable with that

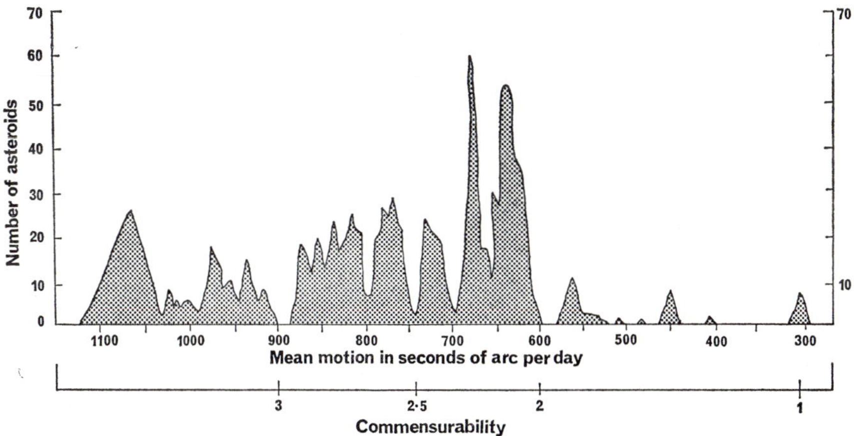

Fig. 36. Groupings of the minor planets.

of Jupiter; that is to say, there seems to be no mean daily motions near the values 598″, 748″ and 898″, which are respectively 2, 2½ and 3 times that of Jupiter. Such mean motions are nearly commensurable with that of Jupiter, and it is shown in celestial mechanics that when they exist the perturbations of the planet by Jupiter will be very large. It was therefore supposed that if the commensurability should be exact the orbit of the planet would be unstable. But it is now known that such is not the case, and that the only effect of even an exact commensurability would be a libration of long period in the mean motion of the asteroid. The gaps cannot therefore be accounted for on what seemed to be the plausible supposition that the bodies required to fill these gaps originally existed but were thrown out of their orbits by the action of Jupiter. The fact can now be more precisely stated by saying that we have not so much a broken series as a tendency to an accumulation of orbits between the points of commensurability and it is probable that the grouping had its origin in the original formation of these bodies.

The broad groups of the asteroids according to Porter[26] are:

[25] The 'true nature' is still in doubt; see Hunter, R. B., *Monthly Notices of R.A.S.*, *136*, 267 (1967), who postulates an asteroid belt between the orbits of Jupiter and Saturn.
[26] Porter, J. G., *J.B.A.A.*, *61*, 4 (1951).

The Hecuba group with a mean daily motion of 600″, a little in excess of twice that of Jupiter.

The Hilda group with a mean daily motion of 450″ at 2/3 of Jupiter's motion.

The Minerva group with a mean daily motion of 750″ at 2/5 of Jupiter's motion.

The Hestia group with a mean daily motion of 900″ at 1/3 of Jupiter's motion.

The Flora group with a mean daily motion of 1050″ at 2/7 of Jupiter's motion.

The Trojan group with the same mean daily motion as Jupiter.

The Trojan group offers us an example of the Lagrangian problem that arises when three bodies, a primary planet, and an asteroid of small mass move about the Sun in the same plane in substantially circular orbits with equal periods of revolution. If the bodies are placed 60° apart then that spacing remains substantially inviolate. The problem was initially considered by Lagrange in 1772 as a mathematical curiosity but today we see the application of this within the Trojan group of the asteroids.

The first asteroid of the group was discovered by Wolf in 1906 by photography. A second and third were discovered by Kopff in 1906 and 1907 and it was Palisa who suggested naming them Achilles, Patroclus and Hector. These three bodies all had mean daily motions almost identical with that of Jupiter, 298″, 299″ and 306″ respectively. Eleven more had been added to the group by 1961 and orbits assigned to them. The list according to Nicholson[27] is given below:

No.	Name	Photographic magnitude	Inclination to ecliptic	Orbital eccentricity	Mean daily motion	Dist. from Jupiter	Diam. miles
588	Achilles	16·0	10°·3	0·15	298″	+44°	35
617	Patroclus	15·8	22°·1	0·14	299″	−63°	38
624	Hector	15·2	18°·3	0·02	306″	+70°	48
659	Nestor	16·3	4°·5	0·11	296″	+74°	30
884	Priamus	16·5	8°·9	0·12	298″	−82°	28
911	Agamemnon	15·4	21°·9	0·07	305″	+69°	44
1143	Odysseus	16·0	3°·1	0·09	300″	+66°	35
1172	Aeneas	16·0	16°·7	0·10	300″	−76°	35
1173	Anchises	16·6	7°·0	0·14	308″	−45°	25
1208	Troilus	16·3	33°·7	0·09	303″	−60°	29
1404	Ajax	16·8	18°·1	0·11	302″	+85°	23
1437	Diomedes	15·8	20°·6	0·04	304″	+45°	38
1583	Antilochus	16·5	28°·3	0·05	293″	+33°	29
1647	Menelaus	18·5	5°·6	0·03	299″	+71°	11

[27] Nicholson, S. B., Astron. Soc. of the Pacific leaflet No. 381, 1961.

According to the latest Palomar-Leiden Survey there are fifteen members of the group with assigned numbers. There are a further fifteen members unnumbered and for details one should consult Gehrels paper in the *Astronomical Journal*, *75*, 659 (1970), in which he estimates that there are possibly some 1700 members within the group.

The disposition of these bodies is such that they take up a lenticular path about the 60° points (Fig. 37). This path derives from a number of

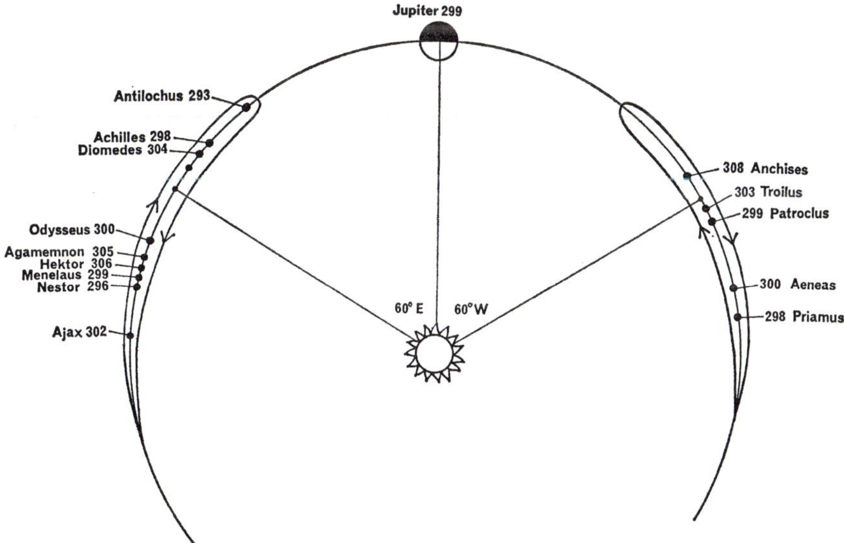

Fig. 37. The Trojan asteroids, 1960 June 20, with their mean motion per day.

facts; their orbits are not circular; Jupiter's orbit also is not circular ($e = $ 0·048); perturbations are set up within them by Saturn, and the orbits are not co-planar. All of which tends to distort the ideal mathematical situation studied by Lagrange. It is not to be wondered at that these small bodies wander far from the stable positions of the 60° points; indeed Diomedes may go as far as 40° beyond the 60° point on the side away from Jupiter and 24° from it on the side toward him.

The asteroids are all very small bodies. Ceres is estimated to be of 770 km. diameter, Pallas 490 km., Vesta 390 km.[28] and Juno 190 km. The rest all come below about 200 km. in diameter. There is little doubt that they all rotate. Eddington once commented on the remarkable fact that everything in the cosmos rotates and revolves—who will wish to dispute that the asteroids, not withstanding their small size, conform with the behaviour of their more noble companions. Eros, on one of its

[28] See Allen, D. A., 'Infrared diameter of Vesta', *Nature*, *227*, 158 (1970); the mean infra-red diameter is given as 573 ± 6 km. from which he estimates a diameter of 600 km. The value of 390 km. above is that of Barnrad.

close approaches to the Earth (1931), revealed a strange stellar-like image that was believed to stem from a quasi-cylindrical body of 25 km. length and 10 km. diameter with a rotational period of 5 h. 17 m. Three of these bodies are known to come within the Earth's orbit: 1566 Icarus, 1620 Geographos and 1685 Toro, with perihelion distances of 0·187, 0·827 and 0·771 a.u. respectively. In August 1969 Geographos passed within 6×10^6 miles (0·061 a.u.) of the Earth and in June 1968 Icarus came as close as 0·042 a.u. It will be clear that these close approaches to the Earth, and to the other planets, afford an opportunity of measuring the masses of the latter with a degree of precision not generally available except with the most advanced techniques of charting by radar the orbits of artificial satellites placed in close orbit about the planet whose mass is to be measured.

The names of the minor planets have given pleasure to many and caused consternation to some, especially those who deplore seeing unsavoury characters enshrined in the celestial spaces. Nowadays names are largely dispensed with in favour of a more humble numerical designation. The Ephemerides of Minor Planets for 1969 under the editorship of Prof. G. Chebotarev, Director of the Institute of Theoretical Astronomy, Academy of Sciences of the USSR, records 1735 named asteroids and gives the elements of their orbits. The current Palomar-Leiden Survey (see 1970 October issue of *Astronomy and Astrophysics Supplement Series*, vol. 2, part 5) lists 1779 asteroids and these have been assigned permanent numbers and orbits. One of the principal deductions of the Survey is that

$$\log N (m_0) = - 3\cdot30 + 0\cdot390 \, m_0$$

represents the number of minor planets within the apparent magnitude range $(m_0 - 0\cdot25)$ to $(m_0 + 0\cdot25)$. This formula has been observed (or is 'known') to be valid as far as $m_0 = 20\cdot5$. The total is obviously in excess of 50,000.

The correct understanding of the formula

$$\log N (m_0) = - 3\cdot30 + 0\cdot390 \, m_0$$

is as follows, for example:

Let $m_0 = 20\cdot0$, then $\log N = +4\cdot50$ and $N = 28,200$. Therefore, in the magnitude interval from 19·75 to 20·25 there are an estimated 28,200 minor planets. The limit of observation was about 20·5, so that any use of the formula beyond that is an unobserved extrapolation. In the range from 19·25 to 19·75 one finds that there are an estimated 27,000 minor planets. The formula loses its validity at brighter magnitudes, e.g. between 09·75 to 10·25 where it estimates that there are only four minor planets.

198

METEORS, METEORITES AND TEKTITES

Meteors (Greek μετέωρα, literally things in the air) enter the Earth's atmosphere in large numbers every day. From careful observation over many years a number of meteor showers are extremely well known. In any goodly shower the individual meteors diverge from a common focus which is a mere effect of perspective on the great backcloth of the night sky; this focus is termed the radiant. In 1876 the number of radiants reported was 850, but many thousands have been published up to the present time.[29] The radiants of the principal showers are given below:

Meteor showers (northern hemisphere)	Radiant R.A.		Dec.	Duration
	h m °		°	
Quadrantids	15 28 (232)		+ 50	Jan. 1–4
Lyrids	18 08 (272)		+ 32	Apr. 19–23
η Aquarids	22 24 (336)		00	May 1–8
Capricornids	21 00 (315)		− 15	July 10–Aug. 5
δ Aquarids	22 36 (339)		− 17	July 15–
	22 36 (339)		00	Aug. 15
a Capricornids	20 36 (309)		− 10	July 15–Aug. 25
ι Aquarids	22 32 (338)		− 15	July 15–
	22 04 (331)		− 6	Aug. 25
Perseids	03 04 (046)		+ 58	July 25–Aug. 17
κ Cygnids	19 20 (290)		+ 55	Aug. 18–22
Orionids	06 24 (096)		+ 15	Oct. 17–26
Taurids	03 28 (052)		+ 14	Oct. 10–
	03 36 (054)		+ 21	Dec. 5
Leonids	10 08 (152)		+ 22	Nov. 14–20
Geminids	07 28 (112)		+ 32	Dec. 7–15
Ursids	14 28 (217)		+ 78	Dec. 17–24

Meteor showers (southern hemisphere	Radiant R.A.		Dec.	Duration
	h m °		°	
Corona Australids	16 20 (245)		− 48	Mar. 14–18
Ophiuchids	17 20 (260)		− 20	June 17–26
Pisces Australids	22 40 (340)		− 30	July 15–Aug. 20
Phoenicids	01 00 (015)		− 55	Dec. 5

The August Perseids have been seen more frequently than any other shower and the radiant has been fixed over 250 occasions.

Historical records according to Denning supply the following dates of abundant meteor displays.

A.D.		A.D.	
902	Oct. 13	1101	Oct. 17
931	Oct. 14	1202	Oct. 19
934	Oct. 14	1366	Oct. 23
1002	Oct. 15	1533	Oct. 23

[29] See *Mem. R.A.S.*, *53*, 203, W. F. Denning's Catalogue of the radiant points of meteoric showers and of fireballs and shooting stars obtained at more than one station.

A.D.	A.D.
1602 Oct. 28	1833 Nov. 13
1698 Nov. 9	1866 Nov. 13
1799 Nov. 12	1867 Nov. 14
1832 Nov. 13	1868 Nov. 14

The meteor shower of November 13, 1866 (The Leonids) was particularly prolific and as far back as A.D. 902 we find a reference to this famous shower. Conde in his *History of the Dominion of the Arabs in Spain* mentions that on the death of King Ibrahim bin Ahmad (902, mid-October, old style calendar) 'an infinite number of stars were seen during the night scattering themselves as rain to the right and left and that year was known as the year of the stars'. Humboldt witnessed a similar display in South America on November 12, 1799. The most recent display of this renowned shower was seen at Kitt Peak,[30] South Arizona, when on the night of November 17, 1966, a rate of 150,000 meteors per hour was recorded.

Radar methods have shown the presence of a number of extensive showers in the daylight hours, particularly in summer months. These showers have small eccentric orbits as also do the Geminids and the November Taurids—more in keeping with the orbits of the minor planets. Many estimates suggest that the average meteor is seen at a height of 70 miles above the Earth and terminates its ephemeral life at a height of about 50 miles.

The meteorites offer to the discerning student of the history of science an outstanding example of that remarkable attribute of the human intellect that enables it to take no cognizance of those facts that do not appeal to the myopic framework in which it chooses to operate. That such an aberration should afflict a body as diverse and as cultivated as the French Academy of Sciences[31] must be unique. Indeed the almost total scepticism in the early nineteenth century shown toward what had been widely acclaimed in earlier times, viz. the fall of stones from the sky[32] was only relieved, prior to Biot's report on the meteoric shower at L'Aigle in 1803, by the labours of Chladni.[33]

It was said in 1962 by Mason that the only adequately observed orbit for a meteorite is that of the Pribram chrondrite of April 7, 1959, which showed conclusively that it had travelled in an elliptical orbit within the

30 See Milon, D., 'Observing the 1966 Leonids', *J.B.A.A.*, 77, 89 (1967).
31 See Paneth, F., 'Science and Miracles', *Durham University Journal*, 10, 49 (reprinted in *Chemistry and Beyond*, Wiley, 1964). It is said that scientists in other countries were anxious not to be considered as backward compared with their famous colleagues in Paris and many curators of museums threw away their precious meteorites.
32 Acts 19[35] record that Ephesus was the guardian of the image which fell down from Zeus. The Kaaba of Mecca is also thought to be a meteorite. Both Livy and Pliny report showers of stones from the sky.
33 Chladni, *Ueber den Ursprung der von Pallas gefundenen und anderer ihr aehnlicher Eisenmassen* (Riga 1794).

asteroid belt prior to its demise as an aerial body. The largest meteorite to fall in Britain recently is the Barwell stones[34] that fell December 24, 1965. The many meteorite falls throughout the world are skilfully catalogued and the student may with confidence turn to the *Catalogue of Meteorites*[35] by G. T. Prior (3rd ed., 1966, published by the British Museum).

The classification of meteorites is complex and most laymen will be content with the scheme proposed by Prior.[36] A more detailed classification due to Rose-Tschermak and Brezina may be discovered in the *Proceedings of the American Philosophical Society, 43,* pp. 211–47.

PRIOR'S CLASSIFICATION OF THE METEORITES (FIGURES IN PARENTHESES ARE THE NUMBERS IN EACH CLASS)

Group	Class		Principal Minerals
Stony meteorites	Chondrites (commonest of all meteorites 85 per cent)	Enstatite (11)	Enstatite, nickel-iron
		Olivine-bronzite ⎱(~900)	Olivine, bronzite, nickel-iron
		Olivine-hypersthene ⎰	Olivine, hypersthene, nickel-iron
		Olivine-pigeonite (12)	Alivine, pigeonite
		Carbonaceous (17)	Serpentine
	Achondrites rare falls	Aubrites (9)	Enstatite ⎫ calcium
		Diogenites (8)	Hypersthene ⎬ poor
		Chassignite (1)	Llivine ⎭
		Ureilites (3)	Olivine, pigeonite, nickel-iron
		Angrite (1)	Augite ⎫ calcium
		Nakhlites (2)	Diopside, olivine ⎬ rich
		Eucrites and howardites (39)	Pyroxene, glagioclase) ⎭
Stony-irons minor group 4 per cent	Pallasites (40)		Olivine, nickel-iron
	Siderophyre (1)		Orthopyroxene, nickel-iron
	Lodranite (1)		Orthopyroxene, olivine, nickel-iron
	Mesosiderites (22)		Proxene, plagioclase, nickel-iron
		Nickel	
Irons second largest group	Hexahedrites (55)	4–6+	Kamacite
	Octahedrites (487)	6–14	Kamacite, taenite
	Ni-rich atazites (36)	> 12	Taenite

The average composition (in percentage by weight) of the chondrites from almost one hundred analyses is

SiO_2	38·04
MgO	23·84
FeO	12·45
Al_2O_3	2·50
CaO	1·95

[34] See *J.B.A.A., 76,* 331, 147, 192 (1966).
[35] A useful catalogue based on the famous Rose-Tschermak-Brezina system of classification is that by Leondard, F. C., and de Violini, R. *A Classificational Catalog of the Meteoritic Falls of the World* (University of California Press 1956).
[36] Prior, G. T., 'The classification of meteorites', *Mineral Mag., 19,* 51 (1920).

Na_2O	0·98
K_2O	0·17
Cr_2O_3	0·36
MnO	0·25
TiO_2	0·11
P_2O_5	0·21
Fe	11·76
Ni	1·34
Co	0·08
FeS	5·73
Total Fe	25·00

The elemental composition of meteoric matter is of profound importance in understanding the origin of the Solar System and the values for the major elements in parts per million according to Levin *et al.* 1956 are are follows:

Atomic no.	Element	ppm	Atomic no.	Element	ppm
3	Li	3·2	47	Ag	0·5
4	Be	0·09	48	Cd	2
5	B	2·6	49	In	0·2
8	O	346,000	50	Sn	20
9	F	40	51	Sb	0·4
11	Na	7000	52	Te	0·4
12	Mg	139,000	53	I	1
13	Al	14,000	55	Cs	0·08
14	Si	178,000	56	Ba	7
15	P	1600	57	La	200
16	S	20,000	58	Ce	2
17	Cl	800	59	Pr	0·8
19	K	900	60	Nd	3
20	Ca	16,000	62	Sm	1
21	Sc	5	63	Eu	0·3
22	Ti	700	64	Gd	1·6
23	V	80	65	Tb	0·5
24	Cr	2500	66	Dy	2
25	Mn	2000	67	Ho	0·6
26	Fe	256,000	68	Er	1·7
27	Co	900	69	Tm	0·3
28	Ni	14,000	70	Yb	1·6
29	Cu	40	71	Lu	0·5
30	Zn	20	72	Hf	0·8
31	Ga	8	73	Ta	0·3
32	Ge	40	74	W	17
33	As	70	75	Re	0·0018
34	Se	9	76	Os	1·1
35	Br	22	77	Ir	0·6
37	Rb	8	78	Pt	3
38	Sr	22	79	Au	0·26
39	Y	5	80	Hg	0·009
40	Zr	90	81	Tl	0·14
41	Nb	0·5	82	Pb	2
42	Mo	5	83	Bi	0·16
44	Ru	2	90	Th	0·2
45	Rh	0·6	92	U	0·05
46	Pd	0·5			

The abundances of the rare earth elements in the chondrites and achondrites in parts per million are given by Schmitt as follows:

	A. Allegan (olivine-bronzite chondrite)	2. Richardton (olivine-bronzite chondrite)	5. Nuevo (pigeonite-plagioclase achondrite)	4. Murray (carbonaceous chondrite)	3. St. Marks (enstatite chondrite)	6. Pasamonte (pigeonite-plagioclase achondrite)
La	0·33	0·32	0·25	0·39	4·03	3·21
Ce	0·54	0·48	0·66	1·04	10·7	8·08
Pr	0·12	0·12	0·13	0·15	1·47	1·26
Nd	0·65	0·61	0·36	0·62	8·0	5·10
Sm	0·24	0·20	0·14	0·21	2·19	1·90
Eu	0·087	0·080	0·045	0·072	0·75	0·68
Gd	0·34	0·34	0·15	0·27	2·45	2·69
Tb	0·049	0·053	0·036	0·049	0·59	—
Dy	0·39	0·34	0·21	0·32	4·14	3·06
Ho	0·082	0·068	0·046	0·079	0·84	0·69
Er	0·22	0·21	0·14	0·21	2·87	1·65
Tm	0·043	0·033	0·024	0·037	0·48	0·30
Yb	0·20	0·19	0·14	0·19	2·34	1·67
Lu	0·038	0·033	0·027	0·030	0·30	0·50
Ca(%)	1·24	1·54	0·87	1·37	7·43	7·34

TERRESTRIAL METEORITE CRATERS AND ASTROBLEMES

Some twelve craters are known on the Earth to be associated with meteoric material, the largest of which is the Barringer crater, Arizona. Another thirty-two sites are thought to be the remains of meteorite impact sites. In these no meteoric material is to be found and the meteorite is believed to have completely disintegrated leaving no trace of itself. The presence of coesite is said to be evidence of a high pressure event. Dietz has suggested that the Sudbury basin in Ontario may be an 'astrobleme', that is, an impact structure made by a meteorite or an asteroid. Dietz and his co-workers believe that about 17 billion (17 × 10^9) years ago the Sudbury structure may have been created by a small asteroid or a large meteorite colliding with the Earth. The energy liberated at impact is calculated to be of the order of 3 × 10^{29} ergs. The meteorite is calculated to have been travelling at 15 kilometres/sec. (9·3 miles/sec.) and to have possessed a diameter of some 4 kilometres (2·48 miles). The crater which the missile generated is believed to have been initially of the order of 48 kilometres (30 miles) in width and of 3·2 kilometres (2 miles) in depth—the meteorite is wisely postulated to have been an iron-nickel body with an appreciable copper content. The case formulated by Dietz is impressive, but not wholly acceptable to all geologists.

Dietz argues for the Sudbury structure as an astrobleme on the following points. His argument carries the inescapable corollary that our wealth in this area may have an extra-terrestrial origin.

(*a*) At impact the meteorite exploded, excavating a great crater. Shock waves spread from the centre and caused severe brecciation and shatter coning. A thick collar of rocks was thrown up around the crater.

(*b*) The meteorite was partly liquefied and forced against the crater wall into stellate cracks to form the ore either by the conversion of a metallic nickel-iron body or by direct replacement from a sulphide body. Some of the native rock also was liquefied and splashed against the crater wall. The native rock cooled as quartz-diorite. Liquid quartz-diorite and meteoritic matrix intermixed in all proportions. Both liquids provided matrix substances for the breccias[37] found.

(*c*) A magma[38] welled up in the crater bowl from the rocks of the Earth's crust. The exposed pool of magma laid down a thick crust of welded tuff.[39] Upon solidification a saucer-shaped complex of igneous material remained.

(*d*) The incompletely filled crater filled up with water, and sediment poured into this declivity from the surrounding crater wall. This is thought to be the genesis of the Onwatin slate and Chelmsford arkose of the area.

(*e*) The original crater of circular form in plan later suffered distortion to produce the present lenticular shape.

Dietz believes that the Sudbury breccia is remarkable and provides evidence of having been subjected to shock, since it occupies fractures that are disposed radially or concentrically to the basin. A reconnaissance of the area has revealed the presence of shatter cones which further suggest that intense shock waves have at some time been in operation at the site. Shatter cones are a mode of conical fracture induced by intense shock waves and are already known from two meteorite craters of the quaternary era. Elsewhere Dietz[40] has marshalled the evidence for shatter cones as a criterion for deciding upon meteorite impact and he is fully aware of the difficulties in attempting to explain the genesis of the complex Sudbury ores solely by this means. The case which he puts forward in support is too lengthy for presentation here, but mention must be made of the high copper content of the Sudbury ores and the composition of known meteorites. The Cu/Ni ratio over the Sudbury district as a whole is approximately 1:1 with only a small excess of

[37] A composite rock of angular fragments of stone, etc., cemented together by some matrix in contradistinction to conglomerate in which the fragments are rounded or water worn.
[38] One of two or more supposed strata of fluid or semi-fluid matter lying beneath the solid crust of the Earth.
[39] Any light porous cellular rock, more often volcanic tuff.
[40] Dietz, R. S., 1960, 'Meteorite impact suggested by shatter cones in rock', *Science*, vol. 131, June 17, 1960, pp. 1781–4. See Millman, P. M., 'The Space Scars of Earth', *Nature*, *232*, 161 (1971); see also Krinov, E. L., *Giant Meteorites* (1966).

nickel,[41] whereas in typical nickel-iron meteorites the copper averages only 0·02 per cent.

*Meteoritic Irons**

Iron	90·8 per cent approx.
Nickel	8·5 per cent approx.
Cobalt	0·59 per cent approx.
Phosphorus	0·17 per cent approx.
Sulphur	0·04 per cent approx.
Carbon	0·03 per cent approx.
Copper	0·02 per cent approx.
Chromium	0·01 per cent approx.

* from Nininger, H., *Out of the Sky*, Dover Pub. Inc. 1952.

The matter of the composition is by no means clear since little is known of the numbers of varieties of these cosmic wanderers in space. Nininger[42] reports the Coppery Eaton Meteorite which fell in Colorado in 1931. The weight of that meteorite was 29½ grammes (a little above one ounce), and it was analysed by Hawley. It consisted mainly of copper with additions of zinc and lead. The specimen bore flight marks and, as with iron and nickel in meteorites, the metals are free from the chemical combinations in which they are found in the Earth. Copper was first reported as a free element in the Richardton meteorite by Quirke in 1919. It is not beyond the bounds of possibility that the Sudbury astrobleme may have been caused by a sulphide meteorite, as suggested by Anders.

Baldwin,[43] the leading authority on the case for the meteoritic origin of the lunar craters, has examined the ballistics of the problem. The Sudbury event is summarized by him as follows:

Impact velocity 15 km./sec. (9·3 miles/sec.)
nickel-iron siderite dia. 4·07 km. (2·5 miles)
original crater 57 km. (36 miles) wide
depth—3853 metres (12,640 ft.)
rim height—1234 metres (4050 ft.)
true depth—2620 metres (8600 ft.)
major brecciation—13 km. (8·2 miles) below the surface.

In a consideration of damage done to the surface of the Earth by the fall of extra-terrestrial bodies, it is of interest to recall the greatest of the meteorite craters. The Barringer crater is near Flagstaff, Arizona (Highway 66), in West Longitude 111 deg. and North Latitude 35 deg. The

[41] Hawley, J., 1962, 'The Sudbury ores and their origin', *Canadian Mineralogist*, vol. 7, part 1, p. 207.
[42] Nininger, H., *Popular Astronomy*, vol. LI, No. 5, May 1943.
[43] Baldwin, R. B., 1963, *The measure of the moon*. Chicago, University of Chicago Press, p. 488; Baldwin, R. B., 1949, *The Face of the Moon*, University of Chicago Press, p. 238.

crater is 1280 metres (4200 ft.) in diameter and 173 metres (570 ft.) deep. The rim rises 49 metres (160 ft.) above the plain which contains it. It has been calculated that the explosion at impact excavated 62 × 10^6 cubic metres (82 × 10^6 cu. yd.) of the Earth's surface material, weighing 3024 × 10^8 kilogrammes (300 × 10^6 tons). Numerous nickel-iron specimens were collected from the area prior to 1902 when Daniel Moreau Barringer, a mining engineer of Philadelphia, turned his attention to winning iron from it. Iron sold at that time for \$80 a ton and the prospect of several millions of tons of iron embedded in the crater was a vast attraction to him. Barringer and his associate, B. C. Tilghman, filed a claim to the land in 1903.

The first drillings were made in the central area and he expected the iron body not to be below 426 metres (1400 ft.), since no Supai sandstone of that level had been ejected. The drillings were unsuccessful and he began to dig trenches in the outer slopes, only to find small meteorites which had fallen with the main mass. Barringer now believed the meteorite had not fallen vertically and he attempted to determine the direction of its flight path. His observations favoured a north-south path with the mass under the southern rim. He gathered funds and, with the help of a mining company, began drilling in 1920. At the 418 metres (1376 ft.) level the drill hit an impenetrable object and the cable broke. The hole was abandoned in 1923. He began another drilling operation some years later but ran out of funds. When he returned to Philadelphia in 1929 the stock market crashed. He did not live to realize his dream. Since Barringer's day many surveys have been made and many meteorites gathered from the site. The best known surveys have been made by Nininger and Rinehart, the latter conducting an expedition in 1956 for the Smithsonian Astrophysical Observatory. The consensus is that the parent mass vaporized on impact and that it was initially of 12 × 10^6 kilogrammes (12,000 tons) weight. Its flight path is believed to have been from the north west. Assuming a nickel-iron sphere of specific gravity 7·703, its radius would be about 31 metres (103 ft.).

The Moon offers the best readily observable site for the examination of large craters and maria with extensive walls possibly cut by fast moving projectiles. The idea that certain craters are of meteoritic origin is now widely held. The meteoritic theory was first advanced by Gilbert[44] in 1893. Urey believes that the entire Mare Imbrium for example was the scene of a cataclysmic event which liberated the kinetic energy equivalent to 4·6 × 10^{11} atomic bombs (4·15 × 10^{32} ergs). The largest terrestrial earthquake liberates no more than about 10^{22} ergs and the Imbrium collision is consequently an event 10^{10} times as great. The missile probably produced Sinus Iridum and hence passed between the two promontories of Laplace and Heraclides. The distance between

[44] Gilbert, G. K., *Bull. Phil. Soc. Wash. 12*, 24 (1893).

On comets, minor planets, meteors, meteorites and tektites

these promontories is about 230 km. (143 miles) and this gives one guide
to the minimum diameter of the missile.

When we reconsider the sites at which nothing of the suspected
meteorite remains we should not forget the views of Cowan, Atluri and
Libby,[45] who believe that a case may be made out for the presence of
anti-matter[46] in at least one meteor. They rest their arguments on the
Tunguska meteor of 1908.

The origin of meteorites is a question of complexity. The general view
that they come from the disintegration of a planet is difficult to support.
The mass of material in the asteroid belt is so small that it would not
make a body the size of our Moon and the thermal cycle of such body
cannot be related to the temperature indicated by the Widmanstatten[47]
structure of known meteorites within the age of the solar system. An
attractive theory to explain many of the characteristics of meteorites is
that of Fish, Goles and Anders in which the meteorites are developed
in asteroidal bodies. The asteroids accumulated some $4\frac{1}{2}$ aeons ago and
melted to give an inner core of metal and an outer core of molten
silicates, a sintered mantle and an unconsolidated surface layer. This
type of body has sufficient diversification to explain the many classes of
meteorite found on Earth. Further temperature fluctuations are thought
to cause recycling of material, the production of capillary veins in stone
meteorites and the depletion of the mantle in certain elements which
would explain the anomalies found in the composition of some
chondrites.

Oort argues for a common origin of comets and asteroids and hence
we can link up all three of our interplanetary wanderers, comets, minor
planets and meteorites.

Oort's[48] arguments, somewhat truncated, are as follows:

It seems impossible that the peculiar inner structure of a meteorite could have
been the result of gradual condensation and accretion at low temperatures.
It would be very interesting in this connection, if it could be decided whether

[45] Cowan, C., Atluri, C. R., Libby, W. F., *Nature*, *206*, 861 (1965).
[46] *Anti-matter*. Matter consisting of antiparticles. The duals, or counterparts, of the fundamental particles; each antiparticle has the same mass and spin as its corresponding particle,
but usually carries the opposite charge. Some neutral particles are also assumed to have antiparticles; some are regarded as their own antiparticles. A collision between a particle and its
corresponding antiparticle results in mutual annihilation, with concomitant conversion of
their masses into high-energy light waves, or (in some cases) into lighter particles moving with
greater speeds.
See *Matter and Antimatter*, Duquesne, M., Arrow Science Series (1960); also *Nature*, *224*, 477
(1969); *Nature*, *227*, 475 (1970); *Nature*, *230*, 261 (1971).
[47] The Widmanstatten structure or pattern consists of bands crossing one another in a number
of directions. The banding depends on the orientation of the section. It is well known that
when a nickel-iron alloy is cooled slowly into the two-phase field the alpha-phase field the
alpha phase precipitates preferentially along the octahedral planes of the gamma phase. The
pattern is good evidence for the very slow cooling of the octahedrites within the alpha-gamma
field.
[48] Oort, *B.A.N.*, *11*, 107.

the orbital characteristics of at least part of the known meteorites resemble those of the comets.

It appears far more probable that instead of having originated in these far away regions, comets were born among the planets. It is natural to think in the first place of a relation with the minor planets. There are indications that these two classes of objects belong to the same 'species'. Like the minor planets the comets' nuclei seem to consist of solid blocks of considerable dimension. The known asteroids are all rather larger than this, but there is good evidence that their number increases considerably when we extend the search to fainter limits. The observed difference in size may therefore be an effect of observational selection, only the larger objects being observed as asteroids.

It seems a reasonable hypothesis to assume that the comets originated to-gether with the minor planets, and that those fragments whose orbits deviated so much from circles between the orbits of Mars and Jupiter that they became subject to large perturbations by the planets, were diffused away by these perturbations, and that, as a consequence of the added effect of the perturbations by stars, part of these fragments gave rise to the formation of the large cloud of comets which we observe today.

It can hardly be doubted that at least a certain number of the minor planets must have disappeared in the course of time through the action of Jupiter in exactly the same way as comets of the Jupiter family disappear through this action. Even at the present time there are some minor planets whose orbits cross the sphere with radius equal to Jupiter's mean distance from the sun, or the spheres with radii equal to Mars' or the earth's mean distance. Such asteroids are likely to suffer at some time in the future so large a perturbation by one of the planets that they will be brought into long-periods orbits. After which they will be gradually diffused out of the solar system by small perturbations that bring them into more and more elongated orbits. Most of the minor planets that have in the past been thus diffused outwards through Jupiter's perturbations must have escaped into interstellar space, but not all. During the diffusing process a certain fraction will get orbits with semi-major axes between 25,000 and 100,000 a.u. Practically all the asteroids that happened to get into this range of orbits would, during the 4 to 30 million years they needed to complete such an orbit, have been diverted into orbits that did no longer come near the large planets. In as much as they would generally have been brought into these orbits soon after their origin, the continued action of the stars would afterwards have dispersed their velocities such as to give them a random distribution.

For an asteroid there are thus two types of more or less stable orbits in which it can get: nearly circular orbits between Mars and Jupiter like the known asteroids, or orbits with mean distances of the order of 100,000 a.u. into which it can be brought through combined action of Jupiter and the stars.

At the present time the number of minor planets being transferred into long-period cometary orbits is certainly very small. The main process is now the inverse one, that of a slow transfer of comets from the large cloud into short-period orbits. But at the epoch at which the minor planets were formed,

78. The far side of the Moon—a view taken from the Apollo XI command ship in lunar orbit. The large central crater is IAU 308, longitude 179°E, latitude 5·5°S. The diameter of IAU 308 is about 50 miles (80 km.).

79. Landing site of the Apollo XII mission in the Ocean of Storms (Oceanus Procellarum) just south-east of the Crater Lansberg. Co-ordinates are 23°W and 3°S. To the right is landing site of Apollo XI in the Sea of Tranquility (Mare Tranquilitatis).

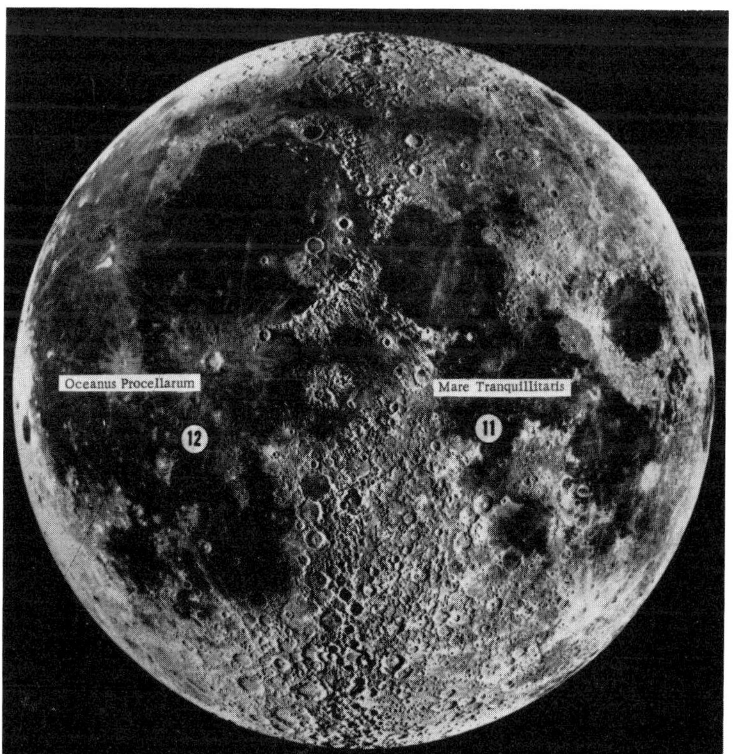

80. The Apollo XII landing vehicle 'Intrepid' seen descending towards the Moon from the mother command ship 'Yankee Clipper'. 'Intrepid' is above Oceanus Procellarum. Photograph by Richard Gordon, 1969, November 19.

81. The Imbrium collision area after Urey. The arrows show the mountainous masses outside the hypothetical collision area and the possible entry direction of the meteorite.

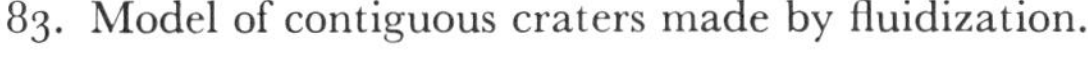

82. The Earth seen from above the Moon on the Apollo XI mission.

83. Model of contiguous craters made by fluidization.

84. Lunar craters Azophi and Abenezra.

85. Pseudo-Thebit, made by fluidization.

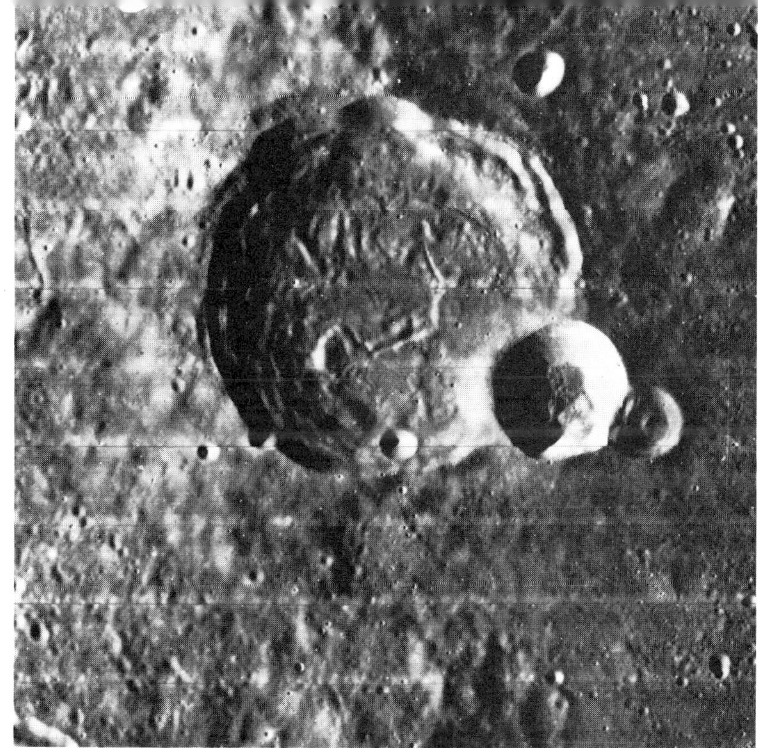

86. Lunar crater Thebit (diameter 48 km.) in the south polar area of the moon.

87. Single crater showing terracing. Compare with, for example, the lunar crater Eratosthenes.

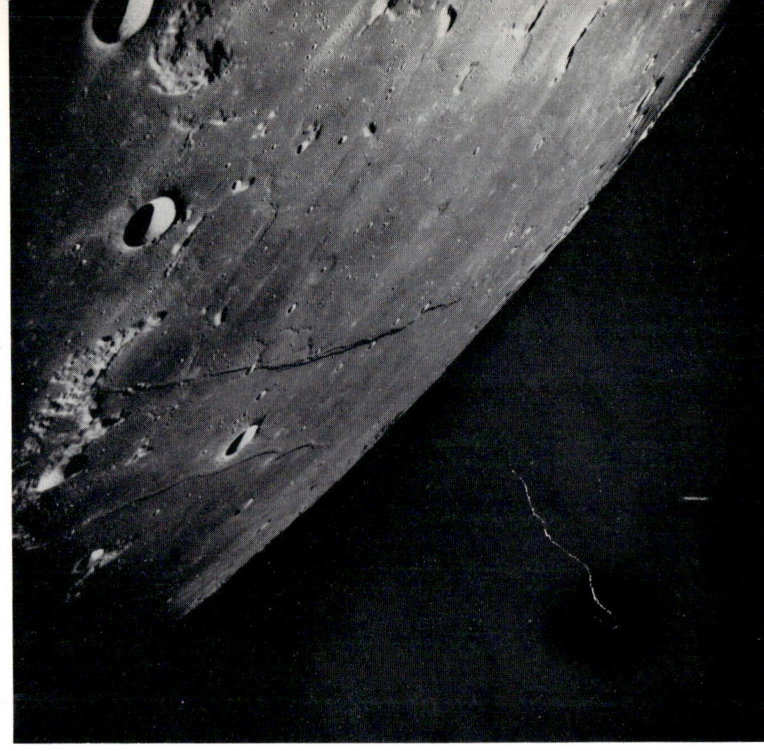

88. Cauchy crater fault and scarp. (Compare with the fault scarps of the Silali Volcano, Kenya.) A view into the Mare Tranquilitatis.

89. Air photograph of the fault scarps of the south-east slope of the Silali Volcano, Baringo district, Kenya. The photograph covers an area four miles across and the scarps are up to 200 ft. high.

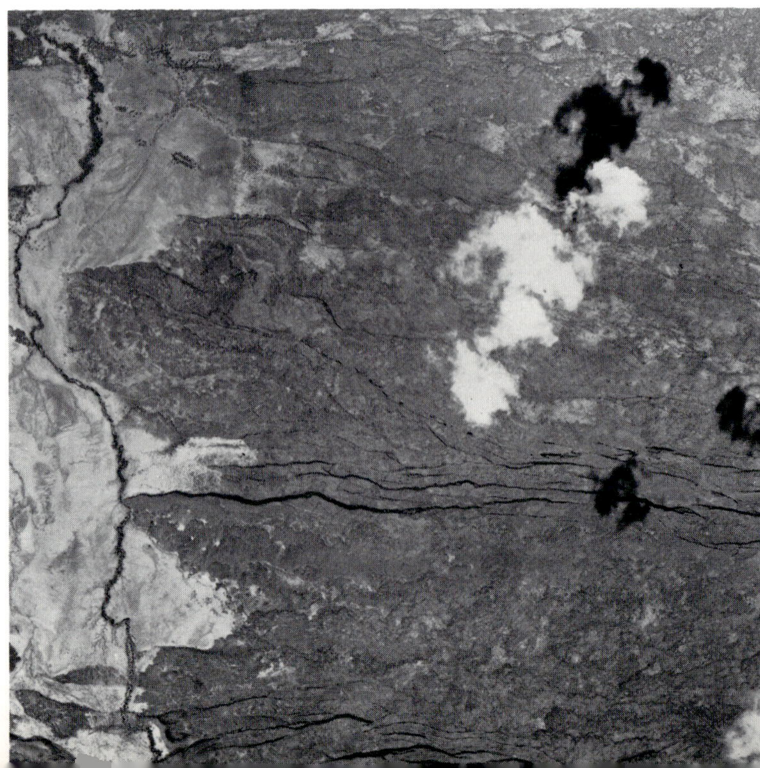

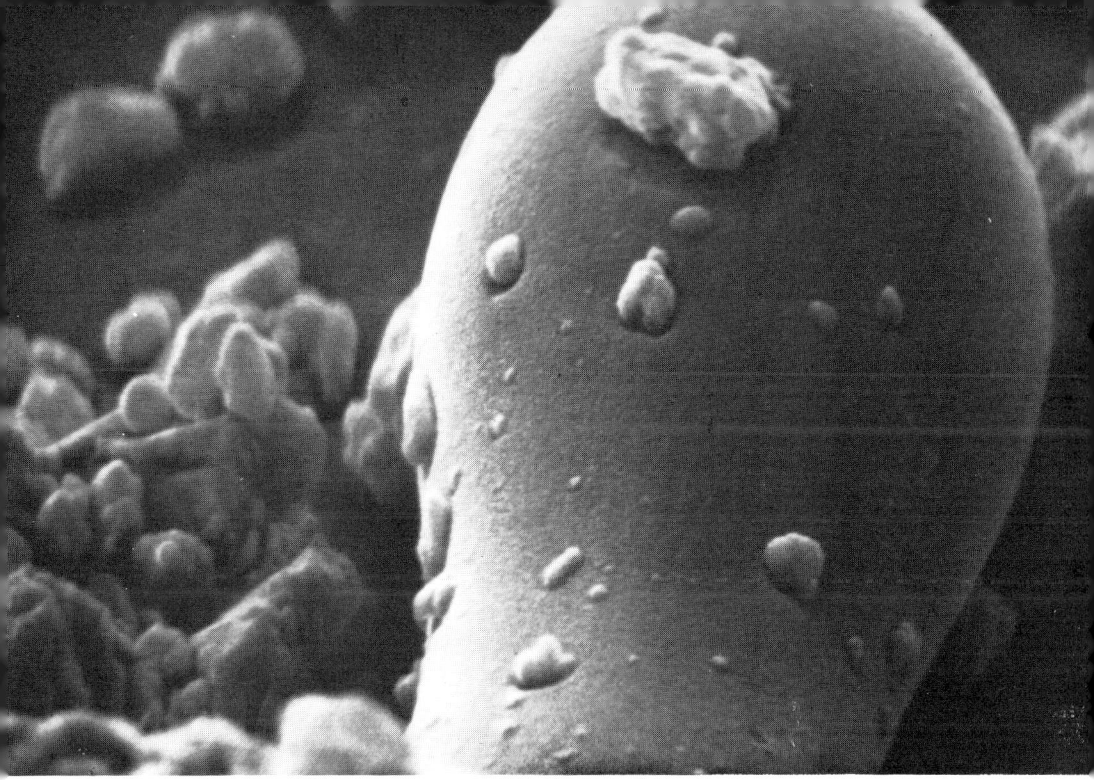

90. Apollo XI fines.

91. Apollo XI fines. Magnification × 9·3 K.

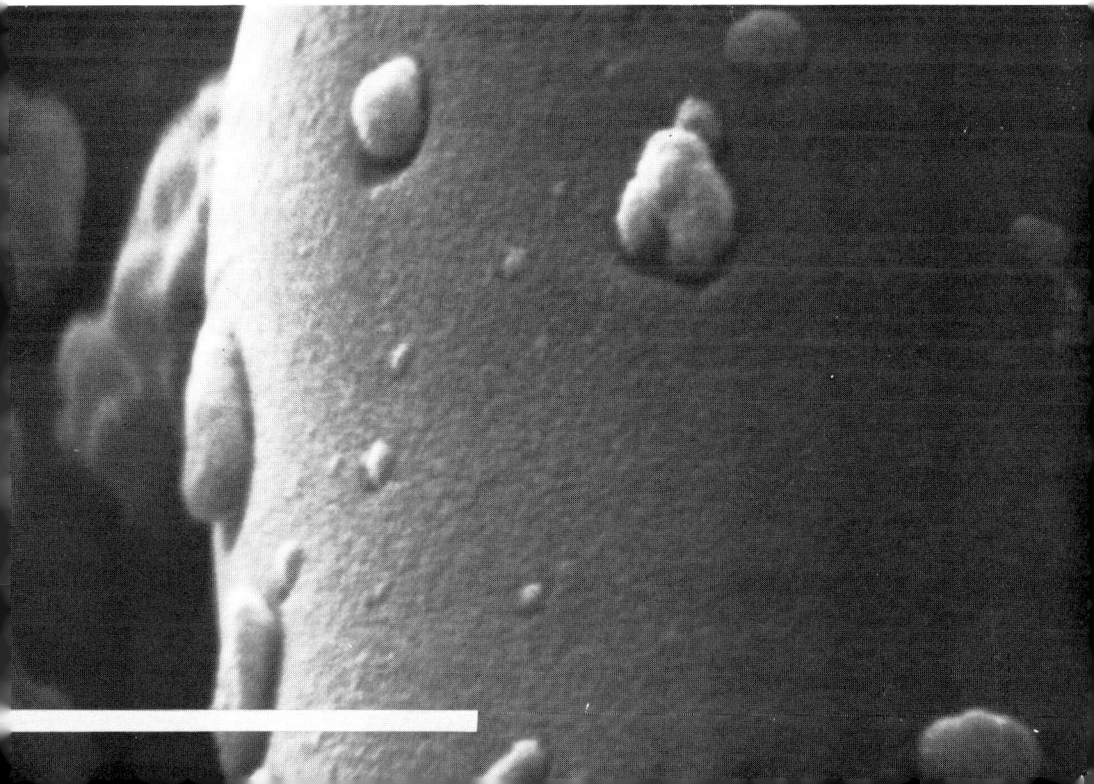

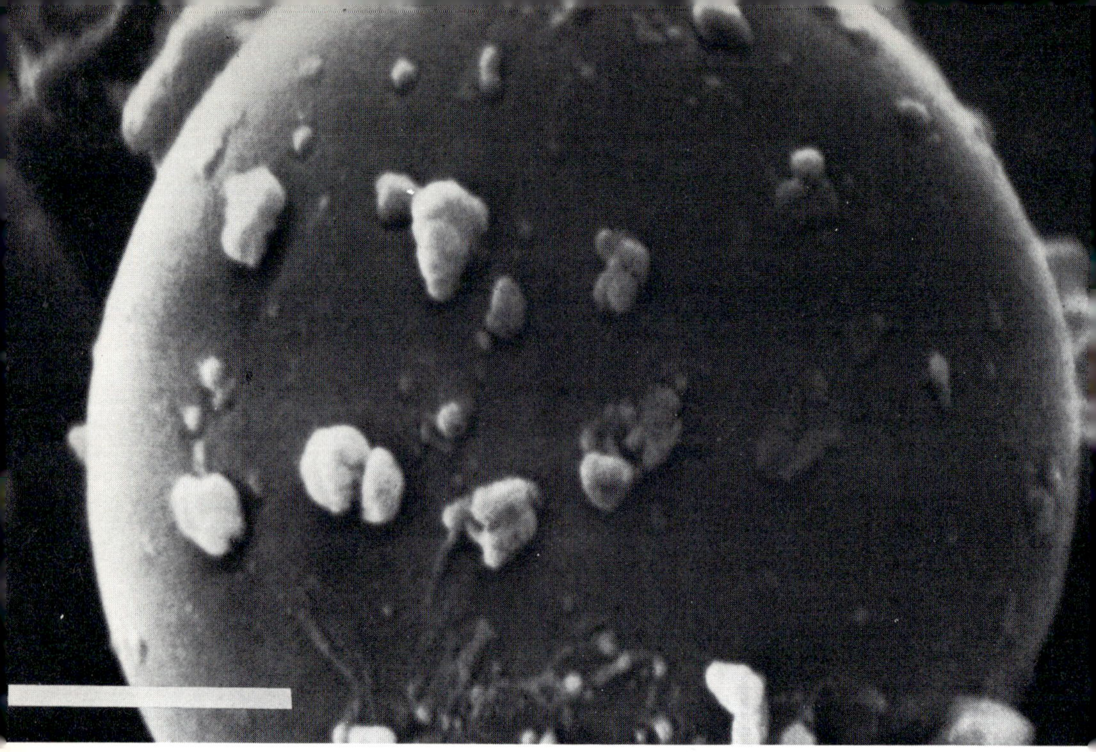

92. Apollo XI lunar sample fines. Magnification $\times$ 4·7 K.

93: (a) Scanning electron micrographs of part of a dumbbell shaped glass bead. (b) Same particle at higher magnification. Note that small particles are partially embedded in the bead surface. (c) Spherical bead showing cratering at the lower edge of the photograph. (d) Spherical beads with partially embedded particles. The horizontal white lines represent 5 μm.

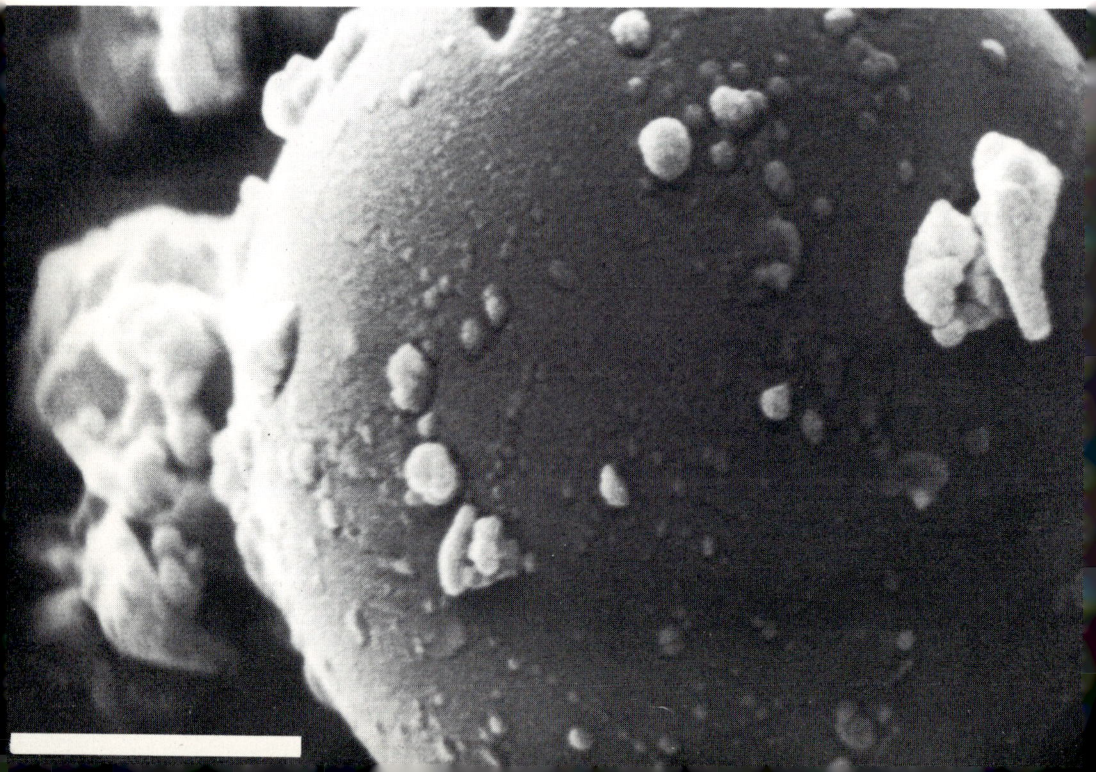

94. Comet Ikeya 1963 I. Perihelion passage, March 21. 4747 U.T. Photograph by Alan McClure, Frazier Mountain 7″ F/7 Triplet., blue sensitised plate, 1963, February 25, 3 h. 13 m.–3 h. 38 m. U.T.

95. Comet Mrkos 1957 V. Perihelion passage, August 1. 4ʰ37ᵐ50ˢ E.T.

96. Comet Arend Roland 1957 III. Perihelion passage, April 8. 03117 U.T.

97. Comet Bennett, 1970 April 7, 3 h. 33 m.–3 h. 41 m. U.T.

98. Comet Bennett 1969 I. The photograph was taken at Archallagan Observatory, Foxdale, Isle of Man, 1970, April.

99. Exploding Andromedid Meteor, Butler 1895, November 23, 2 in. lens.

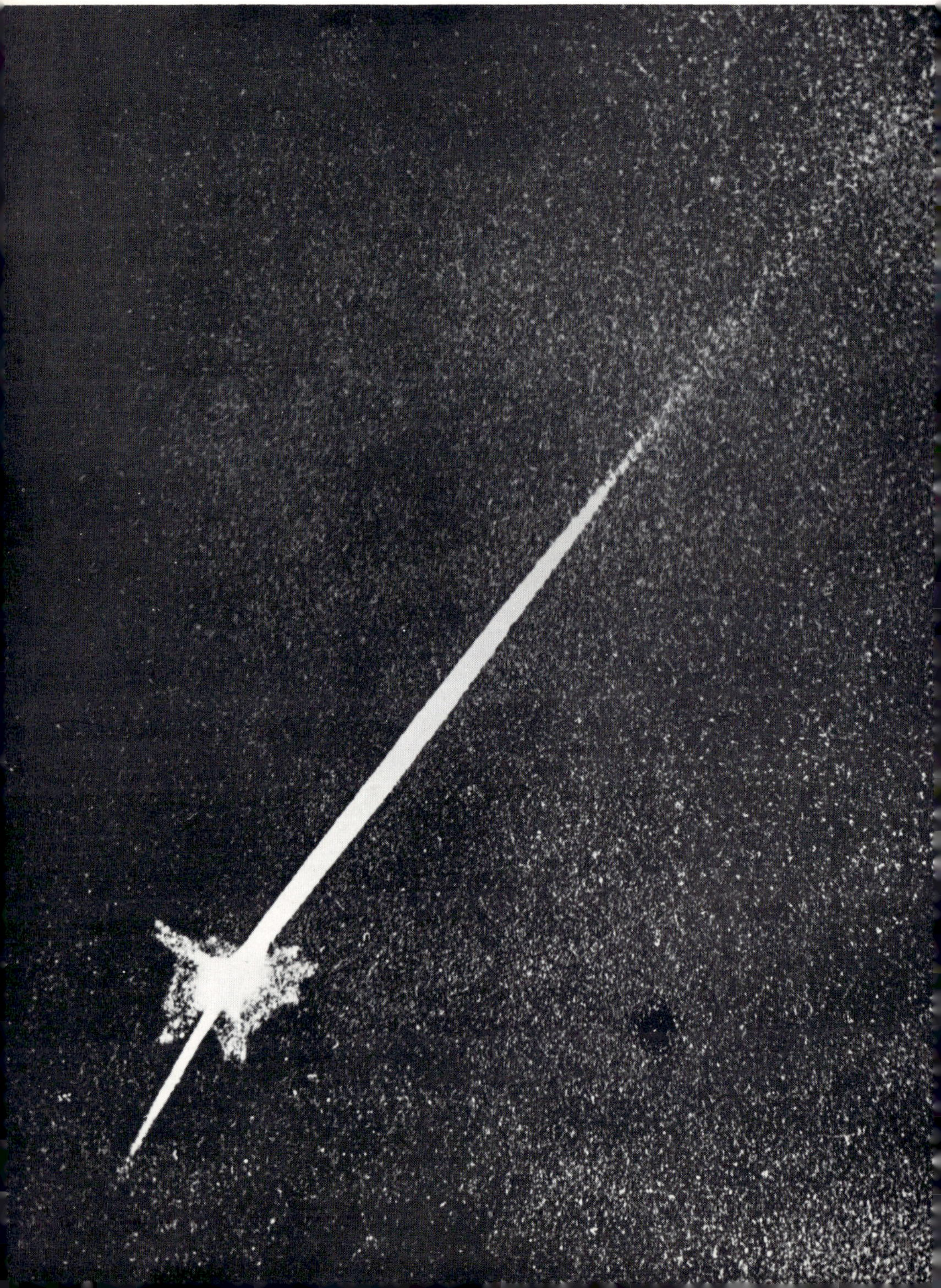

100. Forty-three meteors from the phenomenal Leonid shower at Kitt Peak, 1966, November 17, 12·00 hrs. U.T.

101. The Great Arizona or Barringer Meteorite Crater near Winslow, Arizona; known variously as Arizona Meteorite Crater, Barringer Crater, Meteor Crater, Coon Butte, Meteor Mountain, Canyon Diablo. It is 3 miles in circumference, 4–5 miles in diameter and 600 ft. deep. (See H. H. Nininger, *Out of the Sky*, Dover Books, 1959; R. Gallant, *Bombarded Earth*, London 1964.

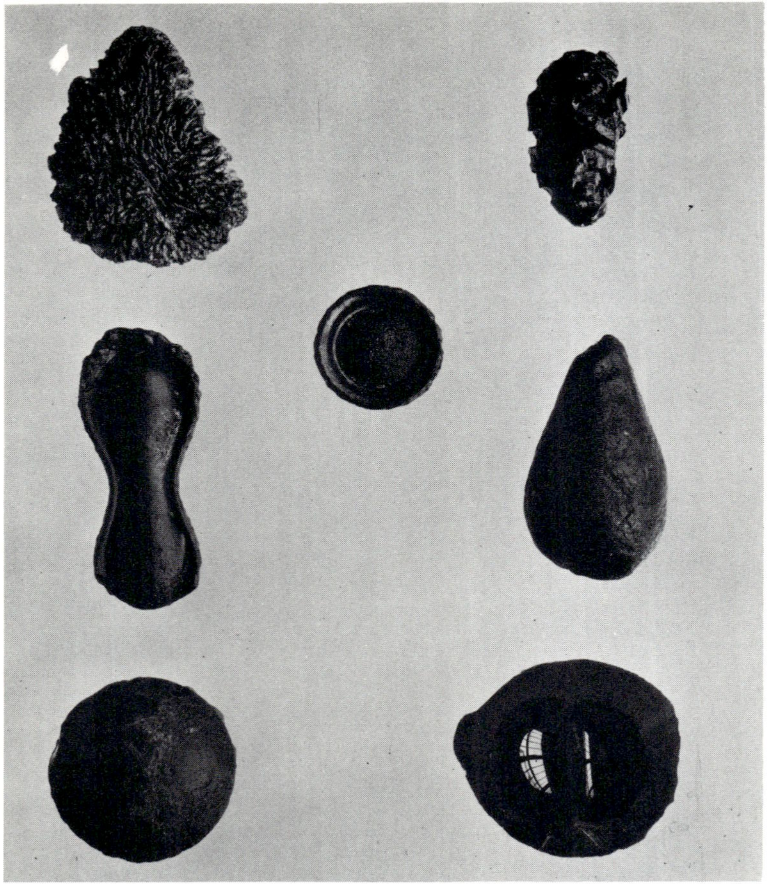

102. Tektites.

103. Shatter cones in mississagi quartzite along the southern margin of the Sudbury Basin, Canada.

when presumably there were a large number of fragments with considerable orbital eccentricities and inclinations, the trend must have been opposite, many more objects being transferred from the asteroid region to the comet cloud than vice versa.

The present data about minor-planet orbits will probably contain no clue by which we could estimate how large a fraction of the fragments once present have escaped by planetary perturbations. This would certainly depend mostly upon the unknown original distribution of the orbital elements. It may well have been a very large fraction. For, the aggregate mass of the present minor planets is very much smaller than would have been expected if there had once been a mass of the order of that of an ordinary planet in the open space between Mars and Jupiter. A reasonable estimate for the mass of all the asteroids, known and unknown, would be $1/1000$ of the earth's mass. It is tempting to suppose that the mass has indeed at the outset been of a dimension comparable to that of an ordinary planet, but that the large majority of the asteroids that were formed from this mass has escaped, the only ones that remained being those whose orbits happened to have small eccentricities and small inclinations.

The difference in general appearance between minor planets and comets can certainly not be taken as an indication of a difference in origin. The fragments that we now call comets having disappeared from the regions near the sun soon after their birth, and having passed their entire further existence at distances where the sun's radiation can have had little or no effect, must have kept the larger part of the gases that were included within them at their origin. The fact that comets can apparently keep up their gas emission until they have got into short-period orbits indicates that it is legitimate to assume that, during the comparable period that would have elapsed between their birth in the asteroid region and their transfer into the cloud of comets, they would not have lost all their volatile constituents. The objects, however, that have remained in the region between Mars and Jupiter will since long have lost their capacity for emitting gases.

TEKTITES

Tektites (Greek tektos—melted) are not meteorites and for that reason we have left them to a short separate section. They consist of a silica rich obsidian glass[49] distinct from terrestrial obsidian. The tektites have the broad composition 70–80 per cent SiO_2 and are found in eight localities from which they derive their names. By far the greater numbers come from the Philippines, Australia, Indo China and Moldavia.

The broad geographical groups are:

Australites	from Australia and Tasmania
Bediasites	from Texas

[49] *Obsidian.* A volcanic glass of granitic composition, generally black with vitreous lustre and conchoidal fracture; occurs at Mt. Hecla in Iceland, in the Lipari Isles, and in the Yellowstone Park, U.S.A. A green silica glass found in ploughed fields in Moravia is cut as a gemstone and sold under the name *obsidian.* True obsidian is used as a gemstone and is often termed *iceland agate.*

The solar system

Billitonites	from Billiton Island, Dutch East Indies
Indochinites	from Indo China
Ivory Coast tektites	from West Africa
Javaites	from Java
Malaysianites	from Malaysia
Moldavites	from Moravia and Bohemia
Rizalites	from Luzon Island, Philippines

The mean chemical composition of these groups is given below with their refractive index and density.

	SiO_2	Al_2O_2	FeO	MgO	CaO	Na_2O	K_2O	TiO_2	MnO
Moldavites	79·60	11·02	2·45	1·30	1·92	0·51	2·90	0·80	0·11
Bediasites	75·64	14·59	4·37	1·29	0·05	1·36	1·85	0·81	0·01
Australites	73·33	12·75	4·70	1·98	3·01	1·25	2·05	0·67	0·16
Indochinites	73·14	12·48	5·02	2·00	2·51	1·45	2·40	0·94	0·13
Javaites	71·83	11·89	5·24	2·79	2·77	1·76	2·22	0·77	0·17
Philippinites	71·21	12·57	5·18	2·90	3·19	1·52	1·93	0·89	0·11
Ivory Coast tektites	71·05	14·60	5·62	3·29	1·67	1·71	1·53	0·73	0·08
Billitonites	70·50	12·53	5·84	3·07	3·14	1·75	2·28	0·84	0·16

	Refractive index	Mean density
Moldavites	1·480–1·496	2·337
Bediasites	1·488–1·512	2·374
Australites	1·498–1·520	2·410
Indochinites	1·506	2·427
Javaites	1·509	2·442
Philippinites	1·511–1·513	2·442
Ivory Coast tektites	1·499–1·518	2·451
Billitonites	1·513	2·456

With the exception of moldavites, which are dark green, the tektites are usually jet black in reflected light. Their form is varied and remarkable since they exhibit a morphology associated with an aerodynamic event at some stage of their formation. Baker[50] has explained their unusual form by appeal to a primary phase in which the buttons, spheres, oblate and prolate spheroids, dumb-bells and apiods were formed in an extra-terrestrial environment with subsequent modification by aerodynamical ablation during high speed flight and kinetic heating subsequent to arrival on the Earth and general erosion.

Chapman[51] has shown from wind tunnel experiments that the australites must have entered the upper atmosphere at a velocity of 12–13 km./sec. and is much persuaded that they may have originated from the Moon. But the origin of tektites is presently unresolved and we will content ourselves with an excellent analysis given by no less an authority

[50] Baker, G., *Mem. Nat. Museum Victoria*, *23*, p. 5–313.
[51] Chapman, D. R., *Nature*, *188*, 353 (1960).

than Urey,[52] which although it is now thirteen years old does not appear to have suffered very much from intellectual erosion. I have taken the liberty of condensing it a little.

If a large mass of glassy material arrived and broke up in the atmosphere, the parts should have distributed themselves over an area some tens of kilometres in dimension, as is observed for certain meteorite showers. The distribution over all southern Australia, for example, would be impossible. Hence, the compact-mass arrival is excluded. If the tektites arrived as a swarm, its mean density must have been greater than 10^{-6} gm. cm.$^{-3}$, for otherwise solar gravitational forces at the Earth's distance from the Sun would break up the swarm and tektites would become distributed over the entire Earth. Yet a swarm of this density some 10^8 cm. in diameter would pile up tektites to a depth of 100 gm. cm.$^{-2}$ over southern Australia, and this is not the case. Arrival from the Moon raises similar objections. I know of no answer to these arguments.

The approach of the comet Arend-Roland (1957) stimulated a little further thought on this problem. What would happen if a comet collided with the Earth? A comet head consists of a very loose aggregate of small particles, and recently it has become evident that these particles consist of, or are mixed with, chemical compounds having large amounts of energy. In fact, the material is a high explosive which 'burns' quietly under excitation by high-speed particles from the Sun only because of its very loose structure and low density. Since comets move on nearly parabolic orbits, their velocity at the Earth's distance from the Sun is 42 km. sec.$^{-1}$. The heads are estimated to be 10–70 km. in diameter and their density may be 0·01 gm. cm.$^{-3}$. The kinetic energy of a spherical object 10 km. in diameter of this density and velocity would be 5×10^{28} ergs, which is equivalent to 5×10^7 atomic bombs of the old-fashioned fission type, or a half-million hydrogen bombs. The chemical energy would only be a fraction of this, perhaps 10^{27} ergs, or even somewhat more.

If an object of this kind entered the atmosphere at this high velocity, compression and heating of the material would occur. It would explode in a chemical sense, and most of its mass would be converted to gases at high temperature. The amounts of silicates expected to be present would be volatilized or scattered as fine dust. Such compaction would be effective at 60–100 km. above the Earth's surface and the explosive reaction should begin at this altitude. The high-temperature mass would continue to move towards the Earth, heat its surface quite easily to the melting point, produce a very compressed region of gas, and this would propel terrestrial material in all directions at high velocity. It is difficult to predict the type of crater to be expected. However, the explosion would resemble the high air burst of an atomic or hydrogen bomb; but the effects would be very much greater and somewhat different because of the great mass of gas both from the detonation of the comet head and from the compressed and heated air, and the much lower temperatures distributed throughout the mass. One would expect the general effect to be more that of a propellant than a detonating type of

[52] Urey, H. C., *Nature*, *179*, 556 (1957).

explosion. The mass would be stopped within one second and the maximum pressure would be about 40,000 atmospheres. Probably a very broad area would be involved and no deep penetration in a limited area would result.

If they are a by-product of a cometary collision, several puzzling features of tektites are explained. (1) Their material substance comes from the surface of the Earth, and hence varying compositions are to be expected; even the nearly pure silica of Libyan glass, and in fact the americanites, which are not usually included as respectable tektites because of their obsidian-like composition, could also owe their existence to this process acting upon granitic rock. (2) The high temperature required for melting is provided. (3) They were melted quickly and briefly and hence did not lose their alkalis, which are known to be very easily volatilized from melts. Also the presence of melted quartz particles, that is, lechatelierite, is understandable. (4) A mechanism for scattering over wide areas is available. (5) The flow marks could have been produced by the blast of hot gases over more slowly moving objects. (It has always appeared surprising to me that a small glass object, 1–2 cm. in diameter, would travel at high velocity through the atmosphere, keeping one orientation as is necessary in the case of the more 'button-shaped' tektites. A blast of high-temperature gas could produce this effect very quickly, and seems a more probable mechanism.) (6) Since more basic materials will crystallize rapidly even during cooling, it can be expected that only very acidic tektites will remain glassy and thus be observed. (7) The residual crater may be very difficult to identify; but it might well be looked for while keeping some flexible ideas as to what its properties may be.

Note: Meteors from Comet P/BIELA. Dr. L. Kresak of the Astronomical Institute, Slovak Academy of Sciences records (IAUC 2362, B.A.A. Circular 537) that: The present orbit of P/Biela (Marsden and Sekanina 1971 Astron J. 76, also IAUC 2347) indicates that the part of the meteor stream moving in the vicinity of the comet may encounter the Earth with appreciably different characteristics from those observed at the past great displays. The nearest approach of the Earth to the perturbed comet orbit occurs 12 days earlier, and the apparent radiant is situated as much as 20° south of the earlier position. The predicted radiant coordinates and no-atmosphere velocity are: $\alpha = 26°.2 + 0°.13\ t$, $\delta = + 24°.6 + 0°.13\ t$, $V\infty = 20.0 + 0.20\ t$ (km/s), where t is measured in days from the date of maximum, 1971 Nov 17.0 U.T. At the point of closest approach the earth passes 0.05 AU from the comet orbit, and it is within 0.10 AU from November 6 to December 1; this distance is smaller than for many major meteor showers associated with other comets though still much larger than for the past great Bielid (Andromedid) showers. Although there is very little hope for a conspicuous display, the possible presence of a diffuse component of the stream (Southworth and Hawkins 1963, *Smithsonian Contr. Astrophys.* **7**, 261) makes observations desirable. The predicted maximum coincides with that of the Leonid meteors, except that the radiant culminates 8.4 hours earlier.

Chapter 12

The scale of the solar system and the law of planetary and satellite distances

Do not the same magnitudes appear larger to your sight when near, and smaller when at a distance? They will acknowledge that. And the same holds of thickness and number; also sounds, which are in themselves equal, are greater when near, and lesser when at a distance. They will grant that also. Now suppose happiness to consist in doing or choosing the greater, and in not doing or in avoiding the less, what would be the saving principle of human life? Would not the art of measuring be the saving principle; or would the power of appearance? Is not the latter that deceiving art which makes us wander up and down and take the things at one time of which we repent at another, both in our actions and in our choice of things great and small? But the art of measurement would do away with the effect of appearances, and, showing the truth, would fain teach the soul at last to find rest in the truth, and would thus save our life. Would not mankind generally acknowledge that the art which accomplishes this result is the art of measurement?
PLATO

Now listen to the rule of the last inch. The realm of the last inch. The job is almost finished, the goal almost attained, everything possible seems to have been achieved, every difficulty overcome—and yet the quality is just not there. The work needs more finish, perhaps further research. In that moment of weariness and self-satisfaction, the temptation is greatest to give up, not to strive for the peak of quality. That's the realm of the last inch—here the work is very, very complex but it's also particularly valuable because it's done with the most perfect means. The rule of the last inch is simply this—not to leave it undone. And not to put if off— because otherwise your mind loses touch with that realm. And not to mind how much time you spend on it, because the aim is not to finish the job quickly but to reach perfection.
SOLZHENITSYN

The distance of the Earth from the Sun is fundamental to any understanding of the solar system and indeed, it enters into almost any calculation of distances, of masses, of sizes and densities, either of planets or their satellites. The same is true when we come to study the more remote stars.

In all ages the Sun's distance from the Earth has been the object of enquiry and in the last two centuries it has attracted the attention of astronomers owing to the inherent difficulties of determining it to an accuracy of one part in a thousand. The Sun is a difficult object for measurement by the well known system of taking its horizontal parallax since under the best conditions its large size and the effect of heat on the instruments used reduces the accuracy of the measurement. If the distance of the Sun could not be determined otherwise than by direct observation it would be a vain hope to expect an accurate answer. Fortunately, Kepler's life work had shown the harmony of the solar system and in his famous third law governing the motions of the planets, which he expressed in the following terms: 'The periodic times of any two planets are to each other exactly as the cubes of the square root of their median distances' lay the possibility of using the measured distance of any single object in the solar system from which to infer the distance to the Sun. The periodic times that the planets take to revolve about the Sun can be measured with little difficulty hence their relative distances can be inferred and one single distance suffices to fix the scale of the entire system.

In earlier times Aristarchus of Samos attempted to obtain the distance of the Earth from the Sun by measuring the angle subtended by the Earth when the Moon was seen at first quarter. He measured the angle Moon Earth Sun when the Moon's face was half illuminated and obtained a value of $87°$, and then calculated that the Sun was nineteen times as far from the Earth as the Earth from the Moon: a result that was too small by a factor of twenty. The angle he measured, viz. $89°51'$, was in truth close to a right angle, and this would place the Sun at an infinite distance and consequently puts the suitability of this method in serious doubt.

Ptolemy, using the observations of Hipparchus for eclipses of the Moon, determined the Moon's distance with considerable precision, and then, combining this value with the researches of Aristarchus, obtained a distance of the Earth from the Sun of 5×10^6 miles; a result that was widely accepted up into the sixteenth century. This value stood unchallenged until Kepler, from his labours with the vast sets of observations gathered so assiduously by Brahe, came to the conclusion that the distance from the Earth to the Sun (the astronomical unit a.u.) must be much greater than 5×10^6 miles. He estimated that the distance was between 3,500 and 4,000 times the Earth's radius, i.e. in the range 14×10^6 to 16×10^6 miles; a distance now seen to be in error by a factor of between six and seven.

Huygens (*c.* 1660) by an ingenious method based on false premises concerning the size of the Earth and the use of Kepler's third law, constructed a plan of the solar system with the Sun at 100×10^6 miles from the Earth; a singularly satisfactory figure in view of the earlier distances arrived at by other philosophers.

The two planets that appeared to offer the best scope for measuring the Sun's distance by *direct* observation were Mars and Venus. Cassini estimated the distance to Mars in 1672 by parallactic methods and placed the Sun at between 82,000,000 and 91,000,000 miles, a value accepted for about a century.

The utility of the transits of inferior planets in furnishing an accurate method for determining the solar parallax was first pointed out by James Gregory, the mathematician, in his *Optica Promota* published in London in 1663. It is usual to ascribe this idea to Halley, but this is a mistake, since Halley's attention was first directed to the subject at the transit of Mercury in 1677; that is to say fourteen years after the publication of Gregory's work. In fairness to Halley, it must be argued that it was his weight of authority nevertheless which induced the British Government to finance the voyage of Cook in 1768 for the observation of the transit of Venus in 1769. Halley also noted with clarity the main difficulty of the method, which lay in ascertaining the longitudes of the stations from which the transit times of contact would be observed. Hence he insisted on comparisons of the *observed duration* of transit reckoned from first to second internal contact as *seen from different stations*. Some stations were chosen as far north as possible, others as far south so as to displace Venus to the south and north of the Sun's disc. This was the genesis of the voyage of 1768, the bicentenary[1] of which was celebrated recently. In view of the wide acclaim for Cook's achievements we will deal with his observations in a little more detail.

To reach Tahiti[2] in good time Cook crossed the line on October 25 and on November 13 put in at Rio de Janeiro for provisions. Here the suspicious envoy was acquainted of Cook's mission to the southward *'to observe the transit of Venus, a very interesting object to the advancement of navigation of which phenomenon he (the envoy) appeared to be totally ignorant'*. Cook then doubled the Horn in clement weather *'with as much ease as if it had been the North Foreland on the Kentish coast'* and stood off Tahiti on April 10, 1769. *Endeavour* came to anchor in the superb bay of Port Royal three days later on the 13th. The next few weeks were devoted to exploration and the building of a fort. On the first day of June the instruments were mounted and the morning of the 3rd was awaited with no little excitement. Cook's personal account from Admiralty Records reads:

---

[1] See the catalogue of the British Museum, *An exhibition to commemorate the bicentenary of Captain Cook's first voyage round the world*, King's Library, 19 July–27 October 1968.

[2] Tahiti (Society Islands) called by the natives Otaheite, discovered by Pedro Fernandez Quiros 1607; re-discovered by Samuel Wallis 1767 and named King George III's Island. Visited by Louis de Bougainville 1768, named La Nouvelle Cythère; visted by Cook, April 1769.

The solar system

The day proved as favourable to our purpose as we could wish; not a cloud
was to be seen the whole day, and the air was perfectly clear: so that we had
every advantage in observing the whole of the passage of the planet Venus
over the Sun's disk. We very distinctly saw an atmosphere, or dusky shade,
round the body of the planet, which very much disturbed the times of the
contact, particularly the two internal ones. It was nearly calm the whole day,
and the thermometer, exposed to the sun about the middle of the day, rose
to a degree of heat we have not before met with.

Charles Green, the coadjutor of Dr Bradley, the Astronomer Royal,
assisted Cook is his observations of the actual transit and found the times
of external and internal contact to be as follows:

Morning $\begin{cases} \text{The first external contact, or first appearance of Venus} \\ \text{on the Sun, was 9 hours 25 min. 4 sec.} \\ \text{The first internal contact, or total immersion, was 9} \\ \text{hours, 44 min. 4 sec.} \end{cases}$

Afternoon $\begin{cases} \text{The second internal contact, was 3 hours 14 min. 8 sec.} \\ \text{The second external contact, was 3 hours, 32 min. 10} \\ \text{sec.} \end{cases}$

Latitude of the observatory, $17° \ 15' \ 29''$ S.
Longitude, $149° \ 32' \ 30''$ W. of Greenwich.

Unfortunately the readings obtained by the other observers differed
from those of Cook more than had been expected and the event proved
to be inconclusive in yielding the information that was desired.

The history of the transits of the planet Venus across the face of the
Sun are of singular interest. The orbit of Venus about the Sun is not in
the plane of the ecliptic, indeed her inclination is $3° \ 23' \ 39''·6$ and the
Earth passes through the line of nodes about June 7th or December 7th
of each year, the date of actual passage changing by about a day later in
each successive century. Thirteen revolutions of Venus in her orbit are
pretty nearly equal to eight of the Earth and accordingly Venus transits
the Sun's disc in pairs eight years apart. After such a pair of transits there
is an interval of more than a century (to be exact an interval of $105\frac{1}{2}$ or
$121\frac{1}{2}$ years) before another pair of transits, and these take place at the
opposite node. The period of 243 years gives an extremely close
approximation to recurrence of the event. December transits occured
in 1631, 1639, 1874, 1882 and will occur in 2117 and 2125. This series
continues unbroken, according to mathematicians, for more than a
thousand years from the present age. June transits occurred in 1761,
1769 and will occur in 2004 and 2012 until 2733 when to our eyes,
Venus just misses the Sun's disc, to cross it again centrally in 2741. We
are indebted to Kepler[3] for asserting that although Venus might some-

[3] Kepler, *Ad Vitellionem Paralipomena*, 1604 (B.M.C. 75 c 13).

times be interposed directly between the Earth and the Sun, the instances are very rare. He announced prematurely that no transit of the planet would occur during the whole of the seventeenth century, but upon completion of his celebrated Rudolphine Tables in 1627 he was enabled to predict the transit of 1631 and to announce that in his view no other transit would occur before the year 1761. In this the illustrious mathematician was wrong and it was left to two young Englishmen cultivating the art of astronomy in the seclusion of Lancashire to be the witnesses of an event that was not known to any other person living. The transit of 1639 was computed by Jeremiah Horrocks and William Crabtree. Horrocks' achievement is one of the most celebrated in practical astronomy, (he died in his twenty-second year; 1619–41) for not only did he compute the event but he contrived to observe it with the slender equipment at his disposal. Returning to his apartment from divine service (the event fell on a Sunday)

and having closed the windows against the light I directed my telescope previously adjusted to a focus, through the aperture toward the Sun I received his rays at right angles upon the paper. The Sun's image exactly filled the circle and I watched carefully and increasingly for any dark body that might enter upon the disc of light . . . About fifteen minutes past three in the afternoon 24th Nov. OS. (4 Dec. N.S.)[4] . . . I then beheld a most agreeable spectacle, the object of my sanguine wishes, a spot of unusual magnitude and of a perfectly circular shape, which had already fully entered upon the Sun's disc to the left.

In this way Horrocks records[5] his labours and became the first mortal to see this sublime event.

The results of the transit of 1769 gave the first fairly reliable determination of the Sun's distance from the Earth by yielding a parallax of some eight seconds of arc. Encke later derived the value of $8''\cdot5776$ from a close examination of the results of the transits of 1761 and 1769 and this was the value used in the Nautical Almanac from 1836–69. The method lost favour however at the carefully observed transit of 1874 when the inability of defining the internal and external contact times owing to what has descriptively been called the black drop phenomenon upset many of the results. Nothing seemed more simple than the precise measurement of the exact instants when the small dark disc of Venus would be in contact with the limb of the Sun's bright disc, but instead of a clean meeting and parting of the two discs a clinging evanescent black drop obtruded and vitiated all the results. It is doubtful whether Venus will ever again be used to find the Sun's distance. I have already

[4] Old Style calendar, relinquished by Great Britain in 1752 by omission of eleven days September 3 being reckoned September 14.
[5] Much to England's shame Horrocks's *Venus in Sole Visa* was forgotten for twenty years. It was Huygens who happened to be in London who caused it to be sent to Hevelius for publication in Dantzig in 1662. An excellent translation into English by Whatton appeared in 1859.

said that Encke reduced all the anomalous results of the earlier observed transits and his work in 1882 put the Sun at 95,250,000 miles from Earth. Prof. Grant says of Encke's work

when we consider the ingenuity of the method employed . . . and the refined nature of the process . . . we cannot refrain from acknowledging it to be one of the noblest triumphs which the human mind has ever achieved in the study of physical science.

This value derived by Encke was not to satisfy future enquiry. In 1854 P. A. Hansen detected an anomaly from his study of the Moon's motion and postulated that his lunar theory could only be reconciled by the distance of the Sun from the Earth being reduced appreciably.

It was agreed in conference at Paris in 1896[6] to accept the value of solar parallax at 8″·8 and in conjunction with Hayford's value of 6378·388 km. (3963·339 m.) for the Earth's equatorial radius this gives an astronomical unit of 149,504,000 km. (92,897,000 miles).

Sir Harold Spencer Jones in 1930 working on the Moon's motion from 1672 to 1908 derived a value of 92,940,000 miles with an uncertainty of ± 40,000 miles and from the motions of the minor planet Eros derived, some ten years later, the improved value 93,005,000 miles ± 9,000 miles (approximately 149,670,000 km.). This corresponds to a parallax of 8″·790 ± 0″·001. Rabe, working between 1926 and 1945 on comparisons of the mass of the Sun with that of the Earth/Moon system, found a parallax of 8″·7984 ± 0″·0006 or an astronomical unit of 149,530,000 km. The most modern technique for obtaining the distance to the Sun lies in sending a radio wave to a planet and receiving the reflected signal. This requires an accurate knowledge of the velocity of the electromagnetic radiation and yields a relation between the astronomical unit and the kilometre. The first successful attempt was made with the inferior planet Venus in 1961. The resultant value of the astronomical unit was between 149,596,000 and 149,601,000 km. This value with some vexation lay between the values of Jones and Rabe, achieved by their very different methods. The difficulties were happily resolved by the discovery of a conceptual error in the analysis performed by Rabe on the 1926 to 1945 figures. It was accepted at Hamburg in 1964 that the distance is 149,600,000 km. (92,957,209 miles) and the solar parallax[7] 8″·79405. The combined mass of the Earth/Moon system is 1/328,899 when compared with the mass of the Sun as unity. Analysis of trajectories of Ranger probes yield an astronomical unit (a.u.) of 149,597,700 ± 400 km. The matter of the astronomical unit is still not

6 Something of the history of the solar parallax is to be found in *Quat. J. Roy. Astron. Soc.*, 5, 23 (1964); 6, 70 (1965).
7 Solar parallax: If π'' is the parallax in seconds, and R is the Earth's equatorial radius the mean distance to the Sun is $206265R/\pi''$. π'' has been determined by several independent methods. The best value to date is 8″·79405, putting the Sun at 92,957,209 miles from Earth.

wholly settled and we will close with a pertinent and valuable contribution by Schubart and Zech.[8]

The dynamical method of determining the astronomical unit is based on a relation between the astronomical unit A expressed in km. and the ratio m of the mass of the Earth/Moon system to the solar mass. According to this relation $m\,A^3$ equals a constant which is known with sufficient accuracy for all practical purposes. The effects of the attraction of the Earth/Moon system on bodies in close approaches permit the determination of m and thus of A. Various authors have tried to determine the mass of the Earth/Moon system from observations of the Minor Planet (433) Eros which comes near to the Earth. The first determination of this kind was performed by Witt and the most recent one was published by Rabe[9]. An independent determination of the astronomical unit became possible using radar echoes from Venus, but the radar result differs significantly from Rabe's result. Considerable effort was made by Eckstein[10] and Marsden[11] to reconcile the two differing values using numerical material previously published by Rabe.

Recently, Schubart tried to correct the orbit of the Minor Planet (1221) Amor which approaches the Earth as close as 0·10 a.u. He used 227 observations (obtained from 1932 to 1964) to derive the equations of condition without formation of normal positions. For the computation of the values $O-C$, a reference orbit was used which was based on Newcomb's value of the mass of the Earth. It turned out that after the differential correction of the orbit, the representation of the observations was unsatisfactory. When in a new attempt the correction of the mass of the Earth/Moon system was added as a seventh unknown to the corrections of the six orbital elements of Amor, a corrected value of the mass resulted, which was in fair agreement with the radar value. The representation of the observations of Amor is still not quite satisfactory, which may result from neglecting the direct attraction of Mercury in the computations, or from the uncertainty of other constants used in the numerical integration of the reference orbit.

The successful application of this method to Amor suggested a revision of the results obtained by Rabe. Zech applied this method to Eros. For this purpose he made use of the thirty-seven normal positions from 1926 to 1945 given by Rabe. In 1931 Eros made an approach to the Earth with a minimum distance of 0·17 a.u. The result for the reciprocal mass of the Earth/Moon system in units of the solar mass is $m^{-1} = 328894 \pm 30$ (mean error). This is in excellent agreement with Schubart's result obtained from Amor ($m^{-1} = 328895$) and with the radar results. On the basis of the radar results the true value of m^{-1} is believed to lie between the limits 328890 and 328922.[12] The mean error found by Zech is only about half as large as Rabe's mean error (± 64). A preliminary comparison of the coefficients of the equations of condition derived by Rabe and by Zech shows strong differences in the case of the coefficients for the correction of m^{-1}. Almost all the coefficients of m^{-1}

[8] *Nature, 214*, 900 (1967).
[9] Rabe, E., *Astro. J., 55*, 112 (1951).
[10] Eckstein, M. C., *Astro. J., 68*, 231 (1963).
[11] Marsden, B. G., *Bull. Astro., 25*, 225 (1965).
[12] *Trans. Intrn. Astro. Union, 12* B, 593 (1966).

published by Rabe turn out to be considerably greater in absolute value as compared with Zech's result. It can be assumed that the different result for m^{-1} found by Rabe is caused by the deviations in the coefficients. This assumption is supported by the following fact: if Rabe's coefficients of the equations of condition are assumed to be correct in case of the six orbital elements of Eros, if the coefficients for m^{-1} are replaced by Zech's values, and if the right-hand sides of these equations are left as in Rabe's paper, the least-squares method applied to these equations for seven unknowns leads to the result $m^{-1} = 328874 \pm 31$. Our results demonstrate that it is possible to reconcile the dynamical value of the astronomical unit with the radar measures.

The numerical computations were performed on the IBM '7094' of the Deutsches Rechenzentrum, Darmstadt, and on the Siemens '2002' computer of the Astronomisches Rechen-Institut, Heidelberg.

Those persons who marvel mainly at the lack of accuracy in the results may care to consider the remarks of Sir John Herschel in 1858. 'The superficial reader may think it strange and discreditable to science to have errors by nearly 4 million miles (the then new determination brought the Sun nearer by that amount). But such may be reminded that the error in the Sun's parallax on which the correction turns, corresponds to the apparent breadth of a human hair at 125 ft.'

As we have already seen, in view of Kepler's third law, once the distance of any one body in the solar system is known the scale of that system is given and the distances to *all* the other bodies within its compass are automatically supplied; a feature of such beauty that we should pause to savour its true sublimity. This leads us to a consideration of a remarkable pseudo-series of numbers that has had a profound effect upon many astronomers, mathematicians and philosophers since Titius and Bode made it known in the late eighteenth century. The pseudo-series to which I refer is now generally known as the Bode-Titius 'law' or more correctly the Bode-Titius relation in which each term is double the preceding term (with the exception of the first two terms, where the zero should be 1·5 for the perfection of the series).

The relation is most simply shown below:

0. 3. 6. 12. 24. 48. 96. 192. 384. 768.

The originator of the relation is Johann Daniel Titius of Wittenberg who announced it in 1772. Bode[13] appears to have been responsible for giving the relation general currency and making it clear that the proportionate distances of the several planets of the solar system from the Sun may be represented by adding four to each term of the series to give:

4. 7. 10. 16. 28. 52. 100. 196. 388. 772., etc.

[13] Bode, Johann Elert, 1747–1826, founder of the well-known *Astronomisches Jahrbuch*, fifty-one volumes of which he compiled and issued; one-time director of the Berlin Observatory. See Poggendorff, *Biogliterarisches Handwörterbuch*, Allgemeine deutsche Biographie iii, 1.

If now we divide each term by ten we have

0·4. 0·7. 1·0. 16·0. 2·80. 5·20. 10·0. 19·6. 38·8. 77·2.

If we assign the first term to that of Mercury's distance from the Sun, the second to Venus, the third to the Earth we note that the distance of the Earth to the Sun is unity, i.e. one astronomical unit (1 a.u.) and consequently all the other terms then conveniently give the distances of the planets in terms of the astronomical unit.

When Bode announced this in *c.* 1772 the distances of the known planets (out to Saturn) satisfied the series remarkably well with one unfilled orbit between Mars and Jupiter. The position, using modern values for the planetary distances, is shown succinctly in the table below.

Planet	Distance in a.u.	Bode's number
Mercury	0·3870987	0·4
Venus	0·7233322	0·7
Earth	1·0000000	1
Mars	1·5236915	1·6
—	—	2·8
Jupiter	5·2028039	5·2
Saturn	9·5388437	10·0

We have already seen that the lack of a planet between Mars and Jupiter was highlighted by the series[14] and that Bode and his contemporaries were encouraged by it to search for the missing planet.

The tenuous nature of their enterprise when viewed in cold epistemological terms raises profound difficulties, yet the history of science is able to record a remarkable success for the empirical approach of Bode—for instead of a single planet they found a celestial cohort. Polanyi[15] reminds us that the quixotic attack of the young Hegel on the empirical methods of science and his swift defeat at the hands of the scientists was one of the great formative experiences of modern science. When de Zach and his band of observers in 1800 set out to find the missing planet required by the Bode discontinuity, Hegel[16] poured scorn on an enquiry that was following a numerical rule which, being meaningless, could only be accidental. Arguing that nature, shaped by immanent reason, must be governed by a rational sequence of numbers, he postulated that the relative spacing of the planets must conform to the Pythagorean series[17] 1. 2. 3. 4. 9. 16. 27., which allowed for precisely what the planetary system exhibited, namely a large gap between the fourth and fifth term of that series. Thus any attempt to find an eighth planet was chimerical. In the event Hegel was greatly discomfited, but we are still

---

[14] See chapter 11, page 193.
[15] Polanyi, M., *Personal Knowledge* (London 1958).
[16] Hegel, *Dissertatio philosophica de Orbitis Planetarum*, 1801, *16*, 28 (Werke, Berlin 1834).
[17] Hegel substituted 16 for 8. The true Pythagorean series is 1. 2. 3. 2^2. 3^2. 2^3 3^3.,
i.e. 1. 2. 3. 4. 9. 8. 27.

not clear whether Bode's relation is the outward sign of some darkly hid pattern of nature or merely a mathematical augury.

The discovery of Uranus in 1781 confirmed the Bode relation in one further term but the discovery of Neptune and Pluto have shown no harmony with it. We give below the completion of the table above for the trans-Saturnian planets.

Planet	Distance in a.u.	Bode's number
Uranus	19·181882	19·6
Neptune	30·057900	38·8
Pluto[18]	39·75000	77·2

We should, in discussing the profound influence of the Bode relation, not fail to recall that both Adams and Leverrier made use of it in assigning a distance to the unknown planet they sought by their audacious mathematical attack upon that singularly difficult problem, thereby to reduce a little the colossal labour involved.[19]

In an attempt to develop a satisfactory model for the origin of the solar system von Weizsäcker[20] has shown from dynamical considerations that a primitive Sun surrounded by a rotating shell that is transformed into an equatorial disc may form streams of material that lead to the formation of planets placed at distances from the Sun that conform to a relation $D = a + b \cdot 2^n$ which can be shown to agree with the Bode relation, since we are able to write it in the form

$$D = a + b \cdot c^n$$

where a is 4 b is 3 and c is 2 with n for Mercury $- \infty$ and for the other planets in increasing distance from the Sun 0. 1. 2. 3. 4. 5., etc.

Two recent computer programmes are reported that give some insight into the more recondite nature of the Bode relation. Woolfson[21] proposes a model for the solar system in which tides produced on the Sun by a passing star caused portions of the solar material to break away at regular intervals of time, the intervals being governed by the natural frequency of oscillation of the disturbed sun. He is able to explain on the model *inter alia* the lack of fit of Mercury and the origin of the asteroids as the collision between two planets with some harmony in the perihelion distances between those of the model and those observed in nature. Hills[22] notes that the Bode law is obeyed by the inner satellites of Jupiter, Saturn and Uranus as well as by the planets *per se*. He makes a special point, however, of stating that the inital conditions in these satellite systems are not likely to have been the same as those in the

<hr>

[18] We should not attach too much significance to Pluto's distance in that there is some evidence to suggest that it may be a detached satellite.
[19] See chapter 5 (page 85).
[20] Von Weizsäcker, C. F., *Zeit für Astrophys.*, *22*, 319 (1944).
[21] Woolfson, M. M., *Nature*, *187*, 47, and *188*, 1175 (1960).
[22] Hills, J. G., *Nature*, *225*, 840 (1970).

planetary system. This suggests to him that Bode's law may have resulted from a process of dynamical relaxation. That an epoch characterized by strong dynamical encounters occurred before the planets relaxed into stable orbits is, he believes, suggested by the surprisingly large differences between the inclinations of their angular momentum vectors of rotation and those of orbital revolution.

To test his theory eleven planetary systems each with a central star of one solar mass but with widely different planetary mass functions were evolved from random orbits on an IBM 360/67 computer by numerically integrating the equations of motion. Appeal is made to the work of Roy *et al.*,[23] who showed that commensurate orbits are more frequent in the solar system, particularly among the Jovian planets and their satellite systems, than is to be expected from chance alone. The programme discloses a predilection for commensurabilities in a narrow range between 9/4 and 8/3 and a conformation with the Bode relation.

Orowan[24] proposes a genesis of the planets from the welding together initially of heavy metal particles to form a core to which subsequently a brittle silicate mantle is added. If iron and nickel with other heavy metals are the nucleating agents and the thickness depends on the initial local concentration of metallic particles related to the local vapour pressure of iron and nickel in the inital gas cloud, then the temperature of the gas around the Sun would depend on the distance from the centre, but the gravitational potential would be the governing factor and the concentration of the heavy elements would decrease with distance. The constitution of the cloud may in this way inhibit the nucleation of a planet and lay a basis for the Bode relation.

A substitute for Bode's law was presented in a paper of considerable erudition by Miss Blagg[25] in 1913. She makes the following observations:

1. It is a familiar fact that Bode's Law is only approximately fulfilled. Thus the late Professor C. A. Young writes in his *General Astronomy*, Art. 488.

'The resulting numbers, divided by 10, are pretty nearly the true mean distances of the planets from the sun, in terms of the radius of the earth's orbit. In the case of Neptune, however, the law breaks down utterly, and is not even approximately correct.'

Professor Young gives a table in support of this statement, from which the first three columns of the subjoined table are quoted. It will be seen that his special condemnation of the law for Neptune arises from his exhibition of the simple *differences* between the actual distances and those given by Bode. But if we take instead the *ratios*, Neptune become less conspicuous; though still notably discordant. It is not three times as discordant as Venus. Moreover, Mercury is only brought into approximate accordance by tampering with the geometrical progression. The term preceding 0·7 in Bode's Law is really 0·55,

[23] Roy, A. E., and Ovenden, M. W., *Monthly Notices of R.A.S.*, *114*, 232; *115*, 296.
[24] Orowan, E., *Nature*, *222*, 867 (1969).
[25] Blagg, M. A., *Monthly Notices of R.A.S*, *73*, 414 (1913).

The solar system

and if we use this, then the ratio is no longer ·968 but ·704, even less than for Neptune.

FROM PROFESSOR YOUNG'S TABLE

Name	Distance	Bode	Difference	Ratio	Residual deviation from mean
Mercury	0·387	0·4	−0·013	·968	+0·012
Venus	0·723	0·7	+0·023	1·033	+0·077
Earth	1·000	1·0	0·000	1·000	+0·044
Mars	1·523	1·6	−0·077	·952	−0·004
Mean Asteroids	2·650	2·8	−0·150	·947	−0·009
Jupiter	5·202	5·2	+0·002	1·000	+0·044
Saturn	9·539	10·0	−0·461	·954	−0·002
Uranus	19·183	19·6	−0·417	·978	+0·022
Neptune	30·054	38·8	−8·746	·774	−0·182
Mean				·956	

To obviate this defect Professor C. V. L. Charlier has recently suggested (*Astr. Nachr.*, 4623) that Mercury really corresponds in the series

$$d = 0·4 + 0·3 \times 2^n$$

(in which $n = 1$ for the earth, and $n = 0$ for Venus) *not* to the value $n = -1$, but to the value $n = -\infty$; so that in between Venus and Mercury we may look for a large number of small planets with values of n between 0 and $-\infty$: corresponding perhaps to Saturn's ring.

2. It occurred to me, towards the end of last year, to look at the matter from a different point of view. Bode's Law is a modified geometrical series, the constant 0·4 being added to remove the chief deviations from a strictly geometrical series, to which the law tends when n is large. Let us omit this modification and return to the notion of a geometrical series, at least in the first instance. If we take logarithms of the distances (as in the 2nd column of the Table below) the differences ought to be constant.

Planet	Distance	Log	Difference	Theoret.	O—C	Number
Mercury	0·3871	9·588		9·551	0·037	1·089
			·271			
Venus	0·7233	9·859		9·789	0·070	1·175
			·141			
Earth	1·0000	0·000		0·026	9·974	0·942
			·183			
Mars	1·5237	0·183		0·264	9·919	0·830
			·240			
Minor Planets	(2·650)	(0·423)		(0·501)	9·922	0·836
			·293			
Jupiter	5·2026	0·716		0·739	9·977	0·948
			·264			
Saturn	9·5547	0·980		0·976	0·004	1·009
			·304			
Uranus	19·2181	1·284		1·214	0·070	1·175
			·195			
Neptune	30·1096	1·479		1·451	0·028	1·067
Mean		0·501	·236			

Note. The 3rd column shows incidentally the flaws in Bode's Law, which demands that the differences shall increase continuously, approaching more and more nearly to ·301 as n grows larger. The values for 'Distance' are taken from the *Annuaire du Bureau des Longitudes*.

From the 3rd column we see that the range is considerable, the departures from the mean value ·236 amounting to 30 or 40 per cent. Let us, nevertheless, construct a perfectly regular series as in the 4th column, starting with the mean value* of the log (distances), viz. 0·501, for the fifth term, and adding or subtracting a constant each time.

3. But instead of the mean value 0·236 we shall adopt the slightly larger constant difference 0·23742 = log 1·7275, a number to which I was brought by much trial and error.

Miss Blagg derived the formula

$$D = A\,(1\cdot7275)^{n}\quad\{B + f\,(a + n\beta)\}$$

where n is a positive integer, D is the distance of a body from its primary, A, B, $\propto$, β are parameters, having different values for the system of the Sun and the systems of Jupiter, Saturn and Uranus. The function f $(a + n\beta)$ is a curve that gives the deviation of the observed distances from the distances given by a simple geometric series. Roy[26] with considerable insight has shown that the Blagg formula is best tested against the new discoveries in the solar system since 1913 when compared with the predictions made by the law itself. Since 1913 six bodies offer a test: Pluto, Jupiter's satellites IX, X, XI and XII and Uranus' satellite V. All six of these bodies according to Roy's calculations fit the Blagg law closely and in his view it must possess some significance.

Schmidt[27] has derived a law of planetary distance from a model of the nascent solar system in which the Sun captures a cloud of interstellar dust particles. The law is written

$$\sqrt{R_n} = a + bn$$

where R_n is the distance of the nth planet from the Sun, a is $\sqrt{R_0}$ and b is the constant difference between successive square roots.

For the large planets b is unity and a is the square root of the radius of Jupiter. For the terrestrial planets b is 0·20 and a is the square root of the radius of Mercury. The agreement obtained by Schmidt with the observed distances of the planets is given in the table below:

	Distance in a.u. observed	calculated
Mercury	0·3870987	0·39
Venus	0·7233322	0·67
Earth	1·0000000	1·04
Mars	1·5236915	1·49
Jupiter	5·2028039	5·20
Saturn	9·5388437	10·67
Uranus	19·181882	18·32
Neptune	30·057900	27·88
Pluto	39·75000	39·44

[26] Roy, A. E., *J.B.A.A.*, *63*, 212 (1953).
[27] Schmidt, O. J., *The Origin of the Earth* (London 1959).

THE ORBITAL PERIOD RELATION

Dermott[28] has examined modifications of the Bode type law $R_n = R_o\,(C)^n$ and in his opinion the loss of simplicity has never been justified.

For the various systems he writes the following laws:

System	Law
Solar	$R_n = R_o(C_o)^n$
Jupiter	$R_n = R_1(C_1)^n$
Saturn	$R_n = R_2(C_2)^n$
Uranus	$R_n = R_3(C_3)^n$

The constants R_0, R_3 and C_0, C_3 are different for each system and he shows that the most important feature of the distribution of the satellites is the preference for commensurability among pairs of orbital periods. Many pairs of satellites, e.g. Mimas and Tethys, Titan and Hyperion, are in a state of dynamical resonance, the ratios of their orbital periods being closely equal to the ratio of two small integers. This feature, which has been statistically proved to be highly significant, must he feels take precedence over all other features.

If we have $R_n = R_o\,(C)^n$ then because of Kepler's third law ($R^3 =$ constant. T^2, in any one system) this can be written $T_n = T_o\,(A)^n$ where T_o and A are new constants, which are related to the old constants, R_o and C, via Kepler's third law. If R_o, C, T_o, A were just any constants then writing Bode's Law in terms of orbital periods rather than orbital radii would add nothing at all and thus would not be worth doing. But what Dermott found is that writing the law in terms of orbital periods revealed that the constants T_o and A are not arbitrary but that T_p the rotational period of the planet determines the scale of the satellite system and thus that Bode's Law is related to the preference for commensurability.

To summarize, Dermott's achievements are:

(1) to have shown that Bode's Law is related to the preference for commensurability among pairs of orbital periods;

(2) to have shown that the constants in Bode's Law are not arbitrary, in particular it is the rotational period of the primary which determines the scale of the system.

The following four diagrams are self-explanatory (Figs. 38, 39, 40, 41). It may be pointed out for any readers drawn to the Pythagorean mould

[28] Dermott, S. F., 'On the Origin of Commensurabilities in the Solar System I, II and III', *Monthly Notices of R.A.S.*, *141*, pp. 349–61, 363–76; *142*, pp. 143–9.

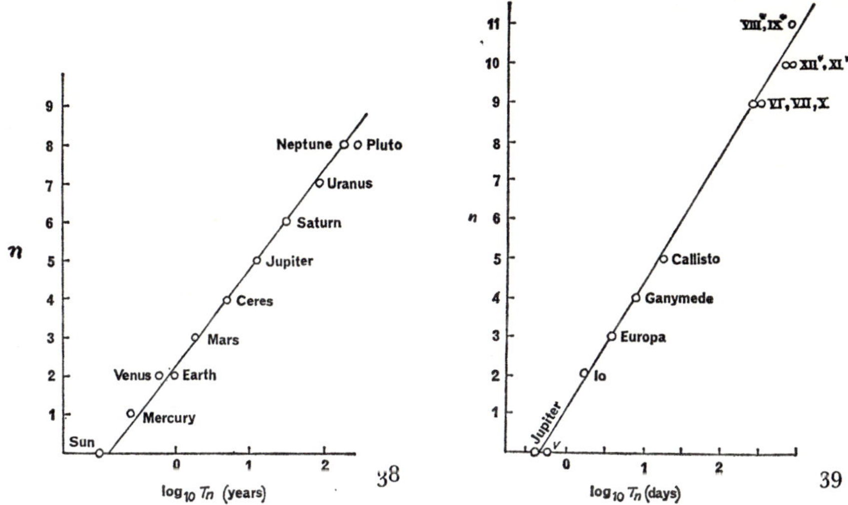

Fig. 38. The orbital period relation $T_n = T_o A^n$ applied to the planetary system. $A^2 = 6$ and $T_o/T_p = 1\cdot442$. The line has been drawn to pass through Jupiter, this being the most massive planet.

Fig. 39. The orbital period relation $T_n = T_o A^n$ applied to the satellite system of Jupiter. $A^2 = 4$ and $T_o/T_p = 1\cdot076$, T_p being the non-equatorial rotational period of Jupiter (this is also shown on the graph). The four Galilean satellites occupy consecutive inner orbitals and lie near the line, the irregular satellites are randomly distributed.

Fig. 40. The orbital period relation $T_n = T_p A^n$ applied to the satellite system of Saturn. $A^2 = 2$ and $T_o/T_p = 1\cdot079$, T_p being the non-equatorial rotational period of Saturn (this is also shown on the graph). The irregular satellites, Iapetus and Phoebe occupy outer orbitals and are randomly distributed ($n = 14\cdot75$ and $20\cdot27$ respectively). These satellites cannot be conveniently shown on the graph.

Fig. 41. The orbital period relation $T_n = T_o A^n$ applied to the satellite system of Uranus. $A^2 = 3$ and $T_o/T_p = 1\cdot076$, T_p being the rotational period of Uranus (this is also shown on the graph). Satellites occupy consecutive inner orbitals and lie near the line, there are no known irregular satellites in this system.

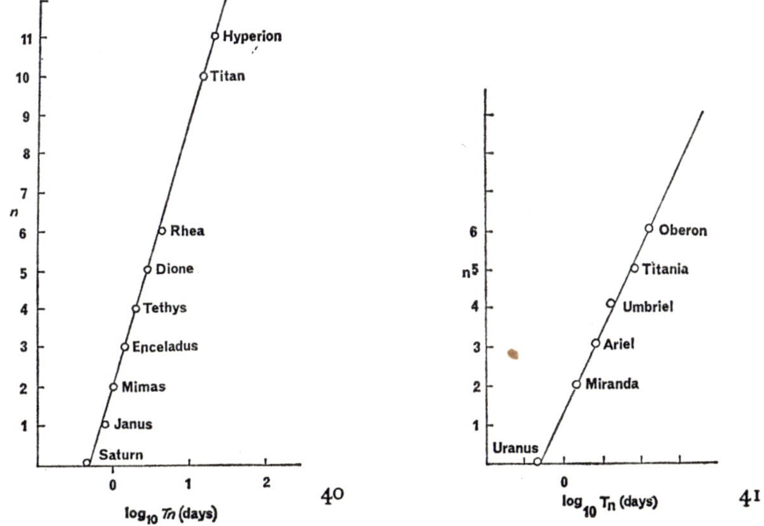

The solar system

that Pythagoras may have interpreted the above[29] as a proof for the existence of the celestial spheres (see chapter 2), the volumes of which are seen to be in a simple numerical ratio and thus in harmony.

[29] For the satellites in any one system we have

$$T_n = 1 \cdot 08 \; T_p \, (A)^n$$

Squaring both sides

$$T_n{}^2 = (1 \cdot 08 \; T_p)^2 \, (A^2)^n. \qquad (1)$$

Let the orbital radius of the nth satellite be R_n.

Kepler's third law states that in general

$$\frac{R^3}{T^2} = \text{constant}.$$

In fact, by Newtonian mechanics we can show that the constant $= \dfrac{GM}{4\pi^2}$, where M is the mass of the planet and G the universal gravitational constant.

$$\frac{R^3}{T^2} = \frac{GM}{4\pi^2}$$

$$\text{or} \quad T^2 = \frac{4\pi^2}{GM} \cdot R^3.$$

Hence, for the nth satellite we have

$$T_n{}^2 = \frac{4\pi^2}{GM} \cdot R_n{}^3.$$

Substituting for $T_n{}^2$ in (1) we obtain

$$\frac{4\pi^2}{GM} R_n{}^3 = (1 \cdot 08 \; T_p)^2 \, (A^2)^n.$$

G, M and T_p are common to all the satellites in the system and therefore

$$R_n{}^3 = \text{constant} \, (A^2)^n.$$

From the orbital period relation, A^2 is always a small integer (I)

$$R_n{}^3 = \text{constant} \, (I)^n.$$

Let V_n be the volume of the sphere of which the radius is R_n, then as

$$V_n = \frac{4\pi}{3} R_n{}^3$$

$$V_n = \text{constant'} \, (I)^n$$

where constant' as a new constant

$$\frac{V_{n+1}}{V_n} \, \text{always} = I$$

The volumes of successive spheres are thus in simple numerical ratio and therefore in 'harmony'!

Chapter 13

The stability of the solar system

Let us now consider what causes the motions of the heavenly bodies. In the first place, if the vast sphere of sky rotates, we must assume that its axis is stabilized and enclosed at both ends by the pressure of extramundane air on each pole.
LUCRETIUS (*c.* 55 B.C.) on '*The movements of the heavenly bodies*'.

For so the God hathe stabeled
The erthe whiche shall not be meued.
MYRR (*c.* 1450), '*Our Ladye*', 297

This subject is one of considerable difficulty, since it is only seen clearly from a deep mathematical insight into the equations of motion of the many bodies that constitute the great concourse of never quiescent matter that we know as the solar system. Further we should with constant humility recognize that we have but a limited knowledge of the behaviour of the system which is currently thought to extend in time back some 5000 million years,[1] for our observations of it, if we extend them to those of Hipparchus (*c.* 146 B.C.), cover but a little over two thousand years, which is a minute fraction of the whole. Today we still do not have precise data on the relativistic, tidal and electromagnetic effects that are essential to a complete analysis of the system. Hagihara[2] has given the most complete treatment of the subject and he places the problem in its anthropocentric perspectives by asking:

Will the present configuration of the solar system be preserved without radical changes for a long interval of time? What can be said about the arrangement of the planetary orbits in the distant past? Were the satellite systems initially very different? A student of the solar system will further note the presence of gaps in the distribution of semimajor axes of the asteroids and in Saturn's rings. Are these gaps due to gravitational causes? Questions such as these constitute aspects of the problem of the stability of the solar system.

The existence of a solution for the three-body problem[3] was proved by

[1] The age of the solar system is put at one-third the age of the Galaxy. The Earth's age is put at 3000 million years. See Professor Holmes's detailed report in *Nature*, *157*, 680 (1946).
[2] Hagihara, Y., *Stability in Celestial Mechanics* (1957).
[3] *Three-body problem.* That problem in classical celestial mechanics which treats the motion of a small body, usually of negligible mass, relative to and under the gravitational influence of two other finite point masses.

The solar system

Sundman, but the solution does not have a form that permits dealing with
questions of stability. At present, celestial mechanics enables us to compute
the positions of the planets and the satellites within the accuracy of modern
observations for a long, but limited, interval of time. The question of the
stability of the solar system is closely related to the form of the solution and
to the behavior of the series employed. The problem can be put as follows:
*What is the interval of time, at the end of which the configuration deviates from the
present by a given small amount?* Present mathematics hardly permits this question to be answered satisfactorily for the actual solar system.

In one recent text that discusses the problem the author presents it and
dismisses it in half a page with a statement of the yearly secular variation
in the mean yearly motion of the planets Mercury, Venus, Earth and
Mars. These are respectively $+ 0''\cdot00000495$
$$+ 0''\cdot00000096$$
$$- 0''\cdot00000403$$
and $+ 0''\cdot00000169$

From this, and in view of the motion of the Earth being $1{,}296{,}000''$ per
year, the secular decrease is seen to be one part in 3×10^{11}. Thus the
tropical year decreases by about 10^{-9} days or 10^{-4} seconds every year.
The author concludes that in 10^{10} years only a relatively small difference
in the Earth's orbit would be apparent.

Newton[4] in *The Principia* observes that the Sun is agitated by a continual motion but never recedes far from the common centre of gravity
of all the planets.

The quantity of matter in the Sun is to the quantity of matter in Jupiter as
1067 to 1; and the distance of Jupiter from the sun is to the semidiameter of
the sun in a proportion but a small matter greater, the common centre of
gravity of Jupiter and the sun will fall upon a point a little without the surface
of the sun. By the same argument, since the quantity of matter in the sun is to
the quantity of matter in Saturn as 3021 to 1, and the distance of Saturn from
the sun is to the semidiameter of the sun in a proportion but a small matter
less, the common centre of gravity of Saturn and the sun will fall upon a point
a little within the surface of the sun. And, pursuing the principles of this
computation, we should find that though the earth and all the planets were
placed on one side of the sun, the distance of the common centre of gravity of
all from the centre of the sun would scarcely amount to one diameter of the
sun. In other cases, the distances of those centres are always less; and therefore, since that centre of gravity is continually at rest, the sun, according to
the various positions of the planets, must continually be moved every way,
but will never recede far from that centre.

Grant,[5] in the beautiful language maintained throughout his celebrated work on physical astronomy, gives an account of the researches

4 *The Principia in Modern English*, Motte's translation, revised by Cajori, p. 419 (California
Press 1947).
5 Grant, R., *History of Physical Astronomy* (London 1852).

of Laplace and Lagrange into certain facets of the question of stability of the system of the world. It is their stability theorems that are still widely used today to show that if two planets revolve in the same direction round the Sun and their orbits have very small inclinations and eccentricities, then the said inclinations and eccentricities will remain small so long as the semi-major axes are not appreciably altered.

'The genius of Lagrange, however, was triumphant in these researches', says Grant, 'and he succeeded in demonstrating, as in the case of the larger planets, that the eccentricities and inclinations would always be confined within very narrow limits. He found that the ecliptic would not be displaced to a greater extent than 5° 23′ by the action of the planets upon the Earth, and that all the planetary orbits would be perpetually comprised within a zone of the heavens whose breadth was 7° 58′. He therefore announced, as the final result of his researches, that the secular variations of the elements were in all cases such as would for ever assure the stability of the planetary system.

The same illustrious geometer extended his researches to the periodic inequalities, which he investigated in two elaborate memoirs communicated by him to the Academy of Sciences of Berlin.[6] He derived the analytical expressions of these inequalities from the periodic variations of the elements, and then computed their numerical values for each planet.

The interesting results obtained by Lagrange relative to the stability of the system were founded upon a knowledge of the masses of the several planets. The computation of the masses of those planets that are accompanied by satellites is not a difficult problem, but it is quite different when the question refers to the other planets of the system. Theoretically speaking, the masses of all the planets may be ascertained by observing the effects of their mutual perturbations, but these effects are generally so very minute that they are almost entirely lost in the errors of observation. Lagrange determined the masses of the planets that have no satellites by combining their volumes with their densities, assuming the latter to vary in the inverse ratio of the planet's distance from the sun. This principle, although naturally suggested by the relative densities of the Earth, Jupiter, and Saturn, was, notwithstanding, gratuitously assumed, and therefore the consequences derived from it could not be altogether free of uncertainty. Lagrange, indeed, shewed the improbability of any minute alteration in the values of the masses affecting essentially the conclusions at which he arrived; but still it was desirable that such valuable truths should be established by an analysis divested of all considerations of a hypothetic character. This important step was made by Laplace.[7] In 1784 he demonstrated that, no matter

6 *Mém. Acad. Berlin,* 1783–4.
7 *Mem. Acad. des Sciences,* 1784.

what might be the relative masses of the planets, the eccentricities and inclinations if once inconsiderable would always continue so, provided the planets were subject to this one condition—*that they all revolved round the sun in the same direction.* This remarkable truth is embodied in two elegant theorems, which the great geometer, just mentioned, was the first to announce to the world. The theorem relative to the oscillations in the form of the orbits may be thus stated: *If the mass of each planet be multiplied by the square of the eccentricity, and this product by the square root of the mean distance, the sum of these quantities will always retain the same magnitude.* Now when this sum is determined for any given epoch, it is found to be small; by the preceding theorem, then, it will always continue so; it follows, therefore, *a fortiori,* that each quantity will continue small, and, consequently, the eccentricity cannot in any case become considerable. The theorem relative to the positions of the orbits is equally elegant. It may be expressed in the following terms: *If the mass of each planet be multiplied by the square of the tangent of the orbit's inclination to a fixed plane, and this product by the square root of the mean distance, the sum of such quantities will continue invariable.* Considerations similar to those we employed in the previous instance enable us to conclude from this theorem that the orbits of the planets will suffer only a very inconsiderable displacement from their mutual attraction.

The laws which thus regulate the eccentricities and inclinations of the planetary orbits, combined with the invariability of the mean distances, secure the permanence of the solar system throughout an indefinite lapse of ages, and offer to us an impressive indication of the Supreme Intelligence which presides over nature, and perpetuates her beneficent arrangements. When contemplated merely as speculative truths, they are unquestionably the most important which the transcendental analysis has disclosed to the researches of the geometer, and their complete establishment would suffice to immortalize the names of Lagrange and Laplace, even although these great geniuses possessed no other claims to the recollection of posterity.

It cannot fail to have occurred to the reader that in these sublime researches the two mighty rivals pressed forward always at an equal pace, insomuch that it would be hardly possible for the most discerning judgment to assign the palm of superiority to either of them. Their investigations of the secular variations were in both cases equally original, and equally entitled to admiration. Laplace's method might be more simple; Lagrange's was more luminous, and had the advantage of being direct. In his researches connected with the mean motion, Laplace displayed a practical sagacity which rarely characterized the speculations of Euler or Lagrange, and, perhaps, this quality was more valuable to him throughout his career than an unexampled command of analysis was to his great rival. In the integration of the differential equations relative to the secular variations of the planets, the genius of Lagrange was eminently conspicuous. Laplace admits that he was compelled to abandon the design of integrating his own equations on account of the difficulties they offered, and was only induced to resume the subject on becoming acquainted with the ingenious method devised for that purpose by his illustrious contemporary.'

These researches in effect introduce the versatile concept of the invariable plane of the solar system, a plane containing the centre of the Sun, which remains absolutely unchanged by any mutual action between the planets and such that forces perpendicular to it mutually cancel out. Laplace expanded the equations of motion of the planets into a set of power series up to the second order of the eccentricities and proceeded to show that the secular terms are really periodic terms of very long period (even millions of years) and could be written as a sine relationship and that the elements associated with them eventually return to the original values. In this sense the solar system is seen to be stable.

Three papers of considerable erudition have appeared in recent years which make an important contribution to the question of stability. Roy and Ovenden[8] show that the occurrence of commensurability between pairs of mean motions of certain bodies in the solar system is more frequent than in a chance distribution. Outstanding examples are Jupiter and Saturn whose mean motions are approximately in the ratio 5 to 2; Tethys and Mimas and Dione and Enceladus where the ratio is 1 to 2, and Hyperion and Titan where the ratio is 3 to 4. In the Jovian system of satellites also not only are the mean motions of Europa and Io, and Ganymede and Europa, very nearly in the ratio 1 to 2 but they are connected by the Laplacian relation

$$n_1 - 3n_2 + 2n_3 = 0$$

where n, n_2 and n_3 are the mean motions of Io, Europa and Ganymede respectively. These authors, however, extended their investigations to forty-six pairs of bodies within not only the broad planetary system of bodies but also the satellite systems of Saturn, Jupiter, Mars and Uranus, including even the retrograde satellites of Jupiter and Saturn. They are influenced by the fact that the well known commensurabilities all involve small integers, the largest being 5. In their work they arbitrarily set the upper limit of the integers concerned in the commensurability ratios at 7. They conclude that the observed distribution may be due either to a property of the mechanism of formation of the solar system or to an inherent stability of commensurable configurations.

Goldreich[9] is critical of some parts of Roy and Ovenden's investigations and in an important analysis of the statellite systems suggests that

[8] Roy, A. E., and Ovenden, M. W., *Monthly Notices of R.A.S.*, *114*, 232; *115*, 296.
[9] Goldreich, P., *Monthly Notices of R.A.S.*, *130*, 159.

the near-commensurable mean motions observed are stable in that their ratios are maintained, even during the evolution of the said satellite systems under the action of tidal forces.

A perusual of these three papers is encouraging to the belief that the preference for near-commensurabilities of mean motion within the planetary system in both primary and secondary bodies may be an index of the stability of these orbits and evidence for a built-in stability function for the solar system as a whole. Indeed this reasoning only reinforces what has been explained by Hagihara,[10] viz. that the stability of a satellite system is, if anything, strengthened by the presence of a commensurability relation in the mean motions. That these commensurabilities exist so widely is a feature of singular interest in these investigations.

[10]Hagihara, Y., *Smithsonian Contr. to Astrophysics*, 5, No. 6.

Chapter 14

The origin of the solar system

*I have been insisting for about
20 years that the claim of finality
for any scientific inference is absurd.*
H. JEFFREYS

Lucretius[1] in his celebrated poem[2] *De Rerum Natura* is one of the first to
apply his mind to the problem of the origin of the world system and
record his thought for posterity. In Book 5 (416–508) he gives us this
pleasing disquisition:

I will now explain in order how that great concourse of matter established the
earth and the sky and the unfathomable ocean, and the sun and the moon and
their courses.

Certainly the primary elements did not intentionally and with acute intelligence dispose themselves in their respective positions, nor did they
covenant to produce their respective motions. In reality, from time everlasting
countless elements of things, impelled by blows and by their own weight, have
never ceased to move in manifold ways, making all kinds of unions and
experimenting with everything they could combine to create; and that is
why, after wandering far and wide during mighty ages of eternity and
experiencing every kind of movement and combination, at last those atoms
have met, which, when suddenly dashed together, often form the foundations
of mighty fabrics—earth, sea, and sky, and the family of living creatures.

At first, it was not possible to see the wheel of the sun soaring aloft with
free-flowing light, nor the stars of the spacious firmament, nor sea, nor sky,
nor earth, nor air, nor indeed anything resembling the things we know. There
was only a newly formed, turbulent mass of primary elements of every kind;
and these were discordantly waging a war that involved constant confusion
of interspaces, courses, interlacements, weights, impacts, concurrences, and
motions, because, owing to the diversity of their shapes and the variety of

[1] Titus Lucretius Carus, 99–55 B.C. I am indebted to the kindness of Mr Martin Ferguson
Smith for permission to quote from his translation of the poem for Sphere Books Ltd. (1969).
[2]   *carmina sublimis tunc sunt peritura Lucreti
        exitio terras cum dabit una dies.
        The verses of sublime Lucretius are destined to
        perish only when a single day will consign the
        world to destruction.*
                Ovid, *Amores*, I. 15. 23–24

their forms, they could not all form lasting unions, or intercommunicate appropriate motions. Then the different parts began to separate, and like elements began to unite with like, thus starting the evolution of the world, the distribution of its members, and the disposition of its vast components.

From the times of Lucretius to the present day some twenty-three separate theories have been advanced in an attempt to explain the origin of the solar system and before one enters this particular intellectual labyrinth it is surely not unreasonable to enquire whether it is necessary to give a detailed account of each one of these theories. A careful study of the position will reveal that there is no wholly satisfactory theory presently available to us and that the theories that have been advanced since 1644 are, by and large, a record of the versatility of the human mind, a contention that will not be challenged by any close student of the human situation.

These reflections stem from my study of a recent short yet detailed survey of the main theories relating to the origin of the solar system by Williams and Cremin.[3] From their analysis it would appear that only two theories are presently acceptable, viz. the theories of Hoyle[4] and McCrea,[5] both of which regard the process for the formation of the planets as independent of the process that formed the Sun; the planets forming after the Sun had become a 'normal' star.

We will now examine these two theories and give a reference and an abstract to each of the theories that has been advanced over the past three hundred years. We should, however, note that the previous chapters of this book seek to give the broad features of the solar system and any valid theory of the origin of that system must ideally explain each and every one of these. In practice the salient features are usually reduced to the following essentials:

1. Regularity of planetary orbits in the sense that the direction of motion around the Sun is the same and in harmony with the direction of the rotation of the Sun itself. This is termed *direct motion*.
2. Similarity of planetary and satellite systems.
3. The differences of mass between the terrestrial and Jovian planets by a factor of about 100 and the difference in composition.
4. The position of the planets and satellites such that they conform to the Bode-Titius law[6] (or the Blagg formula) $r = 0 \cdot 4 + 0 \cdot 3 \times 2^n$.
5. The slow rotation of the Sun.[7] This is a feature of great importance

---

[3] Williams, I. P., and Cremin, A. W., *Quat. J.R.A.S.*, *9*, 40 (1969).
[4] Hoyle, F., *Frontiers of Astronomy*, ch. 6 (London 1955); *Quat. J.R.A.S.*, *1*, 28 (1960).
[5] McCrea, W. H., *Proc. R. Soc. A.*, *256*, 245 (1960). Ciel et Terre, *76*, No. 11.
[6] See chapter 12, pp. 220–6.
[7] This feature is one to be treated with great caution; while all theories of the origin of the solar system consider it, the 'true' rotation of the Sun's core is still not known and the speed of rotation of the Sun at the beginning may have been much faster than now and slowed by the solar wind.

since if the Sun had condensed from a gas cloud in harmony with Keplerian laws concerning its velocity of rotation then its velocity would be 200 times larger than presently observed.

6. The central condensation—the Sun is some 750 times more massive than the whole of the remaining parts.

No theory, as we have said, is satisfactory and the most important points yet to be resolved are, according to a pertinent analysis by Gold:[8]

1. The methods of agglomeration of solid pieces,
2. The magnetohydrodynamics that are involved in transferring angular momentum as far out as necessary by the magnetic field anchored in a body as small as the Sun, and
3. The problem of escape from the Sun of all the extra hydrogen that is believed to be present in the region of the outer planets.

THE THEORY OF THE ORIGIN OF THE SOLAR SYSTEM ACCORDING TO HOYLE

As in the theory of Laplace (1796), in which the cloud from which the Sun is formed has a small amount of angular momentum, Hoyle conveniently starts in the same manner giving it a rotation commensurate with the general galactic rotation. The condensation eventually produces an equatorial disc responsible for the slowing down of the Sun in that there is a steady transference of rotational momentum from the Sun to the disc. The solar condensation first grows its disc at the orbit of Mercury about 3×10^{12} cm. and the pushing outward of the main bulk of the material of the disc explains why the large planets are so far from the Sun. A magnetic torque is now postulated between the disc and the solar condensation. This prevents the ejection of more material and provides a coupling for the transfer of angular momentum across the gap which divides the disc from the solar condensation. The magnetic field strength required is of the order of one gauss. Segregation occurs in the disc and this provides the differences in the composition of the terrestrial and Jovian planets and also explains the large amount of gas in Jupiter and Saturn in contradistinction to the structure of Uranus and Neptune. The theory has many points of agreement with the features observed in the solar system.

THE THEORY OF THE ORIGIN OF THE SOLAR SYSTEM ACCORDING TO McCREA

McCrea commences with a molecular hydrogen cloud of several hundred solar masses at a relatively high density (4×10^{-12} g/cm.^{3}) and a

[8] Gold, J., *Origin of the Solar System*, ed. by Jastrow, R., and Cameron, A. G. W., p. 171 (Academic Press 1963).

temperature of 50°K. The cloud is composed of 'floccules' in random motion with a mean speed of about 1 km./s, their mean free path R being 5×10^{14} cm. Some 10^5 floccules occur in a sphere of radius R. The floccules coalesce by collision and condensations occur within the cloud. A solar mass condenses and captures about 1000 floccules, contra-revolution of floccules cause them to lose momentum and fall on to the main condensation (the Sun), of the remaining floccules, about 200 condense to form protoplanets containing about 20 floccules, thereby giving some 10 proto-planets, which form into major planets by direct condensation. The heavy elements are separated from the hydrogen in the terrestial planets by a disruption within the Roche limit exposing the heavy core which forms in *all* the protoplanets. The major planets lie outside the Roche limit and retain their hydrogen. The theory when viewed in its entirety is extremely subtle and able to satisfy most of the features presently observed in the solar system.

A HISTORICAL SURVEY OF THE MAIN THEORIES OF THE ORIGIN OF THE SOLAR SYSTEM FROM 1644 TO DATE[9]

Note: It is possible to divide all the theories into two broad classes

(1) Monistic or uniformitarian, in which the theory requires *no* interaction with other celestial bodies (M), and (2) dualistic or catastrophic, in which the theory requires interaction with other celestial bodies (D).

1644

Descartes's[10] vortex theory (M) was advanced at a time when only the Sun and six planets were known. Newton's theory of gravitation was not published until 1665. Descartes imagined a whirlpool within the ether and innumerable eddies, with the solid matter tending toward the centre with coarser bodies capturing smaller bodies. Newton showed that this mechanism fails to define the ecliptic plane and does not produce planetary motion in harmony with Kepler's third law.

1734

Swedenborg's[11] nebular hypothesis (M) in which the planets originate from the solar matter and then gradually remove themselves from the Sun and receive a gradually lengthened time of revolution.

1745

Buffon's[12] cometary collision theory (D) is the first of the dualistic

---

[9] The plates illustrating some of the more important theories are reproduced from *Physics Today*, October 1948, by kind permission of the editor, Harold L. Davis.
[10] Descartes, R., *Principia Philosophiae* (Amsterdam 1644).
[11] Swedenborg, *Principia* (1734).
[12] Buffon, G. L. L., *De la formation des planètes* (Paris 1745).

theories. Buffon estimated the mass of the 1680 Comet (comet Kirch, perihelion passage Dec. 18.487 UT) as 28,000 times that of the Earth, that is about one solar mass, and viewed the collision as one not dissimilar to the later tidal theories in which two stars are involved. In the collision matter is torn out of the Sun to form the planets. It can be shown that the shearing forces generated are too strong to allow for condensation by gravitation and that they lead instead to the formation of a solar envelope.

1755

Kant's[13] nebular hypothesis (M) (Fig. 42) is one of the most celebrated of the early phyotheses. He postulates a rotating nebula of gas with the Sun at the centre formed from a concentration process and a subsequent flattening of the nebula into a disc due to the rotation. Secondary mass concentrations form the planets and their satellites. He was unaware of the need to explain the distribution of the angular momentum throughout the nebula and this is its main ground of weakness.

1796

Laplace's[14] nebular hypothesis (M) (Fig. 43) owes nothing to Kant and was formed independently. It postulates a Sun surrounded by a gaseous atmosphere at very high temperature. As it cools it contracts and the equalization of angular momentum causes acceleration and rings of matter are thrown off and from these the planets and their satellites are formed. Unfortunately the Laplacian rings have no ability to form dense planetary bodies.

1878

Bickerton's[15] Star–Sun collision theory (D) is a refinement of Buffon's tidal theory in which the colliding stars produce a nova-like explosion of great violence forming the planets from condensations in the ejected material.

1885

Faye's theory (M) is not dissimilar to that proposed by Laplace.

1901

Chamberlin's[16] Star–Sun encounter theory is a refinement (D) (Fig. 44) of Buffon's according to which the second star passes the Sun in a hyperbolic orbit and raises tides and eruptions from which the planets form out of liquid drops known as planetesimals.

[13] Kant, I., *Allgemeine Naturgeschichte and Theorie des Himmels* (1755).
[14] Laplace, P. S., *Exposition du systeme du Monde* (Paris 1796).
[15] Bickerton, A. W., *Trans. New Zealand Inst.*, *11*, 125 (1878).
[16] Chamberlin, T. C., *Astrophys. J.*, *14*, 17 (1901).

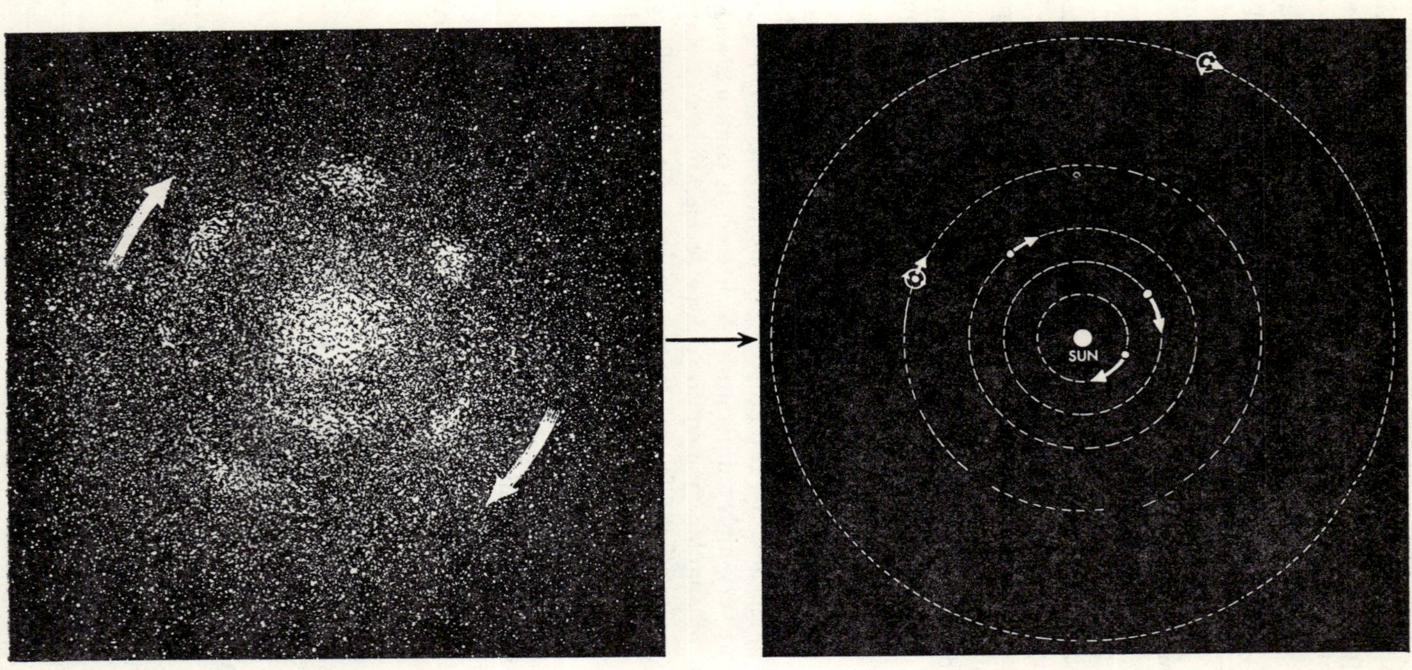

Fig. 42. Kant, 1755. (*Left*) Clotting mass of gas and dust in rotation. (*Right*) Clots grown by accretion to form planets and satellites. Remainder of the nebula contracts to form the sun.

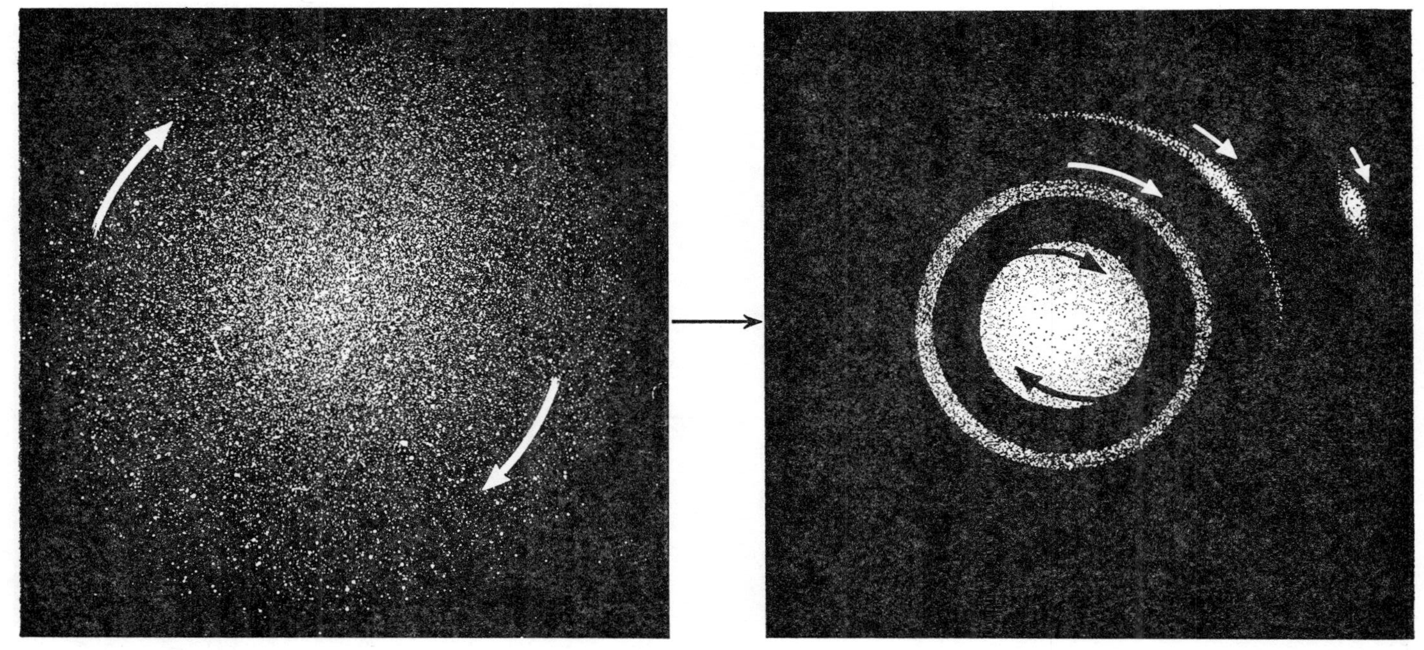

Fig. 43. Laplace, 1796. (*Left*) Rotating nebula of hot gas. (*Right*) Cooling nebula shrinks, spins faster, and is expected to leave rings of gas to condense into planets. The remainder forms the sun.

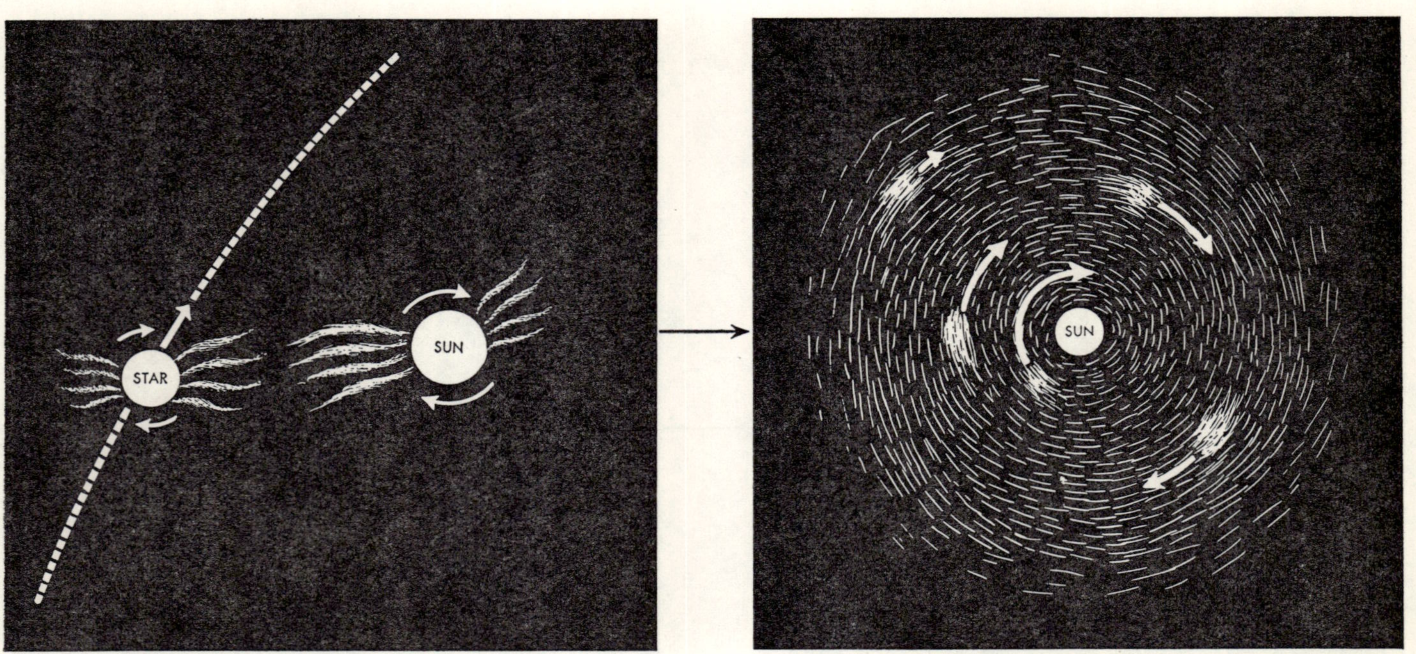

Fig. 44. Chamberlin–Moulton, 1900. (*Left*) A passing star narrowly misses the sun. Huge eruptions were expected to occur on both as they pass. (*Right*) The sun is left with a vast number of plantesimals which condensed from the erupted gases and slowly coagulate to form planets. The intruding star should also have planets forming.

1905
Moulton's[17] Star–Sun encounter theory (D) (Fig. 44) is independent of that of Chamberlin but substantially identical. Chamberlin and Moulton are the first to propose the formation of the planets from the agglomeration of cold bodies rather than from the condensation of hot fluid.

1912
Birkeland's[18] solar magnetic theory (M) introduced a novel mode of genesis. He combined the solar magnetic moment with the ions emitted to produce spiralling particles that form circular concentrations of particles, the radii of which depend on the ratio of the charge to the mass of the particles.

1913
Arrhenius'[19] Star–Sun collision theory (D) requires the two bodies to meet in a head-on collision thereby removing one of the bodies and producing a gaseous filament from which the planets form.

1916
Jeffrey's[20] Star–Sun grazing encounter theory (D) (Fig. 45) produces a filament of matter that condenses into planets.

1917
Jeans'[21] close encounter of the Sun and a Star is similar to that of Jeffreys (D). (Fig. 45.)

1930
Berlage[22] improved Birkeland's theory (M) (Fig. 46) by considering the effect of the solar electric field and ingenously showing how a series of concentric rings may be formed each composed of one particular element or isotope in a descending arithmetic series. From these the planets form and are spaced according to the Bode-Titius relation. The theory is difficult to reconcile with the highly variable solar emission of charged particles which Berlage assumed to be constant.

1935
Russell's[23] Binary Sun theory (D) is of considerable interest for his discussion of the difficulties of the resisting medium and the disagreements which arise between the theory of the formation of the planets and what is observed in nature.

[17] Moulton, F. R., *Astrophys. J.*, *22*, 165 (1905).
[18] Birkeland, K., *Compt. Rend. Acad. Sci.*, *155*, 892 (1912).
[19] Arrhenius, S., *Das Werden der Welten* (Leipzig 1913).
[20] Jeffreys, H., *Monthly Notices of R.A.S.*, *77*, 84 (1916); *78*, 424 (1918).
[21] Jeans, J. H., *Mem. Roy. Astron. Soc.*, *62*, 1 (1917). *Monthly Notices of R.A.S.*, *77*, 186 (1917).
[22] Berlage, H. P., Jr., 'Het ontstaan en vorgaan der werelden' (Amsterdam 1930). Also *Jr. Proc. Koninkl. Ned. Akad. Wetenshap.* Amsterdam, *33*, 614 (1930).
[23] Russell, H. N., *The Solar System and its Origin* (Macmillan, New York 1935).

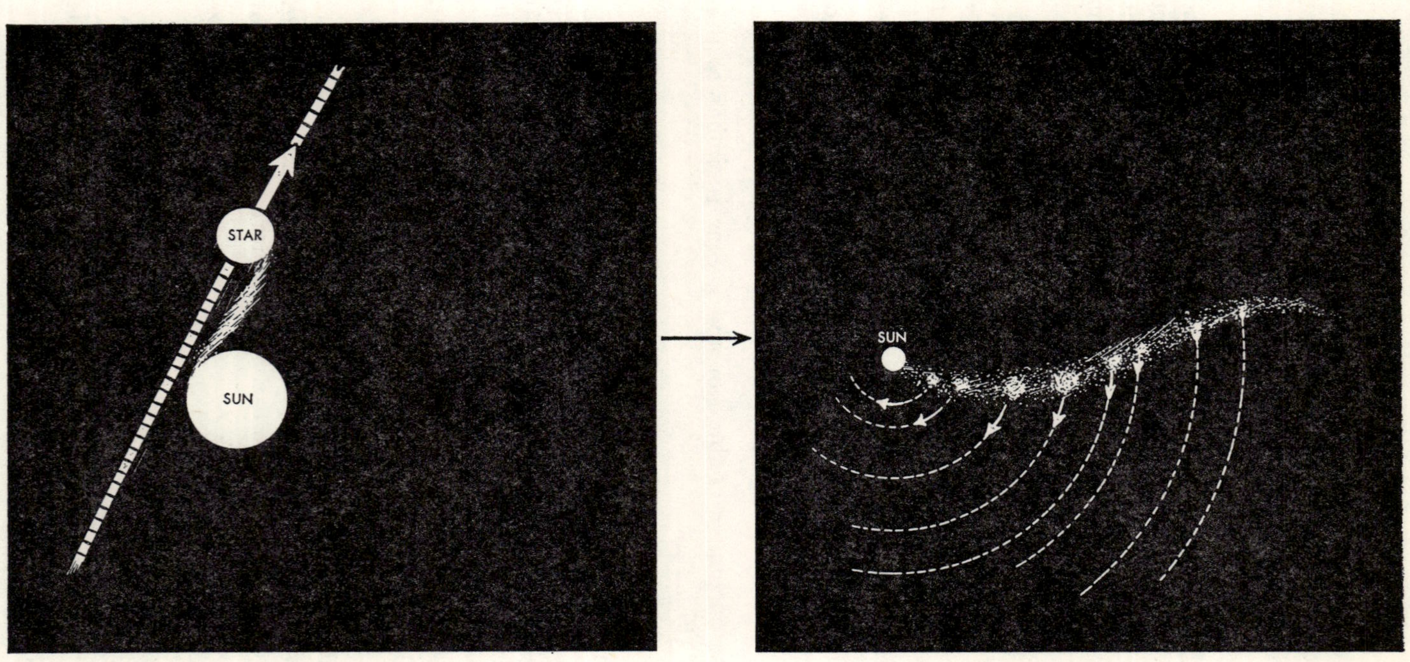

Fig. 45. Jeans–Jeffreys, 1917. (*Left*) A passing star sideswipes the sun, tearing out a long filament of gaseous material. (*Right*) The gas was expected to cool and condense into planets, the largest one in the middle and the smaller ones at either end.

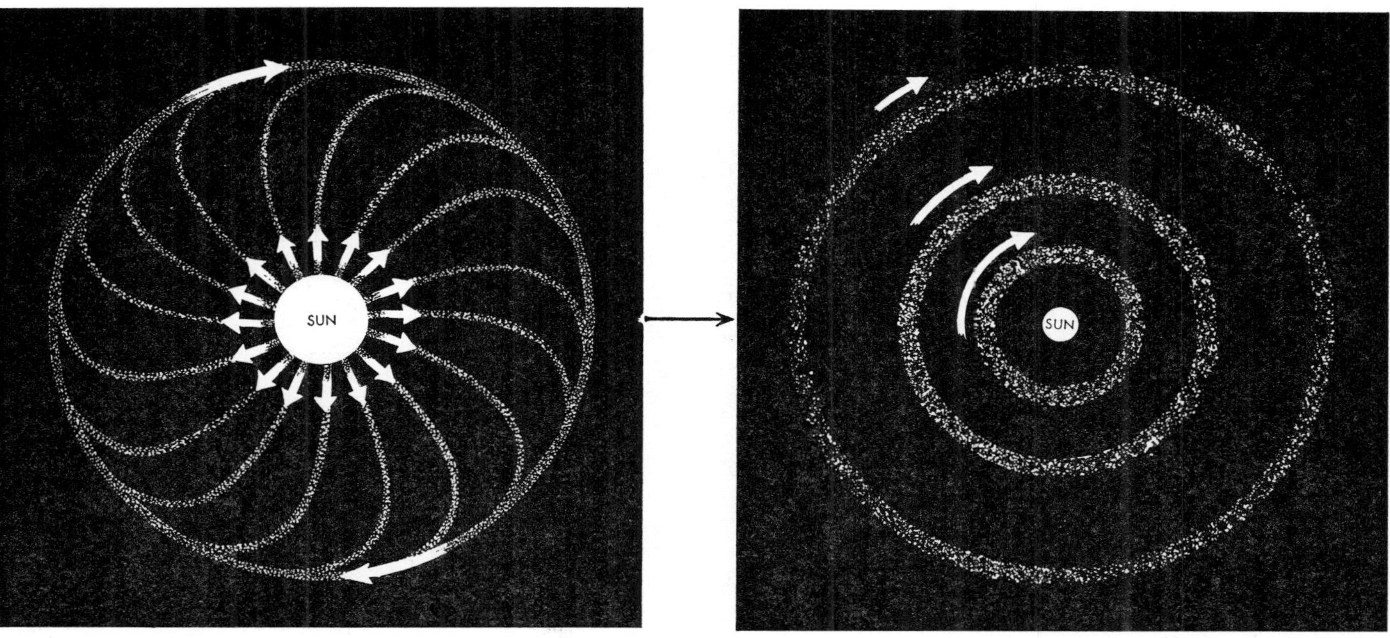

Fig. 46. Berlage, 1930. (*Left*) Electricity charged atoms and molecules shot out of the sun spiral in the solar magnetic field. (*Right*) Rings of gas result, each ring formed of atoms or molecules with the same ratio of charge to mass. Condensation into planets is uncertain.

1936

Lyttleton's[24] triple Sun theory (D) (Fig. 47) requires the binary companion of the Sun to undergo a close encounter with a third star, which results in the disruption of the binary system and the production of a filament of matter from which the planets form.

1942

Alfvén's[25] theory (M) (Fig. 48) in which the Sun's magnetic moment acts on ionized matter. The sun is supposed to meet an interstellar cloud and be immersed in it. Atoms fall in on the Sun and ionization occurs. The ions are forced to move along the magnetic lines of force until they reach an equilibrium position in the equatorial plane of the Sun. The ions form mass concentrations in that plane and this is the origin of the planets. The theory breaks down when one considers the formation of the *inner* planets in that ionization occurs at too great a solar radius. A recent test of Alfvén's theory (Alfvén, H., *On the Origin of the Solar System*, Clarendon Press, Oxford 1954, is proposed by Hauge (see *Nature*, *230*, 39, 1971). He appeals to Alfvén's suggestion that magnetic fields act on the orbits of ions and that this effect was important during the early history of the solar system. It is clear from this that neutral atoms and ions will follow different paths which may cause a separation effect. In Hauge's work, the ionization energy has been used as a parameter and the abundances of the chemical elements in ordinary chondrites, terrestrial magma and lunar rocks have been compared with solar abundances.

This leads to the following tentative conclusions. Ionized matter in magnetic fields may have played an important part in the formation of the Earth and the Moon, which were formed in different ionization conditions.

Hauge assumes that the energy which caused the ionization increases towards the Sun. Then the Moon must have been formed on the inner side of the Earth. This assumption is also supported by the high density of Mercury which may be due to a higher iron content, while the less dense planet Mars should be overabundant on elements with low ionization potentials such as Na, Al, K and Ca.

Considering the ionization potentials only, Hauge cannot give a complete answer to these questions. Many other physical and chemical properties of matter must be taken into account in a complete theory for the solar system, but the analysis is one of considerable encouragement.

1943

Schmidt's[26] theory (D) in which the Sun encounters a swarm of inter-

[24] Lyttleton, R. A., *Monthly Notices of R.A.S.*, *96*, 559 (1936); *98*, 536 (1938); *98*, 646 (1938).
[25] Alfven, H., *Stockholms Obs. Ann.*, 14, No. 2 (1942); No. 5 (1943).
[26] Schmidt, O. Y., *Dokl. Akad. Nauk. SSSR*, 45, No. 6 (1944).

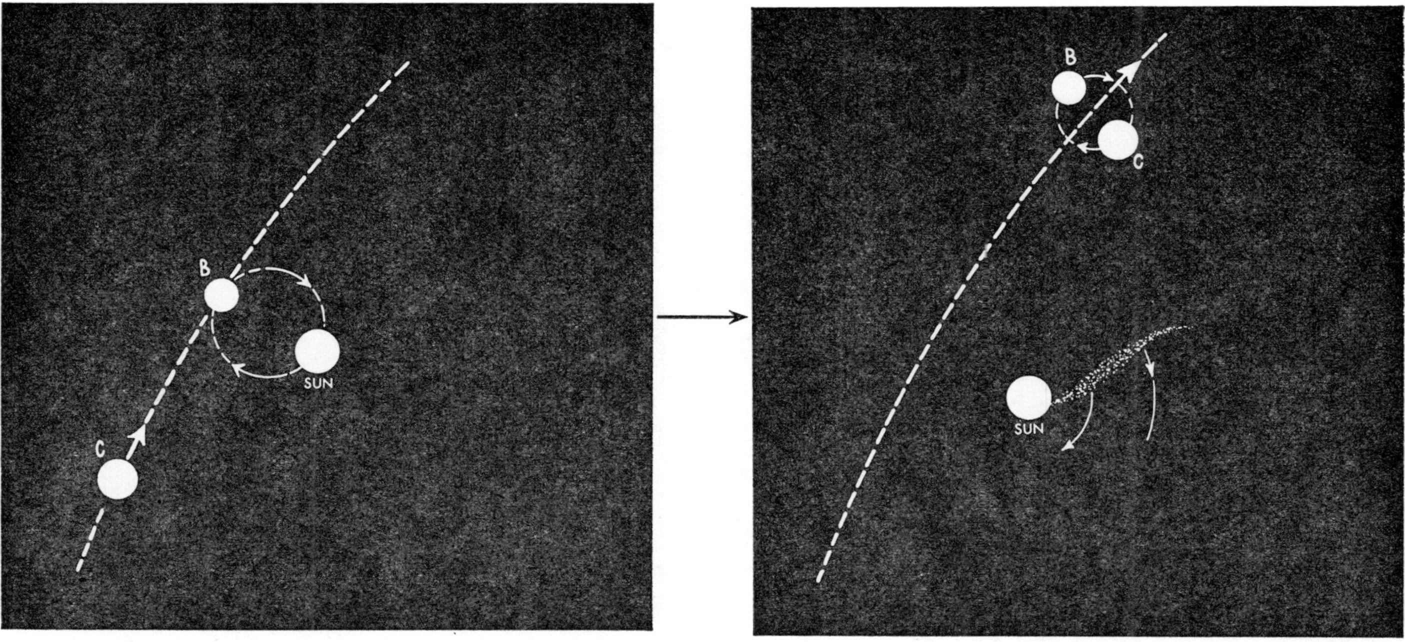

Fig. 47. Lyttleton, 1936. (*Left*) If the sun originally had a close **companion**, *B* spinning around it, a third star *C*, might have sideswiped the companion carrying it away (*right*) and leaving a filament of its gas moving around the sun, with lots of angular momentum.

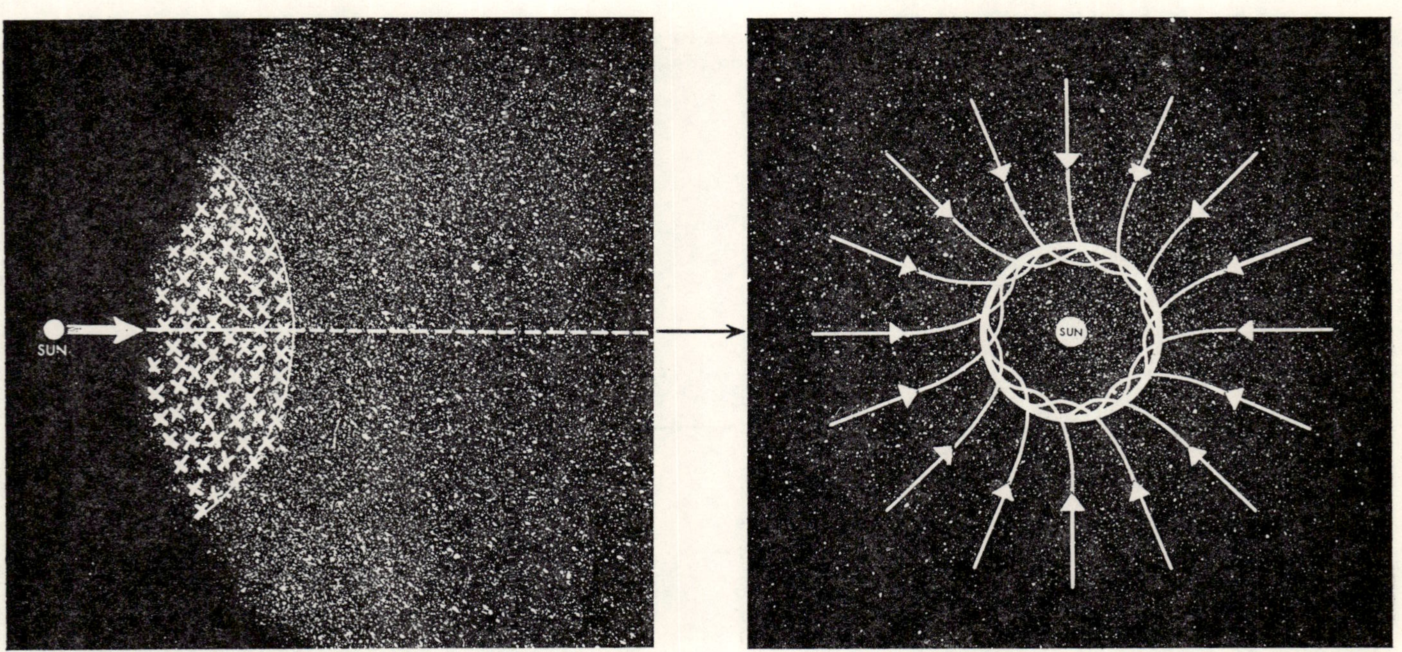

Fig. 48. Alfvén, 1942. (*Left*) The sun, rushing through space at 12 miles per second, passes through a gaseous nebula. Its presence creates electric charges on the atoms of gas. (*Right*) The charged atoms spiral inward to form rings of gas (only one is shown here) which might later condense into planets.

stellar bodies. The planets are formed by collisions and the swarm is changed into a lenticular disc with the planets in an equatorial array.

1944

Von Weizsäcker's[27] theory (M) (Fig. 49) of turbulent eddies is a novel theory. A gaseous disc around the Sun segregates into a system of vortices

Fig. 49. Weizsäcker, 1944. Vortices formed in the equatorial plane of a nebula of gas and dust rotating about the sun. Accretion is expected to take place along the heavy concentric circles to form planets and satellite systems with direct rotation and revolution.

not very dissimilar to the systems postulated by Kant and Berlage before him. He now assumes a system of vortices which gives a Keplerian configuration with secondary eddies called 'roller bearing' eddies from which the planets form at distances in agreement with the Bode-Titius relation.

[27] Von Weizsäcker, C. F., *Z. Astrophys.*, 22, 319 (1944). See, Chernin, A. D., 'Turbulence on the Hot Universe, *Nature*, 226, 440 (1970).

1944
Hoyle's[28] binary Sun and supernova hypothesis (D) (Fig. 50). In this hypothesis the second component disintegrates with a supernova outburst breaking up the binary system and forming the planets by condensation from the material remaining. The results are said to be erroneous owing to the neglect of the exhaustion of the gaseous system generated.

1947
In Whipple's[29] Sun and smoke cloud (D) hypothesis a smoke cloud is supposed that has a radius of 30,000 a.u. and one solar mass. It also has a small angular momentum and contracts to form the Sun. The cloud captures a smaller smoke cloud having its own angular momentum. In the condensations formed the planets have their genesis.

1948
Ter Haar's[30] turbulent contracting solar envelope hypothesis (M) is a modification of Von Wiezsäcker's theory.

1949
Kuiper's[31] gravitational instability theory (M) is also a modification of von Weizsäcker's theory. Kuiper dismisses the regular vortices of von Weizsäcker and introduces instabilities in the solar nebula at which points condensations produce large protoplanets which contract to form the planets with their attendant satellites. The main objection to the theory comes from the long time taken for the protoplanets to lose their non-condensable gases.

1955–60[32]
Hoyle's gas cloud hypothesis (M) was referred to in detail on page 237.

1963
McCrea's stellar cluster hypothesis (M) was referred to in detail on page 237/238.

[28] Hoyle, F., *Proc. Camb. phil. Soc. math. phys. Sci.*, *40*, 256 (1944); *Monthly Notices of R.A.S.*, *105*, 175 (1945); *106*, 406 (1947).
[29] Whipple, F. L., paper read before the American Association for the Advancement of Science, Chicago, December 27, 1947.
[30] Ter Haar, D., *Kgl. Danske Videnskap. Selskab, Mat.-Fys. Medd.*, *25*, 3 (1948).
[31] Kuiper, G. P., *Astrophys. J.*, *109*, 308 (1949).
[32] Hoyle advanced yet another theory at the I.A.U. meeting in Brighton, 1970, August 20. *The Times* reported:
   What he suggests is that at an early stage when the sun was about 40 m. kilometres across it developed arms from which the planets were eventually formed.
   Why did hydrogen and helium become segregated from the other elements in the clouds? The reason seems to be that magnesium, silicon and iron in the form of vapour are expected to condense at about the same temperature. Professor Hoyle has been able to show that this is likely to have occurred at a distance from the sun that corresponds to the earth's orbit.
   Professor Hoyle has a reputation among astronomers for provocative ideas put forward to stimulate the development of the subject. Many astronomers are likely to view this theory of the origin of the solar system in this light.
   It leaves a number of problems unanswered, or only roughly sketched out. Even so his talk yesterday should have the desired effect of causing astronomers to think again about their basic views of the solar system.

250

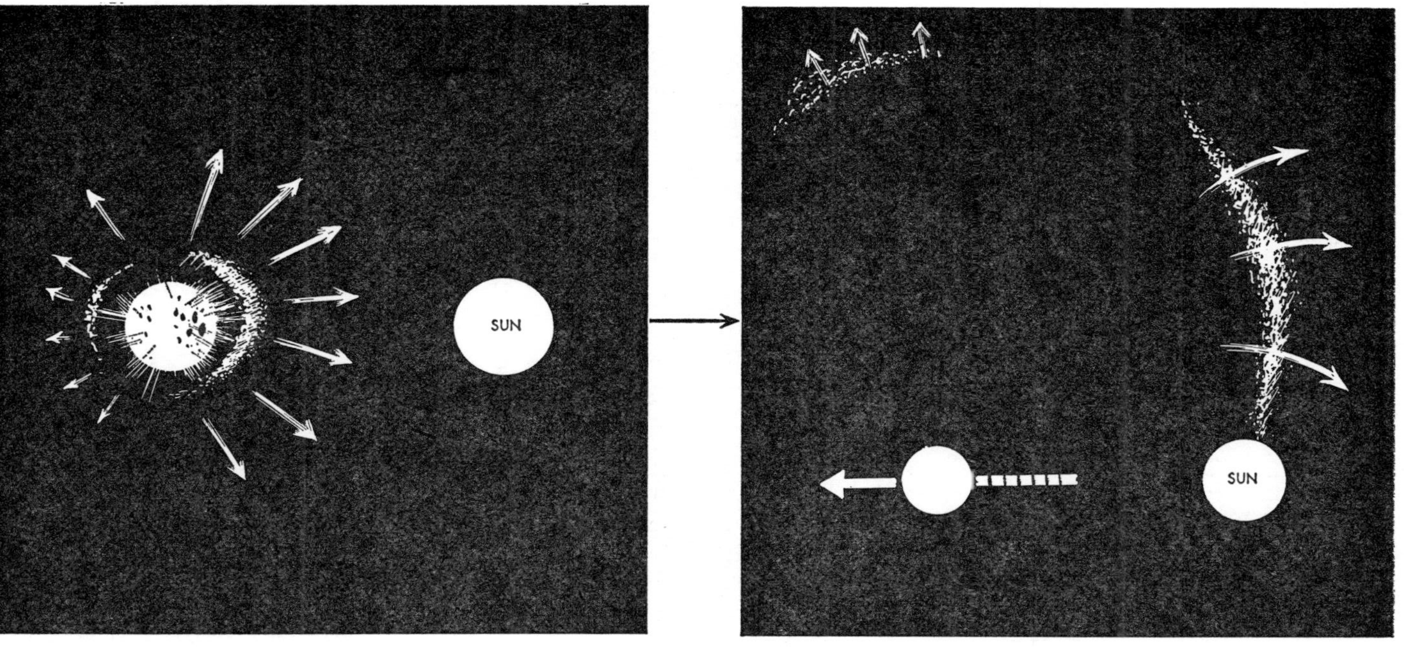

Fig. 50. Hoyle, 1944. A star near the sun might have blown up, throwing off a large shell of material, possibly more in one direction than the others. Such nova explosions are observed frequently. (*Right*) Part of the nova shell could be caught by the suns gravitation, while the nova itself recoiled away from the one-sided explosion.

The solar system

In closing this survey it is my wish to express my debt to the writings of ter Haar,[33] to which any student of this fascinating subject may turn in full confidence not only for the history of the subject but for a profound understanding of the mathematical and physical problems involved.

NUCLEAR CLUES TO THE EARLY HISTORY OF THE SOLAR SYSTEM

The abundances of the nuclei of the isotopes D^2, Li^6, Li^7, Be^9, B^{10} and B^{11} on earth constitute a problem in explaining the origin of the solar system. It is now generally agreed that the heavy elements in the solar system were produced not by the Sun itself but by nuclear processes in countless stellar interiors long before the Sun came into being. The mechanism involves hydrogen 'burning' where hydrogen is the fuel and helium is the ash ($4H^1 \rightarrow He^4$) and later helium 'burning' where helium is the fuel and carbon is the ash ($3He^4 \rightarrow C^{12}$). It is these two fundamental processes that produce the isotopes of the heavier elements, those beyond hydrogen in the periodic table, indeed it can be shown that unstable transuranic elements may be synthesized with hydrogen as the basic building block. Instabilities in the stellar interiors occur and the resultant explosions are the most violent and cataclysmic known to nature; these we witness as supernovae events and it is these that distribute the transmuted materials to interstellar space.

The exception to this form of genesis are the light nuclei other than He^4, viz. the building up of deuterons and the isotopes of lithium, beryllium and boron in terms of thermonuclear reactions in the interiors of stars.

In helium 'burning' the stable isotopes of lithium, beryllium and boron are by-passed. The appearance of these light elements on Earth is currently explained by the formation at some stage of the solar system of planetesimals consisting of H_2O, and oxides of Mg, Si and Fe roughly in the ratio one to one by volume.

The H_2O probably occurred as ice and hydrate. The primaeval sun is thought to have been less luminous than at present since it would then have been at a stage of its evolution before the beginning of hydrogen 'burning'. The temperature of the planetary material near the earth fell in the range $130°$ to $200°K.$, H_2O condensed and froze in the planetesimals, but not NH_3 or CH_4. Thermalization of the neutrons was promoted by 'elastic' scattering in the hydrogen. The neutrons not only produced the required lithium and boron ratios, but also gave sufficient deuterons to produce more than ten times the terrestrial D/H ratio in the material irradiated.

[33] Ter Haar, D., 'Studies on the origin of the Solar System', *Det Kgl. Danske Videnskabernes Selskab. Matematisk—Fysiske Meddelelser*, Band XXV, Nr. 3 (København 1948).
Ter Haar, D., and Cameron, A. G. W., chapter I of *Origin of the Solar System*, ed. by Jastrow, R., and Cameron, A. G. W. (Academic Press 1963).

It is seen from this feature alone that the solar system may have had a most interesting nuclear history and this 'last minute' nucleosynthesis possibly is peculiar to our solar system; but since we have no evidence of any other similar system the claim so made by the experts[34] may seem a little hollow to a philosopher.

The terrestrial-meteoritic abundances of D, Li, Be and B determine the intensity and extent of the break-up and neutron process that have occurred in the planetesimals[35]. In particular the production of radioactive nuclei in the planetesimals such as Pd^{107} and I^{129} leads to an explanation of radiogenic Ag^{107} and Xe^{129} which have been found and considered anomalies in meteorites having high Pd/Ag and I/Xe ratios.

The Pd^{107} decays to Ag^{107} and the I^{129} decays to Xe^{129}. The interval for these events is found to fall in the time range of 10^6 to 10^8 years using Pd^{107} and I^{129} decays as 'clocks'. It is this that entitles us to speak of the 'last minute' nucleosynthesis.

Recent studies on the abundances of Pu^{244} and I^{129} in the early solar system support the continuous galactic nucleosynthesis theory and the agreement between the two 'nuclear clocks' in accordance with the ideas to Kuroda.[36]

[34] See Fowler, W. A., Astron. Soc. of the Pacific Leaflet, No. 411 (1963).
[35] A new hydrogen-deuterium abundance in the promordial Sun is given by Black; see *Nature Physical Science*, *234*, 148 (1971).
[36] Kuroda, P. K., 'Plutonium—244 in the Early Solar System', *Nature*, *221*, 726 (1969). See also Mabuchi and Masuda, 'Possible Clues to the Early History of the Solar System', *Nature*, *226*, 338 (1970); see also Hoffman *et al*, 'Detection of Plutonium-244 in Nature', *Nature*, *234*, 132 (1971).

Note: Since writing this chapter Kumar, S. S., has shown (*Nature*, *233*, 473, (1971) that if Jupiter had originally possessed 40 or 50 times its present mass, it would have made the planetary system unstable over a period much shorter than $4\cdot5 \times 10^9$ yr. The planetary system would also have become unstable if Jupiter's orbital eccentricity were as small as $0\cdot5$. It should be mentioned that the increase of the eccentricity tends to destroy the stability of the orbits of the small planets more effectively than does the increase of the mass of Jupiter. Thus the fact that the planetary system has survived for a period of $4\cdot5 \times 10^9$ yr leads us to the conclusion that the above mentioned properties of the most massive planet (Jupiter) are original properties of the system. If they are original, then any satisfactory theory of the formation of the solar system must take them into account.

Chapter 15

Life in the solar system and beyond

He who through vast immensity can pierce,
See worlds on worlds compose one universe,
Observe how system into system runs,
What other planets circle other suns,
What vary'd being peoples every star,
May tell why Heaven has made us as we are.
But of this frame, the bearing and the ties,
The strong connections, nice dependencies,
Gradations just, has thy pervading soul
Look'd thro? Or can a part contain the whole?
POPE

In this final chapter we will step outside of the confines of the solar system. The reason stems from the widely held view that there *is* life on other worlds[1] and I would like to make a few observations on this problem.

It is clear from recent observations that life in the solar system is extremely difficult to detect by artificial satellite even when it is known to exist. Kilston, Drummond and Sagan[2] have shown that a search for life on Earth at kilometre resolution, using several thousand photographs obtained by the Tiros and Nimbus meteorological satellites, has been undertaken. No sign of life can be discovered on the vast majority of these photographs. Due principally to the small contrast variations involved and the difficulty in reproducing observing conditions at satellite altitudes, no seasonal variations in the contrast of vegetation could be detected. Of several thousand Nimbus 1 photographs of essentially

[1] There are those who may believe that our planet is being visited clandestinely by spacecraft from other worlds. The student should consult two works, one scientific and one psychological. (*a*) *Scientific Study of Unidentified Flying Objects*. Conducted by the University of Colorado under Research Contract Number F 44620–67–C–0035 with the U.S. Air Force, Project Director, Condon, E. U., Bantam Books, 1969 January 8. (*b*) The essay—'Flying Saucers: A Modern Myth of things seen in the skies', C. G. Jung, *The Collected Works*, vol. 10, Civilization in Transition, London 1964, p. 307.
[2] See Kilston, S. D., Drummond, R. R., Sagan, C., 'A Search for life on Earth at Kilometre resolution', *Icarus*, 5, 79 (1966).

cloud-free terrains, one feature was found indicative of a technical civilization on Earth—a recently completed interstate highway—and another suggestive feature was discovered, possibly a jet contrail. A striking rectilinear feature was found on the Moroccan coast; however, it appears to be a natural peninsula. An orthogonal grid, discovered in a Tiros 2 photograph, is due to the activities of Canadian loggers, and is a clear sign of life. It appears that several thousand photographs, each with a resolution of a few tenths of a kilometre, are required before any sign of intelligent life can be found with reasonable reliability. A similar Mariner IV system—taking 22 photographs of the Earth with a resolution of several kilometres—would not detect any sign of life on Earth, intelligent or otherwise.

It was until recently thought that Venus and Mars may be seats of life not dissimilar to that seen on Earth in some form or other, but these views are now hard to sustain in the light of the evidence that Venus has a surface temperature of approximately $700°K$.[3] We should not, however, confine our thinking to the terrestial forms in which a photosynthetic mechanism for CO_2 fixation is predominant. Such a cycle may not be used on Mars for example, and Wolfgang[4] has put forward a proposal for carbon monoxide as a basis for primitive life on other planets.

Outside religious belief, ideas of how living things began are still very much in their formulative stages as far as scientific research is concerned. In 1955 Miller[5] and his contemporaries conducted experiments directed to the simulation of the conditions believed to have existed on Earth at the time of what is now often crudely termed the 'primeval soup', although the precise ingredients and the 'flavour' are very much a matter of speculation.

The Royal Society has given a welcome air of respectability to the subject in a discussion organized by Pirie[6] directed to 'anomalous aspects of biochemistry of possible significance in discussing the origins and distribution of life'. The discussion embraced *inter alia* the occurrence and metabolism of carbon-halogen compounds; silicon compounds in biological systems (mainly in the beautiful Diatomaceae);[7] Vanadium and other metals in ascidians;[8] reversible suspension of metabolism and

<hr>

[3] See Sagan, C., 'Life on the Surface of Venus', *Nature*, *216*, 1198 (1967).
[4] Wolfgang, R., *Nature*, *225*, 876 (1970); see also Postgate's criticisms, *Nature*, *226*, 978 (1970).
[5] Miller, S. L., *J. Amer. Chem. Soc.*, *77*, 2351 (1955).
[6] Pirie, M. W., *Proc. Roy. Soc. B.*, *171*, No. 1022 (August 1968).
[7] *Diatom*. A member of the genus *Diatoma*, or, in a wider sense, of the *Diatomaceae*, an order of microscopic unicellular Algae, with silicified cellwalls, and the power of locomotion, on which account they were formerly placed by many naturalists in the Animal kingdom. They exist in immense numbers at the bottom of the sea, as well as in fresh water; and their siliceous remains form extensive fossil deposits in many localities.
[8] *Ascidian*. Of or pertaining to the Ascidia (or Ascidiae), a group of animals belonging to the tunicate Mollusca, considered by evolutionists to constitute a link in the development of the Vertebrata.

the origin of life; fringe biochemistry among microbes and the limits of microbial existence. The subject is wholly one of speculation but experiments are being conducted although experts are agreed that there is little likelihood in the near future of synthesizing protein and nucleic acid systems that fully self-replicate. The primeval atmosphere is thought to be a reducing environment containing hydrogen, methane, ammonia, hydrogen sulphide and hydrogen cyanide and the biological molecules are said to have *somehow* arisen from condensation products of these gases and water. The 'somehow' offers plenty of scope for hypotheses but these are very limited as we shall presently see. A number of amino-acids, nucleic acid bases and carbohydrates have been made from simple inorganic molecules by irradiating a mixture of gases, or subjecting them to electric discharges and reacting the products with water. One difficulty has been in giving cogent reasons for sufficient quantities of the right kind of amino-acids in the primeval seas to allow their condensation into proteins to occur.

To overcome this difficulty life is now thought to have had its scientific genesis in fresh water pools subjected to drying by evaporation. Eugster and Jones[9] favour pools close to alkaline springs and there will be many who will see wisdom in this refinement. The gas hydrogen cyanide is thought to have had a key role and Matthews and Moser[10] have shown that some fourteen amino-acids are produced when hydrogen cyanide and ammonia are hydrolyzed with water. Significantly the first products of the hydrolysis were polypeptides, not free amino-acids. This suggests that proteins may arise directly from simple organic gases.

Steinman, Smith and Silver[11] have made the first synthesis of a sulphur-containing amino-acid methionine in prebiotic conditions. Four of the five nucleic acid bases have been synthesized but we do not know completely how they are linked to a sugar moiety or how they are phosphorylated and hence their formation of nucleic acids is unclear.

Progress is being made but the origin of biological molecules is still very much unexplained and the origin of life on Earth remains very much of a mystery to the scientific mind. Whether the experimental work will succeed in providing an explanation is open to question.

Polanyi[12] in his latest essays invites us to consider a very real difficulty. 'Mechanisms', he tells us, 'whether man-made or morphological, are boundary conditions harnessing the laws of inanimate nature, being themselves irreducible to those laws. The pattern of organic bases in DNA which functions as a genetic code is a boundary condition *irreducible* to physics and chemistry'. In an earlier paper[13] he points out that,

[9] Eugster, H. P., and Jones, B. F., *Science, 161*, 160 (1968).
[10] Matthews, C. N., and Moser, R. E., *Nature, 215*, 1230 (1968).
[11] Steinman, G., Smith, A. C., and Silver, J. J., *Science, 159*, 1108 (1968).
[12] Polanyi, M., *Knowing and Being*, p. 225 (London 1969).
[13] Polanyi, M., 'Life Transcending Physics and Chemistry', *C. & E.N.*, p. 55 (August 21, 1967).

104. Meteorite from Canyon Diablo, Coon Butte, Arizona. The polished and etched surface shows the Widmanstatten structure.

105. Ahnighito, the largest of the Cape York irons, weighs $36\frac{1}{2}$ tons, is 10 ft. 11 in. long, 6 ft. 9 in. high, 5 ft. 2 in. wide, and was located by Admiral R. Peary during his polar explorations on the northern shore of Melville Bay, Cape York, Greenland. It was brought to New York in 1894–7.

106. Allende, Chihuahua, Mexico, meteorite. Fall of approximately 01·10 Central Standard Time, 1969, February 8. A fireball approached from the south-south-west producing thousands of fusion-crusted stones, found over a strewn field 50 km. long. The photograph is of a typical large individual, 9·3 kg., that was found near the village of Santa Anna, 1969, February 16. Allende is a Type III carbonaceous chondrite, a rare meteorite type. It is the largest stony meteorite shower on record. The photograph shows a typical fusion-crusted surface.

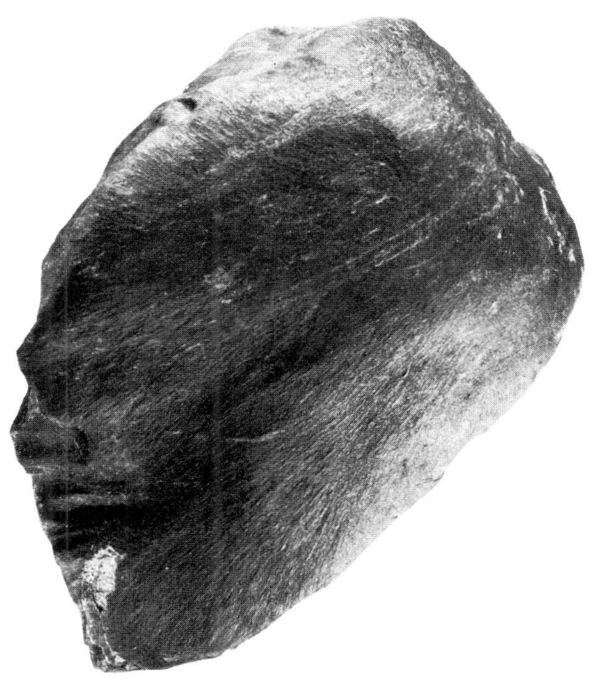

107. Lost City, Oklahoma, meteorite. Fall of 20·14 Central Standard Time, 1970, January 3. This is the largest of four fragments that have been recovered to date, 9·8 kg. Its trajectory through the atmosphere was photographically recorded, and atmospheric debris collections were made. The meteorite is an olivine-bronzite chondrite, and it is being subjected to extensive laboratory investigation. The surface shown in the photograph strongly ablated, the result of oriented flight through the atmosphere.

108. Standing on the slope of the Hadley Delta, David Scott photographs surface features with a 70 mm. camera. In his left hand he holds tongs for picking up rock samples. The Apollo 15 rover vehicle is partially visible in the background. The base of the Apennine Mountains, seen in the distance, is about 17·5 km. away. The view is towards the east.

109. The Apollo 15 mission; Hadley Delta is in the right background. Last crater is to the right of the lunar module. Note the bootprints of the astronauts and the tracks of the lunar rover vehicle.

110. A close study of the floor of a crater termed by the Apollo XV crew as 'relatively fresh'; The Apennine Front is on the left.

111. The Apollo XV mission. Command and service modules in lunar orbit, viewed from the lunar module.

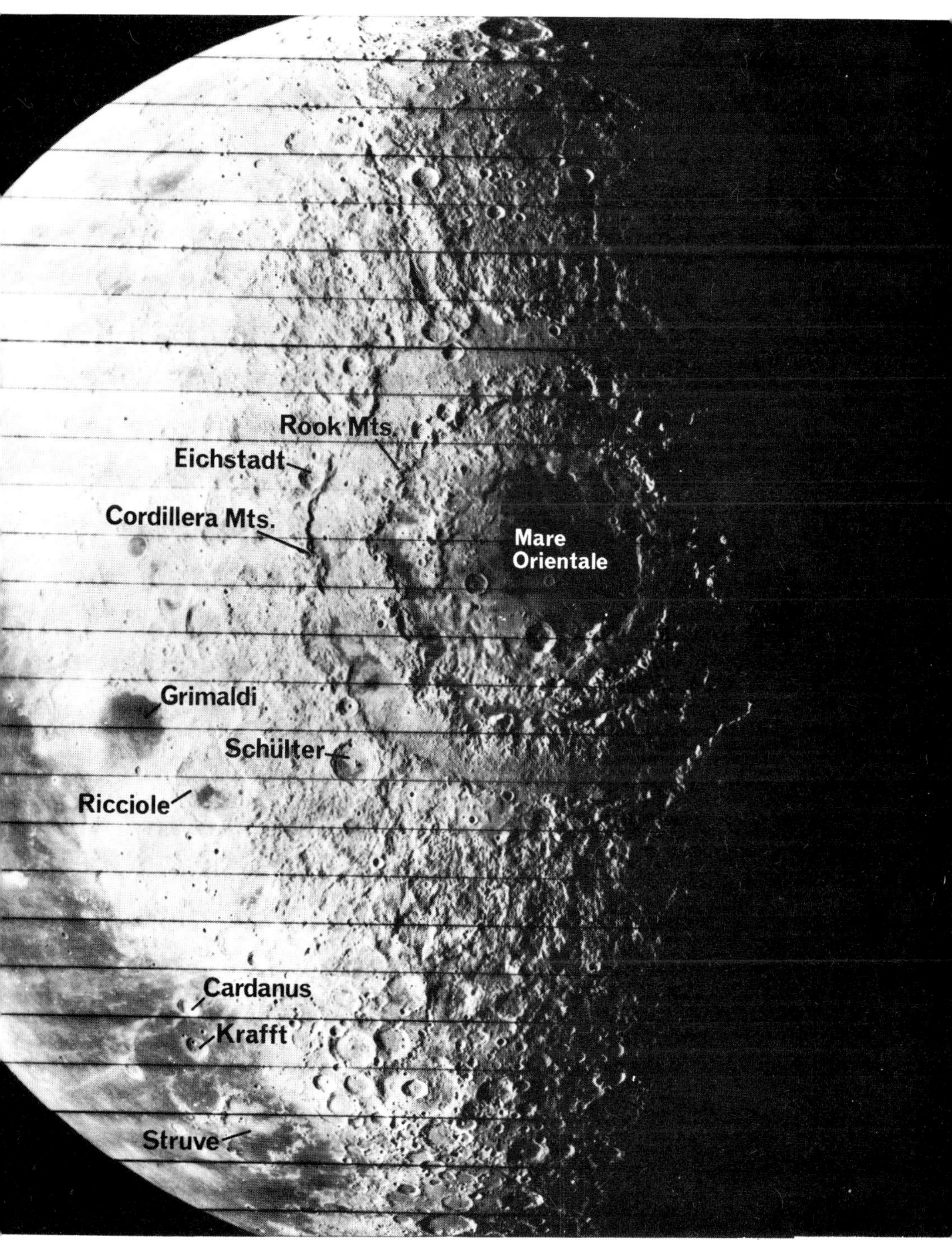

112. The Mare Orientale at the far side of the Moon.

113. The Moon photographed by Apollo VIII (1968 November 24–5).

A. Mare Australe, extended, broken.
B. Mare Crisium, large dark.
C. Mare Fecunditatis, very large.
D. Mare Humboldtianum, gray oval.
E. Mare Marginis, large, dark, irregular.
F. Mare Nectaris, large, very dark.
G. Mare Smythii, large, round, mottled.
H. Mare Struve, small black patch.
I. Mare Undarum, small, broken, dark.

NAMED LUNAR CRATERS
(*Diameters in miles*)
1. Benhaim, 35, central peak.
2. Cannon, 33, bright rim.
3. Cauchy, *8*, bright speck.
4. Condorcet, *46*, grayish.
5. Endymion, *78*, gray floor.
6. Giordano Bruno, *50*, bright ray center.
7. Goclenius, *41*, dark floor.
8. Goddard, *46*, dark floor.
9. Hubble, *45*, dark, bright inner spot.
10. Jeans, *54*, dark, bright-rimmed, above trough at limb.
11. Joliot-Curie, *89*, part of floor dark.

12. Langrenus, *82*, light interior.
13. La Perouse, *48*, gray floor, bright rim.
14. Lomonosov, *37*, very dark.
15. Lyot, *68*, dark floor.
16. Neper, *88*, dark floor, bright center.
17. Oken, *45*, dark floor.
18. Petavius, *110*, central peaks.
19. Proclus, *17*, very bright, ray center.
20. Rheita, *44*, next to Rheita Valley.
21. Stevinus, *46*, prominent, central peak.
22. Tsiolkovsky, *121*, very dark floor with bright "island".
23. Wilhelm Humboldt, *125*, light floor with four dark patches.
floor with four dark patches.

NUMBERED FAR-SIDE CRATERS (*Diameters in miles*)
No. 272, *100*, indistinct rim.
No. 276, *60*, indistinct.
No. 343, *60*, bright floor, in darkish area, irregular dark patch to left
No. 401, *40*, black floor, bright central peak, in Mare Australe.

114. Aratus of Soli, Greek didactic poet, son of Atheno-
dorus and a contemporary of Callimachus and Theocritus.
Born *c.* 315 BC, he is famous for his poem *Phenomena* which
is a hypothetical representation on *lapis lazuli*. The poet
is shown sitting in front of an armillary sphere holding a
pair of compasses and looking up at the Sun, Moon and
stars.

115. Anaximander of Miletus, 610–547 BC, said by some to have invented
the sundial. The picture shows a mosaic from Trier.

116. A representation of Plato. The Vatican Museum, Rome.

117. Aristotle, from a carving at Chartes Cathedral.
118. Socrates, *c.* 471 ?469–399 BC, National Museum, Naples.
119. Thales of Miletus, 640–546 BC, founder of Greek geometry, astronomy and philosophy, Sala dei Filosofi, Museo Capitolino, Rome.
120. Metrodorus, second century BC, Museo Capitolino, Rome.

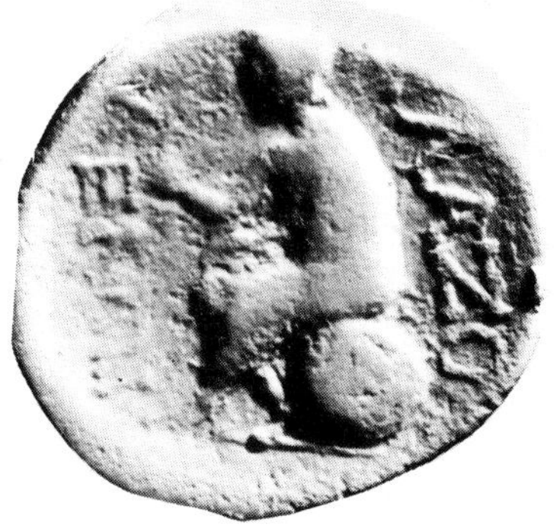

121. Anaxagoras, *c.* 500–428 BC, son of Hegesiboulos and a native of Klazomenai. The representation is from a bronze coin of Klazomenai, *c.* 100 BC.

122. Hipparchus of Nikaia, *c.* 180–126 BC, from a bronze coin of Nikaia. The philosopher is shown seated in front of a globe set on a pillar.

123. Pythagoras, born *c.* 576 BC, the son of Mnesarchus of Samos. The illustration shows a silver coin of Abdera, *c.* 432–425 BC.

124. Eudoxus, *fl.* fourth century BC, son of Aischines, pupil of Archytas. At the age of 23 he attended Plato's lectures. The figure is of a seated man, enveloped in anhimation, bending forward in the act of teaching. National Museum, Budapest.

126. Tycho Brahe, 1546–1601, from an old painting in the observatory at Prague.

125. Nicolaus Copernicus (or Koppernigk), 1473–1543, from an engraving by E. Scriven.

127. Johann Kepler (Keppler), 1571–1630, from a portrait in the possession of the family and preserved at Kremsmünster.

128. Galileo Galilei, 1564–1642, from an engraving by Robert Hart.

129. Sir Isaac Newton, 1642–1727, from an engraving by W. T. Fry.

130. Eise Eisinga, 1744–1828.

131. John Couch Adams, 1819–92, an engraving by Cousins from a portrait by Mogford.

132. Urbain Jean Joseph Leverrier, 1811–1877, French astronomer born at St. Lô in Normandy on 1811, March 11. The discovery of Neptune by mathematical deduction was but an incident in his career. He spent forty years elaborating the scheme of the heavens traced out by Laplace in the *Mécanique Céleste* and *inter alia* reorganised the Ecole Polytechnique and the Paris Observatory. (See Adam's address, 'Monthly Notices of the R.A.S.', *36*, 232.)

134. Professor J. Challis, 1874.

133. Johann Gottfried Galle, 1812–1910, possibly the only astronomer to have seen Halley's comet at its returns in 1835 and 1910. He discovered Neptune 1846, September 23.

135. Edmund Halley, from a portrait by Thomas Murray, 1712, in the Bodleian Library, Oxford.

136. Joseph Louis Lagrange, 1736–1813.

137. Pierre Simon Laplace, 1749–1827.

138. George Biddell Airy, 1801–1892, from an engraving of 1868.

when he says that life transcends physics and chemistry, he means that biology cannot explain life in our age by the current workings of physical and chemical laws. The spectacular work of Watson and Crick[14] is challenged by Commoner,[15] the Watson Crick heretic who points out that the Watson–Crick theory proposes to explain in chemical terms a process which in nature is uniquely associated with life, that is, inheritance. Inheritance involves the reappearance in the offspring of those features of the parent organism which distinguish it from other forms of life, that is, its biological specificity. This is a process of self-duplication, for the specificity of the new organism is derived solely from the specificity of its parents. While other agents, such as suitable nutrients, are also necessary for the process of self-duplication, they do not contribute to the specificity of the product.

The Watson–Crick theory relates the origin and transmission of biological specificity to the biochemical specificity of nucleic acids (that is, the particular sequence of nucleotides that distinguishes one nucleic acid from another) and of proteins (that is, the particular sequence of amino-acids that distinguishes one protein from another). According to the theory, an organism's biological specificity (or at least that part mediated by protein enzymes) is derived solely from the nucleotide sequence of its DNA complement, and the organism's capability for self-duplication is derived from the capability of DNA to act as a 'self-duplicating molecule'.

In 1964 Commoner concluded, from evidence then available that (a) the biochemical specificity of a newly synthesized DNA molecule (that is, its nucleotide sequence) arises in part from the nucleotide sequence of a pre-existing DNA molecule, but also from the enzyme involved— DNA polymerase; thus DNA is not itself capable of self-duplication. He also proposed that (b) the biochemical specificity of protein (that is, its amino-acid sequence) is derived in part from DNA (through mRNA) and in part from the amino-acid-tRNA complexes involved in protein synthesis. Thus the biochemical specificity of protein, and the resultant biological specificity resulting from the enzyme activity of such a protein, is not derived solely from DNA. (c) Thus transmission of biochemical specificity within the cell is fundamentally circular rather than linear and the total system rather than any single constituent is responsible for the biochemical specificity which gives rise to biological specificity.

Commoner's attack upon this widely held theory is a powerful one and worthy of very close study. His final words are thought-provoking and poetic.

Biologists have confronted successively—like a nest of Chinese boxes—levels

[14] Crick, F. H. C., 'Cold Spring Harbor', *Symp. Quant. Biol.*, *31*, 3 (1966).
[15] Commoner, B., 'Failure of the Watson–Crick Theory as a Chemical Explanation of Inheritance', *Nature*, *220*, 340 (1968).

of complexity ranging from the ecosystem to the internal chemistry of the cell. The last box has now been opened. According to the Watson–Crick theory, it should have contained the single source of all the inherited specificity of living organisms—DNA. It is my view that we now know that the last box is empty and that the inherited specificity of life is derived from nothing less than life itself.

It is often forgotten that the idea of life throughout the solar system and beyond into outer space is not a modern scientific view but one that goes back into the early part of the sixteenth century.

Swedenborg[16] (1688–1772) showed that freshness of approach that is the prerogative of the mystic unfettered by the cramping need for evidence. In his survey of the earths in and beyond the solar system he 'encountered' many celestial inhabitants. Let us share his vision. On the Moon some spirits appeared to him and he tells us:

Their faces appeared not unhandsome, but longer than those of other spirits. In stature they were like boys of seven years old, but of more robust frame; so that they were dwarfs. I was told by the angels that they were from the Moon. The one who had been carried by the other came to me, applying himself to my left side under the elbow, and from thence he spoke, saying, that when they utter their voice they thunder in this manner; and that by so doing they strike with terror the spirits who would do them harm, and put some to flight, so that they go safely wherever they please. In order that I might know for certain that the sound they make was of this kind, he retired from me to some others, but not quite out of sight, and thundered in like manner. They showed me, moreover, that their voice, being sent forth from the abdomen after the manner of an eructation, thus resounded like thunder. It was perceived that this arose from the circumstance, that the inhabitants of the Moon do not, like the inhabitants of other earths, speak from the lungs, but from the abdomen, and thus from some collection of air therein; the reason of which is, that the Moon is not surrounded with an atmosphere of the same kind as that of other earths. I was informed that the spirits of the Moon, in the Grand Man, have relation to the ensiform or xiphoid cartilage, to which the ribs are attached in front, and from which descends the *linea alba*, which is the point of attachment of the abdominal muscle.

Closer to our own age scientific men of undoubted ability have made some remarkable use of the scientific evidence available to them. Proctor tells us that the elder Herschel's[17] ideas of the sun's fitness to be an abode for living creatures were fanciful in the extreme:

'The sun appears to be nothing else,' he wrote, 'than a very eminent, large, and lucid planet, evidently the first, or, in strictness of speaking, the only primary one of our system, all others being truly secondary to it. Its similarity

[16] Swedenborg, E., *Earths in our Solar System*, from the Latin, Swedenborg Society (London 1962).
[17] Herschel, Sir Wm., 'The Scientific Papers of', vol. I, 479 (1912), *Phil. Trans.*, pp. 46–72 (1795).

258

to the other globes of the solar system with regard to its solidity, its atmosphere, and its diversified surface; the rotation upon its axis, and the fall of heavy bodies, lead us on to suppose that it is most probably also inhabited, like the rest of the planets, by beings whose organs are adapted to the peculiar circumstances of that vast globe. Whatever fanciful poets might say in making the sun abode of blessed spirits, or angry moralists devise in pointing it out as a fit place for the punishment of the wicked, it does not appear that they had any other foundations than mere opinion and vague surmise; but now I think myself authorised *upon astronomical principles* to propose the sun as an inhabitable world, and am persuaded that the foregoing observations, with the conclusions I have drawn from them, are fully sufficient to answer every objection that may be made against it.'

Proctor continues with an illustration of the insight on these matters afforded to Herschel's son

Sir John Herschel he tells us was one of the last to abandon the belief in Nasymth's willow-leaves. For a time he even entertained the strange fancy that the willow-leaves are organisms, and may even possess some form of vitality. 'These flakes,' he said, 'are evidently the *immediate sources of the solar light and heat*, by whatever mechanism or whatever processes they may be enabled to develope, and as it were elaborate, these elements from the bosom of the non-luminous fluid in which they appear to float. Looked at in this point of view, we cannot refuse to regard them as *organisms* of some peculiar and amazing kind; and though it would be too daring to speak of such organisation as partaking of the nature of life, yet we do know that vital action is competent to develope at once heat and light and electricity.' Regarded as solar inhabitants whose fiery constitution enables them to illuminate, warm, and vitalise the whole solar system, the solar willow-leaves would not want that evidence of might which gigantic size affords. Milton's picture of him who on the fires of Hell 'lay floating many a rood' seems tame and commonplace compared with Herschel's conception of these floating monsters, the least covering a greater space than the British Islands.[18]

Lovejoy[19] explains that many Englishmen, and perhaps most English clergymen, of the early eighteenth century derived their general notions of astronomy largely from William Derham's *Astro-Theology, or a Demonstration of the Being and Attributes of God from a Survey of the Heavens* (1715). The book appeared under royal patronage, and its author was a Canon of Windsor and a Boyle Lecturer, as well as a Fellow of the Royal Society. It thus presumably represented a position officially approved by the orthodoxy of the time, theological as well as scientific. Derham unequivocally supports the infinitist cosmography, which, under the name of the 'New System', he carefully distinguishes from the Copernican.

[18] See *Monthly Notices of R.A.S.*, *25*, 152.
[19] Lovejoy, A. O., *The Great Chain of Being*, Harvard University Press 1936, Harper Torch Books 6th printing 1965. The William James Lectures on Philosophy and Psychology 1933.

It is the same as the Copernican as to the Systeme of the Sun and its Planets. . . . But then whereas the Copernican Hypothesis supposeth the Firmament of the Fixt Stars to be the Bounds of the Universe, and to be placed at an equal Distance from its Center the Sun; The *New Systeme* supposeth that there are many other Systemes of *Suns* and *Planets*, besides that in which we have our residence: namely that every Fixt Star is a Sun, and encompassed with a Systeme of Planets, both Primary and Secondary, as well as ours. . . . In all probability there are many of them (Systemes of the Universe), even as many as there are Fixt Stars, which are without number.

And Derham holds that all the planets (including the moon) in our own system, and all those in the infinity of other solar systems, are 'Places, as accommodated for Habitation, so stocked with proper Inhabitants'. This 'New System' he thinks 'far the most rational and probable of any'; and his first and chief reason for this opinion is the usual theological one:

(This System) is far the most magnificent of any; and worthy of an infinite CREATOR: whose *Power* and *Wisdom*, as they are without bounds and measure, so may in all probability exert themselves in the Creation of many Systemes, as well as one. And as Myriads of Systemes are more for the Glory of GOD, and more demonstrate his *Attributes*, than one, so it is no less probable than possible, there may be many besides this which we have the Privilege of living in.

And the moral which is drawn from the 'New System' is precisely that which the medieval writers and the early anti-Copernicans had drawn from the Ptolemaic:

From the consideration of the prodigious Magnitude and Multitude of the Heavenly Bodies, and the far more noble Furniture and Retinue which some of them have more than we, we may learn not to overvalue this world, not to set our hearts too much upon it, or upon any of its Riches, Honours, or Pleasures. For what is all our Globe but a Point, a Trifle to the Universe! a Ball not so much as visible among the greatest part of the Heavens, namely the Fixt Stars. And if Magnitude or Retinue may dignify a Planet, *Saturn* or *Jupiter* may claim the preference; or if Proximity to the most magnificent Globe of all the Systeme, to the Fountain of Light and Heat, to the Center, can honour and aggrandize a Planet, then *Mercury* and *Venus* can claim that dignity. If, therefore, our World be one of the inferior parts of our Systeme, why should we inordinately seek and desire it?

Derham, moreover, adds the pleasant suggestion that among the principal advantages of 'the Heavenly State' will be improved facilities for astronomical observation—or exploration.

We are naturally pleased with new things, we take great Pains, undergo dangerous Voyages, to view other countries: with great Delight we hear of new Discoveries in the Heavens, and view these glorious Bodies with great Pleasure through our Glasses. With what pleasure, then, shall departed, happy Souls survey the most distant Regions of the Universe, and view all those glorious Globes thereof, and their noble Appendages, with a nearer View.

The truly revolutionary theses in cosmography which gained ground in the sixteenth and came to be pretty generally accepted before the end of the seventeenth century were five in number, none of them entailed by the purely astronomical systems of Copernicus or Kepler. In any study of the history of the modern conception of the world, and in any account of the position of any individual writer, it is essential to keep these distinctions between issues constantly in view. The five more significant innovations were: (1) the assumption that other planets of our solar system are inhabited by living, sentient, and rational creatures; (2) the shattering of the outer walls of the medieval universe, whether these were identified with the outermost crystalline sphere or with a definite 'region' of the fixed stars, and the dispersal of these stars through vast, irregular distances; (3) the conception of the fixed stars as suns similar to ours, all or most of them surrounded by planetary systems of their own; (4) the supposition that the planets in these other worlds also have conscious inhabitants; (5) the assertion of the actual infinity of the physical universe in space and of the number of solar systems contained in it.

The first of these—and, of course, still more the fourth—deprived human life and terrestrial history of the unique importance and momentousness which the medieval scheme of ideas had attributed to them and Copernicanism had left to them. The theory of the plurality of inhabited worlds tended to raise difficulties, not merely about the minor details of the history included in the Christian belief, but about its central dogmas. The entire moving drama of the Incarnation and Redemption had seemed manifestly to presuppose a single inhabited world. If that presupposition were to be given up, how were these dogmas to be construed, if, indeed, they could be retained at all? Were we, as Thomas Paine afterwards asked, 'to suppose that every world in the boundless creation had an Eve, an apple, a serpent, and a Redeemer'?

This indeed is the present position with many Christians who see the Earth, the stage on which the drama of the fall and the redemption of man through Christ was set, as a unique place and science can have little or no part in this. They see as with Oupnekhat 'Et is similis spectatori est, quod ab omni separatus spectaculum videt'.[20]

But what of the thoroughly hard-going pragmatic scientific approach? Prejudice in scientific matters is deeply ingrained and never more so than in the answers that are given to the question 'Is there life on other worlds?' The evidence for and against is meagre and the answer is

[20] 'And he is like a spectator, because, separated from everything he beholds a drama.'

always an opinion or assertion, not a statement of scientific fact.[21] The position today is such that in my view an unrealistic note of absolute dogmatism pervades the utterances of practically all of those who answer the question in the affirmative. The landing of men on the Moon (1969) has engendered a surge of rash speculation with statements over the mass media to the effect that the achievement is the greatest thing to have happened since 'the fish stood up and walked out of the sea', and that the answer to extra-terrestrial life is at hand. This in defiance of the true position, for no one is able to show, outside of fiction, that the fish have ever stood up and walked out of anything or that the problem of extra-terrestrial life is any less intractable since Apollo XI landed men on the Moon.

Puccetti[22], for example, analyses in turn three candidates for the role of persons other than human persons—the supercomputer, the organic artifact, and the extra-terrestrial being. The first of these he rejects on the grounds that a true machine cannot have feelings, and without feelings it will not qualify as a moral agent. The second candidate, the organic artifact, he rules out for the reason that none, to our knowledge, exist as yet. This leaves the third alternative, the extra-terrestrial being, and the major part of this work is devoted to this particular problem. He might with commendable foresight have saved himself the task of writing his book and his readers also much labour, since a small measure of thought will show that there is no satisfactory evidence for the third alternative either.

Puccetti takes the view from the start that evolution is the sole mode of animal genesis. On Mars for example he tells us that 'the scarcity of oxygen and formidable temperature variations *limit evolution* to the simplest forms of life'. Indeed in every case his appeal is to *evolution* of life. Four extracts will show the position he adopts and this is typical of most other commentators.

(a) About 5 per cent. of all visible stars are both single—hence capable of providing stable planetary orbits—and of the right size to create 'habitable' temperature zones for the spontaneous generation and *evolution* of life.
(b) The large early spectral types (O, B, A, and early F) with broad habitable zones, unfortunately evolve too quickly themselves for *biological evolution* to take place on whatever planets they might have.
(c) Even without sunlight, this internally produced energy should be sufficient for chemical evolution and photosynthesis of a distinct sort. Thus life could be generated on such crusted stars and self-heating distant planets; and if so, why should not intelligent beings *evolve* there?

[21] 'There is something fascinating about science. One gets such wholesale returns of conjecture out of such a trifling investment of fact.'
MARK TWAIN, *Life on the Mississippi*
[22] Puccetti, R., *Persons. A study of possible moral agents in the Universe* (London 1968).
(For a more recent study of the problem see Dole, S. H., *Habitable Planets for Man*, 2nd ed., 1970.)

(d) About 5 per cent. of all single stars are of the right size to have planetary systems in which one or more planets would fall in the 'habitable' temperature zone for long periods of time; and life could generate spontaneously and *evolve* into intelligent forms during that period of time by the same means as obtained on the surface of the Earth.

A not dissimilar case is advanced by Sullivan[23] in a popular work. He relies primarily on a statistical case in which the development of complex organic systems is the outcome of random interactions by appeal to the oft cited illustration of the monkey on the typewriter who will eventually, given the time, write *Hamlet*. (Some commentators go as far as to say *all* the books in the British Museum.) This is an unsatisfactory hypothesis when one considers that a monkey instead of following a true random sequence over an infinite period will no doubt choose certain keys or sequences of keys and may well produce a lifetime of gibberish.[24] We should reflect that the solar system has not been in existence for more than 10^{18} seconds.

[23] Sullivan, W., *We are not alone* (Penguin Books 1969).
[24] The problems of the monkey on the typewriter is of perennial interest and I am indebted to Dr R. L. Tweedie (*Personal communication from Statistical Laboratory, University of Cambridge,* 1970 April 10). for the following analysis.

Suppose the monkey is seated before, not a typewriter, but a linotype machine, and that this supposedly immortal monkey is randomly hitting the keyboard, and continues to randomly hit the keyboard.

Suppose further that the keyboard has sixty elements, fifty-two letters, lower and upper case for convenience on separate keys, seven punctuation marks (. , ; : ' ! ?) and a space bar. This is probably sufficient for *Hamlet*, and if one wants brackets, or any other symbols, one can use sixty-five or seventy if desired.

Then let us take a standard edition of *Hamlet*—say that in *W. Shakespeare, The Complete Works*, edited by Charles Jasper Sisson (Odhams, 1960). This has, I calculate, of the order of 260,000 symbols in it—it can then be shown that the *average* length of time needed for the monkey hitting the machine at the rate of one symbol per second, to produce this copy of *Hamlet* would be $60^{260,000}$ secs, or $10^{460,000}$ secs. This is, of course, a number of incredible size: the earth has been cool, I believe, for some 5000 million years, and this is rather less than 10^{18} seconds.

What conclusion can one draw from this result? The tempting thing to say is that this proves more or less conclusively that Shakespeare was not a random number generator, but that *Hamlet*, since it has been produced in rather better than average time, must have another factor entering its composition, say intelligence: the flaw in this argument is, of course, that it is we who judge *Hamlet* to be coherent—objectively, one should say that Shakespeare would take just as long to produce the first 260,000 monkey-letters, whatever they might be, and so one has to judge *a priori* which of the two sequences is more valuable.

One might also argue that, given an alphabet the monkey might just as well be asked to dip into a dictionary of, say, 3,000 words (which I believe to be about the size of Shakespeare's vocabulary), and thus produce *Hamlet*. This would reduce the average time to completion of the play to something like $10^{53,000}$ seconds.

And again, if the monkey is not just randomly scratching with a stick, but has an alphabet, one may suppose he also has a dictionary and that he also has someone standing over him to throw out all the meaningless combinations he has made: this would correspond much more closely to the working conditions of Shakespeare who had, in general, to obey the rules of grammar.

Then one might require a critic, to remove non sequiturs of plot or character, and so on. For *Hamlet* is far from random, in the sense that, if half of it were presented, written, to any literate person one would have far fewer completions possible than if the first six lines were presented, say. Thus one has to beware of two things in using the figures above: firstly, one has to be aware that one has adjudged *Hamlet* non-random in a subjective, not an objective, manner; and secondly, that *Hamlet* was written in English, with all the rules and regulations that implies.

The solar system

The point is better expressed by the Jansenist logician Arnauld of Porte Royale, 'In some cases the likelihood of success is so slight that no matter how great the advantage or how small the expense, good sense advises against risking a wager. It would be sheer folly to bet even ten coppers against 10,000 gold pieces that a child arranging at random a printer's supply of letters would compose the first twenty lines of Virgil's Aeneid.'

Kepler's attack upon the Epicurean view that the star of 1572 was a fortuitous concourse of atoms is noted by Sir David Brewster.[25]

When I was a youth with plenty of idle time on my hands, I was much taken with the vanity, of which some grown men are not ashamed, of making anagrams, by transposing the letters of my name, written in Latin. Out of *Joannes Keplerus* came *Serperns in Akuleo* (A serpent in his sting); but not being satisfied with the meaning of these words, and being unable to make another, I trusted the thing to chance, and taking out of a pack of playing cards as many as there were letters in the name, I wrote upon each, and then began to shuffle them, and at each shuffle to read them in the order they came, to see if any meaning came of it. Now, may all the Epicurean gods and goddesses confound this same chance, which, although I have spent a good deal of time over it, never showed me anything like sense even from a distance.

So I gave up my cards to the Epicurean eternity, to be carried away into infinity; and, it is said, they are still flying about there in the utmost confusion among the atoms, and have never yet come to any meaning. I will tell those disputants, my opponents, not my own opinion, but my wife's.

Yesterday, when weary with writing, and my mind quite dusty with considering these atoms, I was called to supper, and a salad I had asked for was set before me. 'It seems, then,' said I aloud, 'that if pewter dishes, leaves of lettuce, grains of salt, drops of water, vinegar, and oil, and slices of egg, had been flying about in the air from all eternity, it might at last happen by chance that there would come a salad.' 'Yes', says my wife, 'but not so nice and well-dressed as this of mine is!'

Let us now re-examine the case for life on other worlds as it is currently presented. It rests, as I see it, on three clearly separated hypotheses or fictions.

The *first* asserts that there are billions of planetary systems in the cosmos not dissimilar to our own solar system.

The *second* asserts that primary organisms have evolved from inorganic matter contained in a primeval ocean which once covered our planet Earth and that the same phenomenon is therefore to be expected to have taken place in the hypothetical oceans of other hypothetical planets similar to Earth.

The *third* asserts that the primary organisms alleged to have appeared according to the preceding hypothesis, will have evolved into complex

[25] Brewster, D., *The Martyrs of Science*, p. 202 (London 1874).

264

organisms by a process of transformism (i.e. evolution) on the Darwinian model[26] as postulated for the primary organisms of the Earth.

This assembling of fiction upon fiction or of hypothesis upon hypothesis is intellectually unobjectionable provided the tenuity of the nexus in the argument is not lost sight of. What is objectionable is the blatant claim too frequently made today that life *does in fact* exist on other worlds.

FICTION[27]

The first fiction is a statistical abstraction that Shapley[28] gives succinctly, on the basis of an initial guess in our own Galaxy[29] and then extrapolates to embrace the population of the extra galactic nebulae[30] out to the periphery of the known universe. He asserts that there are more than 10^{20} stars, each one capable of maintaining the photo-chemical reactions that are the basis of known terrestrial plant and animal life. At the very lowest he postulates one star in a thousand to possess a planetary

[26] *Darwinian model.* That is to say the non-proven biological theory of Charles Darwin concerning the evolution of species, etc., set forth especially in his works entitled *The Origin of Species by means of Natural Selection, or the preservation of favoured races in the struggle for life* (1859), and *The Descent of Man and Selection in relation to Sex* (1871). The origination of species of animals and plants, as conceived by those who attribute it to a process of development from earlier forms, and not to a process of 'special creation'. Often in phrases *Doctrine, Theory of Evolution.*

[27] The use of the word *'fiction'* is one that has given me much thought. I would have preferred to use *'hypothesis'*—but from a careful reading of Professor H. Vaihinger's seminal work *The Philosophy of 'As If'* (London 1924), I think *'fiction'* is correct. Undoubtedly fiction and hypothesis are externally very similar but hypothesis tries to discover, fiction to invent. One single fact at variance with an hypothesis destroys it. The real difference between the two is that fiction is a mere auxiliary construct, a scaffolding to be demolished, while hypothesis looks to definite establishment. As long as fictions are treated as hypotheses without a realization of their nature, they are *false hypotheses.*

[28] Shapley, H., *Of Stars and Men*, p. 67 (Boston 1958).

[29] *The Galaxy.* The name given to the milky white appearance that encircles the heavens now known to be the belt of stars that form our own system of stars in which the Sun is but a minor star. The milky way—a term used in English since *c.* 1384. Chaucer: 'Se yonder too the Galoxie whiche men clepeth the milky weye. Other remote systems of stars are generally known as the extra-galactic nebulae.

Ten thousand Suns, prodigious Globes of Light
At once in broad Dimensions strike our Sight;
Millions behind in the remoter Skies,
Appear but Spangles to our wearied Eyes:
And when our wearied Eyes want further Strength,
To pierce the Void's immeasurable Length,
Our vigorous tow'ring Thoughts still further fly,
And still remoter flaming Worlds descry.
But ev'n an Angel's comprehensive Thought,
Cannot extend so far as thou hast wrought;
Our vast Conceptions are by swelling brought,
Swallow'd and lost in Infinite, to nought.

[30] The extra galactic nebulae are the nebulae outside of the galaxy or Milky Way system. The galaxy is part of the Local Group of galaxies which comprise some twelve nebulae less than 450,000 parsecs from one another. viz. the two Megallanic Clouds, the elliptical galaxies in Sculptor and Andromeda, the galaxy in Canes Venatici, in Fornax Leo, Ursa Major, and Draco, also the Maffei galaxies. (See *Nature, 231*, 35 (1971) and Journal of the R. Ast Soc. of Canada, *62*, 145, 219 (1968).)

system (he personally would put it at one in fifty), one in a thousand of these to possess a planet at the correct distance from the star to provide a suitable environment for terrestrial protosplasm to exist in, one in a thousand of these planets to be large enough to hold an atmosphere and finally one in a thousand of these with air and water to allow naturally arising complex inorganic molecules to develop into organic molecules. He then summates these four separate one-in-a-thousand chances, each of which is a guess, to obtain the answer that one star in 10^{12} stars meets his criteria. He now divides 10^{20} by 10^{12} and obtains the answer that there *are* 10^8 planetary systems suitable for the second hypothesis to work in.

Any reasonable man would wish to give some latitude to the assumptions he makes and thus arrive at a numerical range for the number of life-bearing planets, rather than a single number; 10^8. This general weakness in scientific inference, exemplified in Shapley's reasoning, has been exposed by Edwards.[31] Shapley's solitary figure of 10^8 is a guess and not even an informed guess, since he had no evidence available to him to lend it any validity.

In 1949 Hoyle marshalled the evidence for life elsewhere in the universe in a radio talk given over the British Broadcasting network. He was keen to form the planetary systems from binary stellar systems in which one component is a supernova. This gave him an ingenious yardstick with which to work out the number of planetary systems in the cosmos. Hoyle states that it is known from observation that supernovae occur in each galaxy of stars at the rate of one every 500 years. To our own galaxy he accords an age of 5×10^9 years; hence 10^7 supernovae will have appeared and thus 10^7 planetary systems will have developed in the galaxy.

There are 10^8 galaxies and thus ($10^8 \times 10^7$) planetary systems (i.e. 10^{15}). Hoyle estimates (in contradistinction to Shapley) that there are 10^6 planetary systems suitable for life in our galaxy and hence 10^{12} planetary systems for life in the known cosmos. This unique planetary-forming idea of Hoyle has not gained general acceptance, indeed the formation of our own solar system is still very much unresolved, and Hoyle has now rejected his own supernovae hypothesis.

We have then two estimates for the number of possible planets in the cosmos available to life 10^8 (Shapley), 10^{14} (Hoyle), one only of these is based but in part on evidence and that evidence is now regarded to be without value in the genesis of planetary systems. Such figures must be treated with the greatest of caution and on no account are to be enthroned in the mind as facts.

Consider now the so-called direct evidence on the question of planetary

31 Edwards, A. W. F., 'Statistical Methods in Scientific Inference', *Nature*, *222*, 1233–7 (June 1969).

bodies in the galaxy. In 1963 Van de Kamp reported the 'discovery' of a new planet after twenty years' assiduous searching of the heavens. The 'new' planet was not observed but postulated from the dynamic aberrations of Barnard's star B. in its celestial path, which is found to deviate from the accepted uniform rectilinear motion. Barnard's star had been photographed by Van de Kamp and his colleagues every year since 1938 to produce 2413 photographic plates from which it was possible to calculate the motion and distance of the star. On the assumption that Barnard's star has 15 per cent of the Sun's mass, the unseen companion is said to be one and a half times the mass of Jupiter.

The broad question of the planetary companions of stars has been reviewed by Van de Kamp.[32] The method of detection is one of long focus photographic astrometry combined with a calculation of the lower limit of the mass of the unseen companion. The evidence to date for the existence of planetary companions is sparse. The possibility of a companion to the star 70 Ophuichi has not been confirmed. Another spurious result is that of the postulated companion to the star Lalande 21185 derived by Mannino in 1951. Investigations concerning the stars Cincinnati 1244 and η Cassiopeia continue but the evidence is not conclusive. The existence of an often quoted companion to the star Krüger 60 is not yet confirmed. An unseen companion of 0·016 solar mass is believed to exist about the well-known double star 61 Cygni. The difficulties in labour, accuracy of observation and calculation are immense. Something of the difficulty is shown by Su Shu Huang.[33] He records that 'to detect the existence of a planet the size of Jupiter in a hypothetical Sun–Jupiter system at a distance of ten parsecs[34] requires an ability to measure an astronometric angular devitation of 0·00005 seconds of arc or an ability to measure by spectroscopy a radial velocity correct to 0·01 km./sec or to measure by photometry a change in luminosity[35] of 0·01 magnitude'.[36]

The difficulty of obtaining any meaningful results in the immediate future is considered by O'Leary.[37] It should be kept clearly in mind that

[32] Van de Kamp, 'Planetary Companions of Stars', *Vistas in Astronomy*, vol. 2 (London 1956).
[33] Su Shu Huang, 'The Problem of Life in the Universe and the Mode of Star Formation', *Pub. A.S.P.*, *71*, 421 (1959).
[34] *Parsec*. The chief unit used in measuring stellar distances, being the distance at which the semi-major axis of the earth's orbit would subtend one second of arc; hence, a star's distance in *parsecs* is the reciprocal of its annual parallax in seconds of arc; one parsec equals 30·857 × 10¹² km, 19·174 × 10¹² miles, 206,265 a.u., 3·2616 light years.
[35] *Luminosity and magnitude*. The measure of the amount of light actually emitted by a star, irrespective of its distance; synonym for *absolute magnitude*.
[36] *Magnitudes*. The scale by which the brightness of stars is measured: (1) *apparent magnitude* is the measure of the brightness on Pogson's logarithmic scale, in which each step of one whole magnitude represents a light ratio of 2·512, and this increases numerically with decreasing brightness; (2) *absolute magnitude* is the apparent magnitude a given star would have at the standard distance of 10 parsecs. This is in harmony with Fechner's Law, which states that equal increases of sensation are produced by equal *ratios* of stimulus.
[37] O'Leary, B. T., 'On the Occurrence and Nature of Planets outside the Solar System', *Icarus*, *5*, 419 (1966).

variations in proper motions[38] are not of necessity to be attributed to an unseen planet but clearly all unseen companions whatever their nature will produce perturbations of some magnitude though whether they will be observed or not is open to doubt. It is wrong to accept uncritically the present view that dynamical anomalies are only to be explained by the presence of superplanets. The perturbing mass is unseen and no one can be certain on the point. To say that a superplanet *is discovered* is one of those statements which show a complete lack of integrity in presenting the evidence.

It cannot be emphasized sufficiently that the statistical argument concerning possible planetary companions to 'near' stars is a statistical abstraction since it makes no appeal to experimental evidence which alone can give it validity. In any statistical argument worthy of consideration the link with reality is the sample which is taken for the ground on which the edifice is to stand. In this argument no sample has been taken and none is yet within our skill. The numbers of planetary systems in the Galaxy is unknown and there is therefore no firm ground from which to make an extrapolation in order to obtain a figure for the number of planetary systems in the cosmos.

FICTION II

The second fiction in the nexus of argument is expressed with erudition by Oparin.[39] During the first quarter of this century there was little or no discussion on the nature and origins of life—a profound point for anyone interested in the philosophy of science. An article by Haldane aroused a flicker of interest in 1929 but it was Oparin's *Origin of Life* published in 1937 which triggered off the many speculations since. Oparin takes the story in several distinct stages. First he forms the Earth out of the Sun's atmosphere on the now discredited theory of Jeans, asserts that carbon made its first appearance on the Earth's surface, not in the oxidized form of carbon dioxide but in the reduced state in the form of hydrocarbons and that nitrogen first appeared in its reduced state in the form of ammonia. He then postulates cozaervation in the helter-skelter accumulation of moving particles to obtain varying concentrations at different points of the aqueous medium.

He finds difficulty in explaining the exceptional energetic activity of

[38] *Proper motion.* That component of a star's own motion in space which is at right-angles to the line of sight, so that it constitutes the apparent change of position of the star on the celestial sphere.

[39] Oparin, A. I., *Origin of Life* (London 1938), republished by Dover publications in paperback form with introduction by S. Margolis (1953); see also the Academic Press Ed. NY (1957). His book of 1937 is still the best source of his general ideas and this is supported by the issue in 1953 of the Dover edition with an erudite introduction by Sergius Margolis, Professor of Biochemistry at the University of Nebraska, and the very recent translation of Oparin's *Genesis of Evolutionary Development of Life*, translated from the Russian by Eleanor Maass (Academic Press 1969), which is the 1937 book brought up to date. (Some inaccuracies in this later work have been noted; see *Nature, 223* 103 (1969).)

enzymes and pathways for the appearance of such a perfect apparatus in the living cell (p. 168).[40] To overcome this difficulty he appeals to natural selection but omits to say how the selective intelligence operates in inanimate structures (p. 175). In this manner (p. 195) he produces 'the simplest primary organisms' and with their appearance the question of the origin of life on Earth is, in his view, closed since to use his own words 'what follows now is a history of the *evolution* of living creatures'. He admits, however, that 'unfortunately the problem of the origin of the cell is perhaps the most obscure point in the whole study of the evolution of organisms.'

More recently this question has been examined by Bernal,[41] one of the world's most distinguished crystallographers. He presents the unusual scientific nature of the case in these words: 'We are here trying to settle a question of a different character from those of the rest of science; it is not merely a description . . . but an attempt to carry out an intellectual reconstruction based on assumptions of inner logic themselves drawn from experimental science here and now.' The inner logic follows some curious pathways since Bernal expresses his dislike of the philosophical emptiness of special creation yet accepts it as axiomatic that life arose on the surface of the Earth. He departs from Oparin's views decisively in thinking that cells are a late stage in the process, preceded by about 2,000 million years of chemical evolution of macro molecules in 'sub-vital areas' not defined by membranous boundaries. Bernal is unable to explain—except by chance—how the key phenomenon of molecular self-replication arose in the course of chemical evolution. As with Oparin—once the intellectual schema has produced the cell from molecules and the molecules from organelles the story is abandoned to the theory of evolution (transformism) to be found elsewhere in the literature on natural history.

One is left with the feeling that he is guided by an assumption and pursues a purpose, that of finding what he presupposes and not unnaturally arrives at a safe intellectual haven. Both Oparin and Bernal allow evolutionary concepts to enter their reasoning which then ceases to be exact. They both require unlimited time for the creation of organisms, and especially for their further evolution to operate in. (The reader may reflect that the reduction of the toe of Eohippus—the alleged progenitor of the parade horse of today—is said to require 60×10^6 years.) Yet time *per se* achieves nothing, and they are compelled from the start to break the rapid synthesis of organic matter. Even chemical processes become evolutionary, in spite of the fact that at present they are repeated nearly instantaneously in any terrestrial laboratory. The problem of life from inorganic compounds has been extensively studied

[40] Pages in brackets refer to the Dover edition (1953).
[41] Bernal, J. D., *The Origin of Life* (London 1967).

more recently in 'The Origins of Prebiological Systems and their molecular matrices' (edited by S. W. Fox, Academic Press, 1965), in which the papers from a conference on the origin of life are gathered together and which have been subjected to a severe criticism by Ulbright[42] who asks some extremely pertinent questions in a review paper. These questions are seldom formulated and are worthy of careful consideration.

Ulbright points out that many of the papers in the Fox compilation describe experiments relating to the synthesis of simple organic compounds such as amino-acids, nucleic acid bases and so on, under 'primitive earth conditions', using heat, high-voltage electrical discharge or ultra violet radiation as the energy source. The chronology of the earth's history accepted at the moment is that from its beginning 4.5×10^9 years ago until 4×10^9 years ago, a strongly reducing atmosphere prevailed (containing hydrogen and the hydrides of the common elements—water, ammonia and methane). From that time until about 1.8×10^9 years ago, the atmosphere is believed to have been neither strongly reducing nor strongly oxidizing, containing neither free hydrogen nor free oxygen (and with nitrogen and carbon dioxide replacing ammonia and methane). The strongly oxidizing atmosphere we know today dates from 1.8×10^9 years ago.

Since the oldest known fossil, a calcareous alga, is dated at 2.7×10^9 years, one may suppose that towards the end of the second period some free oxygen may have been present. For, had oxygen been completely absent, there would have been no ozone in the upper atmosphere and consequently the terrestrial atmosphere would have been transparent to solar ultra violet radiation in the critical 240–300 mg. region (in which nucleic acids, for example, absorb very strongly). If this ozone 'barrier' were now suddenly to disappear, it is doubtful if life could survive on the earth. With this as background, one can try to consider the significance of the experiments reported, in which, for example, amino-acids are formed in gas phase reactions and trapped in water. Vallentyne, in his paper, pointed out that these experiments ignore the time element. The actual concentration that might have occurred on the primitive earth would be the result of synthesis and decomposition. 'The results obtained from experiments run over a few hours may bear no relation whatever to the actual state of affairs resulting after a period of thousands or millions of years.'

There was no comment on this in the discussions, and the decomposition of organic compounds as a result of the intense ultra violet radiation that must have prevailed on the primitive earth was not considered. Calculations by Hull[43] have shown that, as a result of this radiation, the

[42] Ulbright, T. L. V., *Chemistry and Industry*, pp. 43–5 (1966).
[43] Hull, D. E., *Nature, 186*, 693 (1960).

maximum concentration in the ocean of the simplest amino-acid, glycine, might have been 10^{-12} molar (it might have been as little as 10^{-27} molar). The concentration of more complex compounds would have been far less. From thermo-dynamics, the equilibrium concentration of glucose is 10^{-134} at unit activities of the component reactants. Bernal has conceded that the original concept of the 'primitive soup' must be rejected as far as the oceans are concerned.[44]

In his paper, Bernal posed a number of awkward questions. Here is one of them: 'Which of the various synthetic studies that have been made of the formation of elementary molecular compounds is relevant to the question of the origin of life?' This can be considered in conjunction with Lipmann's comment: 'That one finds such compounds (amino-acids, for example, in the experiments mentioned) doesn't tell us that this is a process really related to the origin of life. It means only that what the living organism does effectively in an organized way, can ineffectively be done in an unorganized way outside the living organism.'

An example is the synthesis of adenine nucleotides by irradiation of solutions containing adenosine (10^{-3} molar) and ethyl metaphosphate, described by Sagan, and which has been published. It was stated that 'in primitive terrestrial oceans . . . ethyl metaphosphate was probably not the most abundant phosphorus source'. As this reagent is prepared by the reaction of phosphorus pentoxide with anhydrous ether in chloroform solution and is rapidly hydrolyzed by water, one may agree. Nevertheless, a calculation of the production rate of adenosine triphosphate in these 'primitive terrestrial oceans' was based on the assumption that 'conditions there were similar to those in the present experiments'.[45] In a review in the same journal, this paper was cited as evidence that nucleoside triphosphates could have been produced in significant amounts in the primitive oceans, but the reagent was not mentioned.[46]

Rather appropriately, it was in the discussion of Fox's paper on the thermal synthesis of amino-acids at 1000°C. that the most heat was generated. Sagan criticized this work as irrelevant, because of the high temperature used and of the necessity of absorbing the products in water before they are decomposed at the same temperature. Fox believes that volcanic regions could have provided the high temperatures required and that rain would save the products from decomposition. When the critics were not convinced, he said 'the premise that it does not rain on volcanoes cannot be defended.'

If we grant that simple organic precursor molecules could have been formed on the primitive earth, the question of how they were built up into a living system remains. Even if we suppose that when life first began

[44] Bernal, J. D., *Nature, 186*, 694 (1960).
[45] Ponnamperuma, C., Sagan, C., and Mariner, R., *Nature, 199*, 222 (1963).
[46] Ponnamperuma, C., *Nature, 201*, 337 (1964).

it was much simpler than the simplest unicellular organisms known today, any replicative system we can conceive would appear to require the existence of ordered polymers. The formation of such polymers and of primitive enzymes is usually ascribed to 'prebiological natural selection', i.e. an extension of the concepts of evolution to the world of molecules. This was strongly criticized at this conference by Dobzhansky, and by Mora, who regards 'prebiological natural selection' as a contradiction in terms. 'Molecules are supposed to have accurate and persistent self-copyability, sufficient to resist randomization and yet to have a moderate mutability rate, leading to the "evolution" of the first self-reproductive system', which seems extraordinarily improbable because, as Mora says elsewhere, you cannot get more order out of a system than you put in.

There seems little opportunity to get any direct evidence from the biologists and against the ideas of Oparin and Bernal we can set the equally occult investigations of the mathematicians. In this change of disciplines we may find with Arbuthnot—'that mathematics are friends in as much as they charm the passions, restrain the impetuosity of imagination and purge the mind from prejudice'. Charles Eugene Guye, a Swiss mathematician, calculated the chances of manufacturing a single molecule of some protein like substance and also the quantity of material to be mixed in the experiment. The odds according to his calculations come out at 10^{320} to 1 against, with the quantity of material required to produce one molecule larger than that in the known universe, indeed sextillion sextillion sextillion times in excess. The calculation for the time involved for the experiment to have taken place on Earth is 10^{234} years which is an interesting figure to compare with the present estimated age of the Earth 4×10^9 years. Proteins are very individual substances and the long chains linked by the amino acids cannot 'as we have seen' be put together in any way one chooses, indeed the wrong order of assembly may cause them to be inimical to life. Professor Leathes has calculated that the links in the chain of a single protein may be put together in 10^{48} different ways. When we reflect that the Earth has only been sufficiently cool to accommodate life for some 10^{18} seconds we may marvel at the urgency with which chance would have to operate to show some small measure of success.

More recently a further mathematical boulder has been cast into the evolutionist's pond and the ripples are likely to be wide and persistent provided the event does not fall beneath the notice of astronomers and biologists. Salisbury[47] calculates that a typical small protein might contain about 300 amino-acids, and its controlling gene about 1000 nucleotides (three for each amino-acid). Because each nucleotide in a chain

[47] Salisbury, F. B., 'Natural Selection and the Complexity of the Gene', *Nature*, *224*, 342 (1969).

represents one of four possibilities, the number of different kinds of chains is equal to the number 4 to the power of the number of links in the chain; that is, $4^{1,000}$, or about 10^{600}.

Imagine that the primaeval ocean was uniformly 2 km. deep, covering the entire Earth, containing DNA at a concentration of 0·001 M (about 700 g of DNA/1. of solution), each double stranded molecule with 1,000 nucleotide pairs. Also imagine that each DNA molecule reproduces itself one million times per second, a single nucleotide substitution (a mutation) occurs each time a molecule reproduces, and no two DNA molecules are ever alike. In four billion years, $7·74 \times 10^{64}$ different kinds of DNA molecules will be produced. On 10^{20} similar planets in the Universe, this would be $7·74 \times 10^{84}$ (say 10^{85}) different molecules. If only one DNA molecule were suitable for our act of natural selection, the chances of producing it in these conditions are $10^{85}/10^{600}$ or only 10^{-515}. If 10^{100} different kinds of molecules could each carry out the necessary precursor synthesis, this is equivalent to saying that 166 of the nucleotides might be changed without loss in ultimate activity of the enzyme. Still, only one molecule out of every 10^{500} would be acceptable, and after four billion years on 10^{20} planets, 10^{415} of the first 10^{500} possibilities remain to be synthesized. The chances are, then still unimaginably small (10^{-415}) that a proper DNA molecule would be produced in this time. And if the proper molecule did appear by that fantastic accident, the problem comes up again the next time a precursor becomes limiting.

In the 2 km. deep oceans on the 10^{20} planets during the 4×10^{12} years, the DNA chain can have only 141 nucleotides if all 10^{85} possible kinds are to be produced. This would code a protein chain only forty-seven amino-acids long.

The point of these numbers is that one DNA chain 1000 nucleotides long can be a unique individual in a population of 10^{600} other unique individuals. Numbers of this size have no precedent in anything but the concepts of information theory. Assume, for example, a cubic universe with dimensions of 20 billion light years on each side. In Ångstroms, this would be about 10^{39} Å on a side, with a volume of 'only 10^{117} Å^8. Imagine the number of universes required to contain 10^{600} tightly-packed DNA molecules!

In spite of the wild assumptions, the problem should be apparent. In the evolution of life on Earth, we are dealing with millions of different life forms, each based on many genes. Yet the mutational mechanism as presently imagined could fall short by hundreds of orders of magnitude of producing, in a mere four billion years, even a single required gene.

To compound the problem, consider the fantastic information content of a nucleus. The DNA in man contains about 10^9 nucleotide pairs per nucleus (other organisms from 10^7 to 10^{11} pairs). Written in standard type, this would occupy about 1000 volumes (10^9 bits, 2000 bits/page, 500 pages/volume). Britten and Kohne have shown that certain DNA sequences of higher organisms recur anywhere from a thousand to a million times per cell. Hence much is redundant. Work on amino-acid sequences within a given protein also implies a high redundancy. Yet if only one tenth of the genome in man is relevant, that is still 10^8 bits of information (100 volumes). Could the mutation process account for it?

Special creation or a directed evolution would solve the problem of the complexity of the gene, but such an idea has little scientific value in the sense of suggesting experiments.

This brings me to fiction No. III.

FICTION III

How realistic is it to advance the case for extra terrestrial life on the fiction that neo-Darwinian evolution has occurred on the planet Earth? Practically all biologists and the greater part of the informed lay public accept it as fact. But universal acceptance is no substitute for evidence. Anyone familiar with Schopenhauer's stratagems will recall his eternally valid warning:

But to speak seriously, the universality of an opinion is no proof, nay, it is not even a probability, that the opinion is right.

When we come to look into the matter, so-called universal opinion is the opinion of two or three persons; and we should be persuaded of this if we could see the way in which it really arises.

We should find that it is two or three persons who, in the first instance, accepted it, or advanced and maintained it; and of whom people were so good as to believe that they had thoroughly tested it. Then a few other persons, persuaded beforehand that the first were men of the requisite capacity, also accepted the opinion. These, again, were trusted by many others, whose laziness suggested to them that it was better to believe at once, than to go through the troublesome task of testing the matter for themselves. Thus the number of these lazy and credulous adherents grew from day to day, for the opinion had no sooner obtained a fair measure of support than its further supporters attributed this to the fact that the opinion could only have obtained it by the cogency of its arguments. The remainder were then compelled to grant what was universally granted, so as not to pass for unruly persons who resisted opinions which every one accepted, or pert fellows who thought themselves cleverer than anyone else.

When opinion reaches this stage, adhesion becomes a duty; and henceforward the few who are capable of forming a judgment hold their peace. Those who venture to speak are such as are entirely incapable of forming any opinions or any judgment of their own, being merely the echo of others' opinions; and, nevertheless, they defend them with all the greater zeal and intolerance. For what they hate in people who think differently is not so much the different opinions which they profess, as the presumption of wanting to form their own judgment; a presumption of which they themselves are never guilty, as they are very well aware. In short, there are very few who can think, but every man wants to have an opinion; and what remains but to take it ready-made from others, instead of forming opinions for himself?

Let us look a little more closely at the very heart of transformist (evolutionist) doctrine. I speak of the central idea of evolution theory which is that new species are formed by micro-mutation and natural selection. To those who believe in it this idea provides an ample intellectual dwelling place roomy enough to house all the immense achievements of modern biological research. To those not so convinced, however, the idea is strangely constricted and totally inadequate.[48] To show that the fiction is not wholly satisfactory to biologists I draw on one or two recent expressions of doubt by learned men in the field of biology by whom the normal lip service to Darwin is not paid. Koestler in the Alpbach Symposium[49] draws attention to one of the four pillars of *unwisdom* viz. that biological evolution is the result of nothing but random mutations[50] preserved by natural selection. Polanyi[51] invites us to consider the 'impatience with which most biologists set aside today all the difficulties of the current selectionist theory of evolution because no other explanation than that accepted as scientific appears conceivable. This kind of argument based on the absence of any alternative that is accepted as scientific may be valid, but it seems to me the most dangerous application of scientific authority.'

Medawar[52] in his opening remarks as Chairman of the Symposium held at the Wistar Institute of Anatomy and Biology in 1966 reminded his audience that

the immediate cause of this conference is a pretty widespread sense of

[48] Whatever determines the survival of any particular species does not explain the one-way upward course (in broad psychological terms) from amoeba to man. Every toss of a normal coin is completely determined by the forces acting on the coin in that toss, but it is impossible to deduce from that what the ratio of heads to tails will be in, say, 1,000 tosses. We know from experience that it is approximately unity, and it is the same for any large number of tosses, but that is completely unexplained, and if it were discovered that the ratio was 0·5 in Cambrian times, 0·6 in Silurian, 0·7 in Carboniferous, and so on up to unity now (which is a rough analogue of the biological case) that also would be unexplained by the mechanism that determines the result of each toss and would still more obviously need some other explanation.

[49] Koestler, A., *The Alpbach Symposium* (Hutchinson, London 1969); see also *Encounter*, vol. 23, No. 5, p. 61 (Nov. 1969).

[50] *Mutation*. Sudden and relatively permanent chromosomal change. The most important mutations are those occurring in the gametes, since they can produce an inherited change in the characteristics of the organisms developing from them. The majority of mutations are changes of individual genes (genes-mutations), but some are gross structural alterations of chromosomes or changes in numbers of whole chromosomes per nucleus (e.g. polyploidy). Mutation is normally a very infrequent event, though it can be greatly speeded up by irradiation with X-rays, gamma-rays, neutrons, etc., and by some chemicals (e.g. mustard gas). From the standpoint of adaptation of the organism, mutations are random. The great majority are deleterious, upsetting the balanced mechanism of embryonic development of the organism; and on the whole the larger the change they induce the more deleterious they are.

[51] Polanyi, M., *Knowing and Being* (Routledge & Kegan Paul 1969).

[52] Medawar, P., *Mathematical Challenges to the Neo-Darwinian Interpretation of Evolution*, Wistar Institute Symposium Monograph, No. 5 (Philadelphia 1967).

dissatisfaction about what has come to be thought of as the accepted evolutionary theory in the English-speaking world, the so-called neo-Darwinian Theory. This dissatisfaction has been expressed from three quarters and is not only scientific. First of all, religious: Where once the complaint was that evolution happened at all, now the complaint generally is that it happens without Divine motivation. Many of you will have read with incredulous horror the kind of pious bunk written by Teilhard de Chardin on this subject, if Professor Schützenberger will excuse my putting it that way.

Then, there are philosophical or methodological objections to evolutionary theory. They have been very well voiced by Professor Karl Popper—that the current neo-Darwinian Theory has the methodological defect of explaining too much. It is too difficult to imagine or envisage an evolutionary episode which could *not* be explained by the formulae of neo-Darwinism.

Finally—and these are really, I think, the only objections that should concern us—there are objections made by fellow scientists who feel that, in the current theory, something is missing; and we look forward to hearing their formulation of what, precisely, is missing.

These objections to current neo-Darwinian theory are very widely held among biologists generally; and we must on no account, I think, make light of them. The very fact that we are having this conference is evidence that we are *not* making light of them.

Nilsson[53] in his great work *Synthetische Artbildung* goes further and claims that a slavish acceptance of Darwinian evolution prevents a proper system of biology being built up.

A claim that no valid mechanism for evolution exists is made by the eminent biologist Thompson[54] in a critical essay. He tells us: 'Since no one has explained to my satisfaction how evolution could happen I do not feel impelled to say that it has happened. I prefer to say that on this matter our information is inadequate'.

Kerkut[55] in an outspoken preface to a book of exceptional erudition says:

the present book is concerned with an examination of certain basic assumptions and implications that have become involved in the present-day concept of the evolutionary relationships within the animal kingdom. The majority of books on Evolution either blatantly treat these assumptions as part of an old (and concluded) historic argument or else they avoid discussing the assumptions and instead deal with the more scientific and mathematical parts of Evolution.

If one tries to question this avoiding reaction, the protagonists round on one and say in an accusing tone of voice, 'Don't you believe in the Theory of Organic Evolution? What better theory have you got to offer?'

[53] Nilsson, H., *Synthetische Artbildung*, vol. 2 Lund, Sweden 1953); (there is a 100-page abstract in English).
[54] Thompson, W. R., *Introduction to Darwin's Origin of Species*, Everyman No. 811.
[55] Kerkut, G. A., *Implications of Evolution*, International Series of Monographs in Pure and Applied Biology/Zoology Division, vol. 4 (Pergamon Press 1960).

May I here humbly state as part of my biological *credo* that I believe that the theory of Evolution as presented by orthodox evolutionists is in many ways a satisfying explanation of some of the evidence. At the same time I think that the attempt to explain all living forms in terms of an evolution *from a unique source*, though a brave and valid attempt, is one that is premature and not satisfactorily supported by present-day evidence. It may in fact be shown ultimately to be the correct explanation, but the supporting evidence remains to be discovered. We can, if we like, believe that such an evolutionary system has taken place, but I for one do not think that 'it has been proven beyond all reasonable doubt'. In the pages of the book that follow I shall present evidence for the point of view that there are many discrete groups of animals and that we do not know how they have evolved nor how they are interrelated. It is possible that they might have evolved quite independently from discrete and separate sources. There are only a limited number of chemical elements that are capable of forming stable polymerization compounds and it is not at all surprising that the same compounds have been formed on several occasions. Quite complex materials such as carbohydrates, peptides and even nucleic acids can be formed by irradiating water containing simple salts and gases.

It may be suggested that the problem we are examining here, namely that of the evolution and interrelationship of the basic living stocks is a major problem and one that will test the strength and ability of many hundreds of research workers. If this book merely indicates to some of the readers that certain lines of thought are still open to examination, then I shall consider that it has done its allotted task.

There is, however, a second point that I should like to make, and this concerns not factual material but an attitude of mind. It is very depressing to find that many subjects are becoming encased in scientific dogmatism. The basic information is frequently overlooked or ignored and opinions become repeated so often and so loudly that they take on the tone of Laws. Although it does take a considerable amount of time, it is essential that the basic information is frequently re-examined and the conclusions analysed. From time to time one must stop and attempt to think things out for oneself instead of just accepting the most widely quoted viewpoint.*

This point is amplified in the article already mentioned by Salisbury[56] who points out that

Modern biology is faced with two ideas which seem to me to be quite incompatible with each other. One is the concept of evolution by natural selection of adaptive genes that are originally produced by random mutations. The other is the concept of the gene as part of a molecule of DNA, each gene being unique in the order of arrangement of its nucleotides. If life really depends on each gene being as unique as it appears to be, then it is too unique to come into being by chance mutations. There will be nothing for natural selection to act on.

---

[56] Salisbury, F. B., *loc. cit.*
* *Grateful acknowledgements are expressed to Professor G. Kerkut and Pergamon Press for permission to reproduce this extract.*

The problem was discussed at a symposium of mathematicians and biologists in 1966, but they failed to solve the difficulty. I feel that virtually no one present except Eden and Schützenberger, who outlined the problem, really understood what the commotion was all about. Some years ago I also outlined the problem. My outline begins by overstating the case somewhat, but it provides a background for discussion of modern discoveries which may be pointing toward a solution. I believe that the solution remains to be found.

Let no one be deluded on this point of mutations which is crucial to the present position in the argument for transformation across the species and for a virile system of evolution to exist. It may not be amiss if I give the position as I see it at a little more length than is usual in a work devoted to astronomy.

The student may care to be reminded of the manner in which variations occur and of the vast increase in knowledge obtained on this point since Darwin's day. Since organisms are seen to breed true Darwin advanced the now discredited theory of pangenesis. According to this theory there are hereditary carriers, pangenes, in the blood streams to the sexual organs and the sex cells. By this means a certain organ acquires, on account of changed external conditions, a stronger function so that the pangenes are increased and transmitted to the sex cells in this multiplied form. Hence the progeny obtains this organ more powerfully developed. The acquired characteristics have then become hereditary according to Darwin, yet he had always held that environment did not play a part in the evolutionary process. All this appears farcical now in view of the rediscovery by de Vries of Mendel's work which he rescued from oblivion. These discoveries gave a new aspect to the problem of the origin of species, for they taught us that the individual species pattern is controlled by genes, which are supreme and unchangeable. Thus if two individuals have different hereditary units these genes cannot fuse into a new gene in the offspring. But when two different varieties are crossed there takes place in the offspring a re-combination of the free genes so that new varieties are produced. If the parents differ in 10 characters, not less than 1024 new varieties[57] may be formed. Mendel's greatest achievement lies in his having settled the conflict between heredity and variation. Mendelism introduces us to a new concept, since the forms of variability, which Darwin and others had imagined as successively emerging from one another along the ages in an interminable chain emerge in a single moment from one and the same cross in their millions as a great sphere of variants. These variants can be classified in accordance with their appearance in a series, but they have *not* appeared in that series. We classify everything according to *likeness*, but likeness does *not* prove an evolutionary likeness, *only* a

[57] i.e. 2^{10}.

constitutional likeness. We cannot do better than let the authoritative voice of the late Professor Nilsson take up the story.[58]

Mendelism has thus solved the problem of variation and also solved the problem of heredity, but at the same time its results on the theory of evolution are similar to the result of a boulder falling into a narrow stream. For the main result of Mendel's discovery is this: *Genes are constant*. They are not influenced either by one another or by the environment. Variation is caused by the re-combination of the genes, not by their change. Variation is therefore restricted by the combination possibilities of the genes. And these are limited by the crossing possibilities. Then again, since individuals belonging to different species of plant or animal cannot even be paired, much less produce offspring, the combination of variations is confined to the species. Variants are formed, out-crossed and arise anew in a kaleidoscopic sequence *within* the species. *But the species remains the same sphere of variation*. The various species will remain like circles that do not intersect. *Species are constant*.

With the solution of the conflict between variation and heredity a new conflict has arisen, that between Mendelism and evolution. Is this conflict then a serious one?

So far, we have not at any rate reached exactly and critical point. A remaining possibility is that new hereditary units, new genes, may be formed. And how can this come about? By mutation.

The occurrence of mutants has been observed not only in plants but also in animals. And the most talked of species at present and the one most closely studied with regard to its mutability is the little fruit fly, *Drosophila melanogaster*.

Mutations, that is, forms of variations, which do not appear capable of being classified in the ordinary Mendelian scheme of variations thus occur. They are said to arise spontaneously by a sudden gene change or by a change in the gene-bearing elements—the chromosomes—in the nucleus of the cell. Here we have then an entirely new regeneration factor with respect to variations. Here we have *new possibilities of evolution*.

But—and there is really another but—it is *not enough* that mutations are *formed*. We must be quite certain that they are *caused by gene changes*, and that the change is in a positive, in an *evolutionary direction*.

That the mutations in certain cases are not due to gene changes has been conclusively proved by recent researches. And they arise in an entirely unexpected and quite fantastic manner. It has been found that certain varieties are, we are tempted to say, a kind of double entities. They have also been called chimaeras. Such varieties are found among our common pelargonia. They are recognized by their mottled leaves. If we make an anatomical examination of a plant having white margined leaves we shall see that the outer cell layer or layers are made up of white cells but underneath there are green tissues. The entire plant is like a green frame-work covered with a white mantle. These varieties have therefore been called periclinal chimaeras.

It now happens that in certain cases branches with purely green leaves

[58] Nilsson, H., 'The problem of the origin of species since Darwin', Inauguration address, Lund University, 1934, April 14; *Hereditas*, p. 227–37 (1935).

appear on these white-margined plants. It was formerly thought that a mutation, and in this case a bud mutation, had taken place. But now that we know the nature of the plant the process gets quite a different explanation.

A great sensation has been caused in recent years by the experiments made by the American scientist Muller in which he evoked mutation in *Drosophila* by means of X-ray radiation. Chimaeras (mosaics) are also known in *Drosophila*, indeed they are not rare, and it is quite probable that in this case radiation only caused disturbances in the tissue, by means of which hidden chimaerical constituents were differentiated, in the same manner as in the case of the hyacinth mentioned above. That the mutational process is in general caused by a natural radiation is out of the question, as this does not amount to the quantitative values required to evoke mutants experimentally. *What part of the known mutants is caused by a real gene change and what part is caused by a chimaerical differentiation is at present impossible to say.* But if we assume that at any rate some of them are spontaneous gene mutations there still remain a few questions that call for an answer, viz. whether they run in a positive direction and whether they have an evolutionary selective value.

As regards the first question it is a remarkable fact that the *vast majority* of mutants have not *acquired* any new character by mutation, but have instead *lost* one of the genes of the species. They are loss mutants. To build an evolution on such mutants it would be necessary to assume that a higher differentiation, a development, is associated with a loss of genes. The consequence of this would be that amoeba must be endowed with an enormously greater stock of genes than Homo, and no one will readily take the consequence of such a view. For then vertebrates would be in genetic equipment only greatly thinned amoebae.

A very *small number* of mutants seem, however, to have been enriched with a gene, to be positive new-formations, *progressive* mutants. And for them the final test will therefore be of vital importance, and that is: Have they also selective value, which enables them not only to prevail in the struggle for existence but also to oust the parent species?

As far as viability is concerned mutants are in general poorly equipped. Not less than 60 of the 400 mutants of *Drosophila* have obtained the new character to perish. Chlorophyll deficient and diabetic plants and flies with useless wings or cleft legs are also of course doomed to a rapid elimination. But even the less divergent mutants have a weaker constitution and a shorter duration of life than the parents species. This is clearly and conclusively shown by the fact that *mutants of Oenothera and Drosophila are not found in nature.* They occur of course here and may quite casually be met with, but they are soon eliminated and do not form a component of nature's stock in trade, and much less are they able to oust the present species. *They appear quite simply to be non-viable variants, which we may see only under certain favourable conditions of culture, but which in nature are quickly swept away and disappear without leaving any trace.* And this is true of the dominant mutants as well as of the loss mutants. As a rule it may be said that *the more divergent a mutant is the less viable it is.* To endeavour to build an evolution on mutants is more than difficult. It is like setting out on a stormy sea in a nutshell.

It is obvious that the investigations of the last three decades into the problem of the origin of species have not been able to show that a variational material capable of competition in the struggle for existence is formed by mutation. Further, as it has also been impossible to demonstrate a progressive adaptation by means of the transmission of acquired characters (all the numerous experiments made have yielded negative results), we are forced to this conclusion that *the theory of evolution has not been verified by experimental investigations of the origin of species.*

This comment from a world authority on palaeobiology is now thirty years, old, but does anyone know of evidence to controvert it?

Let each reader examine his own intellectual position with honesty. If his study of the matter is perfunctory let him turn to what is widely held by biologists to be the most erudite readily available work on the subject; I refer to Dr Ernst Mayr's *Animal Species and Evolution* from the Belknap Press of Harvard University Press, Cambridge, Massachusetts, 1965. He will then see the paucity of the evolutionist's case since Dr Mayr neglects to discuss the points made by Professor Nillson yet states (pp. 174–181) that

it must not be forgotten that mutation is the ultimate source of all genetic variation found in natural populations and the only raw material available for natural selection to work on.

As for the evidence that the universal acceptance of the Darwinian position is open to serious challenge I quote from Dr Marjorie Grene's[59] essay 'The Faith of Darwinism':

Great new inventions, new ideas of living, which arise with startling suddenness, proliferate in a variety of directions, yet persist with fundamental constancy—as in Darwinian terms they would have no reason in the world to do. Neither the origin and persistence of great new modes of life—photosynthesis, breathing, thinking—nor all the intricate and co-ordinated changes needed to support them, are explained or even made conceivable on the Darwinian view. And if one returns to read the *Origin* with these criticisms in mind, one finds, indeed, that for all the brilliance of its hypotheses piled on hypotheses, for all the splendid simplicity of the 'mechanism' by which it 'explains' so many and so varied phenomena, it simply is not about the origin of species, let alone of the great orders and classes and phyla, at all. Its argument moves in a different direction altogether, in the direction of minute specialized adaptations, which lead, unless to extinction, nowhere. And the same is true of the whole immense and infinitely ingenious mountain of work by present-day Darwinians: *c'est magnifique, mais ce n'est pas la guerre!*. That the colour of moths or snails or the bloom on the castor bean stem are 'explained' by mutation and natural selection is very likely; but how from single-celled (and for that matter from inanimate) ancestors there came to be castor beans and moths and snails, and how from these there emerged llamas and hedgehogs and lions and apes—and men—that is a question which neo-Darwinian theory

[59] Grene, M., 'The Faith of Darwinism', *Encounter'*, vol. 13, No. 5, pp. 48–56 (Nov. 1959).

simply leaves unasked. With infinite ingenuity it elaborates the microscopic conditions for such macroscopic occurrences; but it provides no conceptual framework in terms of which they can be admitted to exist, let alone an 'explanation' of their descent from 'lower' forms.

Moreover, evolutionists sceptical of the neo-Darwinian synthesis have themselves empirical evidence to support their doubts. For despite the neo-Darwinians' claims, two great biological disciplines, paleontology and embryology, appear to lend their chief weight against the selectionist dogma.

That the evidence from palaeontology was inimical to evolution was advanced as early as 1914 in Spengler's great philosophical work *The Decline of the West*. In volume two he states:

There is no more conclusive refutation of Darwinism than that furnished by palæontology. Simple probability indicates that fossil hoards can only be test samples. Each sample, then, should represent a different stage of evolution, and there ought to be merely 'transitional' types, no definition and no species. Instead of this we find perfectly stable and unaltered forms persevering through long ages, forms that have not developed themselves on the fitness principle, but *appear suddenly and at once in their definitive shape*; that do not thereafter evolve towards better adaptation, but become rarer and finally disappear, while quite different forms crop up again. What unfolds itself, is ever-increasing richness of form, is the great classes and kinds of living beings which *exist aboriginally and exist still, without transition types*, in the grouping of to-day.

Where biologists fall out it is foolish for astronomers to attempt to mediate. We can but note the *two* sides of what is a most complex question and reserve our judgment. If evolution is not yet a settled fiction why should we pass it on incomplete to the stars? We would do well to follow Medawar and say, 'I shall say nothing about the principle of Evolution that I would not be prepared to say about any other attempt to pass off a mere inductive *collage* as a work of philosophic art.'

DIRECT EVIDENCE FOR EXTRA-TERRESTRIAL LIFE

The only direct evidence available of which I am aware is that gathered from a study of the carbonaceous chondrites of which twenty that have fallen on the Earth have been discovered. The main evidence comes from two[60] of these meteoritic stones. A review of the position has been given by Hutchinson *et al.*[61] The evidence is controversial and turns on the so-called 'organized elements' which may be terrestrial contaminants. The research work leaves much to be desired and cannot be seen to be

[60] The Orguei and Murray stones, which fell resp. 14th May 1864 and 20th Sept. 1930.
[61] Hutchinson D. R., Jones, T. J., Potter, R., 'Indications of other life in the Solar System' *J.B.A.A.*, 74, pp. 59–64 (1964).

conclusive.[62] More recently the subject has been reopened by Brooks and Shaw,[63] who claim to have detected sporopollenin in the insoluble part of a small portion of the Orgueil and Murray stones (0·1g. and 0·9g. resp.). The sporopollenin of these two stones has a similar profile (when subjected to infra-red spectroscopy and pyrolysis—gas chromatography) to that of sporopollenin in *Lycopodium clavatum*, *Tasmanites punctatus*, B. carotene and Lilium henryii carotenoids.

The authors feel able to assert from this that since meteorites have an age of 4·5 — 4·7 × 10⁹ years their results imply that there was life in the universe before the time of formation of our planet. This is a conclusion that is wholly unacceptable since the sporopollenin may come from a meteorite which was of terrestrial origin, it being held by some that volcanic action on the Earth may have been the initial force to have ejected the carbonaceous chondrites (they closely resemble Pre-Cambrian sediments) which were then re-captured by the Earth. Until this very real possibility of circular reasoning is removed by incontrovertible evidence for an extra-terrestrial origin of the Orgueil and Murray stones the evidence is inconclusive.

Finally, it should not detain us long to point out that the presence of contemporaneous life in the cosmos is basically unconfirmable. With the discovery by Ole Röemer that light has a finite velocity in the universe we contemplate an old universe. Consider the 'close' spiral galaxy in Andromeda (M31). It is 2 × 10⁶ light years[64] distant. If by some remote chance life was thought to be discovered there tonight it would be 'life' of 2 × 10⁶ years ago. To know whether it is flourishing tonight one has to wait 2 × 10⁶ years for an answer.

Anyone who has followed me this far will see that we are faced not with a scientific case but a philosophical one. It is pertinent therefore to allow a philosopher to speak. Lord Russell writes with more lucidity than most.

In regard to probable opinion, we can derive great assistance from *coherence*, which we rejected as the *definition* of truth, but may often use as a *criterion*. A body of individually probable opinions, if they are mutually coherent, become more probable than any one of them would be individually. It is this way that many scientific hypotheses acquire their probability. They fit into a coherent system of probable opinions, and thus become more probable than they would be in isolation. The same thing applies to general philosophical hypotheses. Often in a single case such hypotheses may seem highly doubtful, while yet, when we consider the order and coherence which they introduce into a mass of probable opinion, they become pretty nearly

[62] Claus, Nagy, *Nature*, *192*, 4803 (1961). Urey, *Nature*, *193*, 4821 (1962). Fitch, Anders, *Scientific American*, *208*, 6·10.
[63] Brooks, J., Shaw, G., *Nature*, *223*, 754 (1969).
[64] *Light-year*. A spatial unit sometimes used to express distances in the stellar universe. It is the distance travelled by light in one year, amounting to 63,240 astronomical units or 9·4607 × 10¹² km. or 5·8786 × 10¹² miles.

certain. This applies, in particular, to such matters as the distinction between dreams and waking life. If our dreams, night after night, were as coherent one with another as our days, we should hardly know whether to believe the dreams or the waking life. As it is, the test of coherence condemns the dreams and confirms the waking life. But this test, though it increases probability where it is successful, never gives absolute certainty, unless there is certainty already at some point in the coherent system. Thus the mere organization of probable opinion will never, by itself, transform it into indubitable knowledge.

The question 'Is there life on other worlds?' is now seen to be an open one. We have no indubitable knowledge concerning it. The three fictions presently put forward for an affirmative answer are shown to have no firm foundations. The evidence available incontrovertibly shows that there is life on Earth and that the solar system is mainly hostile to it. How life arrived and flourished on Earth is a question open to considerable debate. The most frequently advanced answer today, in which evolutionary forces are appealed to, is one of scientific prejudice since the available evidence when viewed dispassionately shows a leavening element of mystery associated with the arrival of complex life *per saltum*. In such a situation demanding the exercise of humility we may close with Melville's powerful lines from 'Clarel'—

> *But how if Nature vetoes all*
> *Her commentators: Disenchant*
> *Thy heart. Look round!*

Appendix A

ABBREVIATIONS USED IN THIS BOOK

Å	Angström unit
Am. J. Math.	American Journal of Mathematics
Anc. Mém. Acad. Sci.	Anciennes Mémoires de l'Academie des Sciences
Ann. Obs. Meudon	Annales de l'Observatoire de Meudon
AP of AE	Astronomical Papers prepared for the use of the American Ephemeris and Nautical almanac
Ap J.	Astrophysical Journal University of Chicago
Astron J.	Astronomical Journal
a.u.	Astronomical Unit
B.A.A.	British Astronomical Association
BAN	Bulletin of the Astronomical Institutes of the Netherlands
BM	British Museum (the numbers and letters that follow these letters are the main catalogue entry)
Bull. Astro.	Bulletin d'Astronomie
Bull. Amer. Phys. Soc.	Bulletin of the American Physical Society
Bull. Inst. Theor. Astr.	Bulletin of the Institute of Theoretical Astronomy
J.B.A.A.	Journal of the British Astronomical Association
J. Geol. Res.	Journal of Geological Research
J. Franklin Inst.	Journal of the Franklin Institute, Philadelphia
M.N. or Monthly Notices of R.A.S.	Monthly Notices of the Royal Astronomical Society
Mém. Acad. Sci.	Mémoires de l'Academie des Sciences
Mem. R.A.S.	Memoirs of the Royal Astronomical Society
Phil. Trans.	Philosophical Transactions of the Royal Society, London
Philos. Mag.	The Philosophical Magazine
Phys. Rev. Letters	Physical Review Letters
Proc. Am. Phil. Soc.	Proceedings of the American Philosophical Society
Proc. I.R.E.	Proceedings of the Institute of Radio Engineers
Proc. R. Soc.	Proceedings of the Royal Society (London)
Pub. A.S.P.	Publications of the Astronomical Society of the Pacific
Quat. J.R.A.S.	The Quarterly Journal of the Royal Astronomical Society
R.A.S.	Royal Astronomical Society
Rev. Mod. Phys.	Review of Modern Physics
Space Sci. Rev.	Space Science Review
Trans. I.A.U.	Transactions of the International Astronomical Union
Zeits. f. Astrophys.	Zeitschrift für Astrophysik

Appendix A

THE ELEMENTS OF PLANETARY ORBITS

2a   The major axis.

a   The semi major axis sometimes called the 'line of apsides'. This gives the size of the orbit.

b   The semi minor axis.

e   The eccentricity $= \dfrac{1}{a} \sqrt{a^2 - b^2}$

This gives the shape of the orbit.

i The inclination, the angle between the orbital plane and a reference plane. In planetary orbits the reference plane is the plane of the ecliptic, i.e. the orbital plane of the Earth about the Sun. Hence by definition i for the Earth's orbit is zero.

Ω Longitude of the ascending node, this gives the position at which the orbit cuts the plane of the ecliptic from South to North.

ω Argument of perihelion.

P   Sidereal period.

T   Time of perihelion passage.

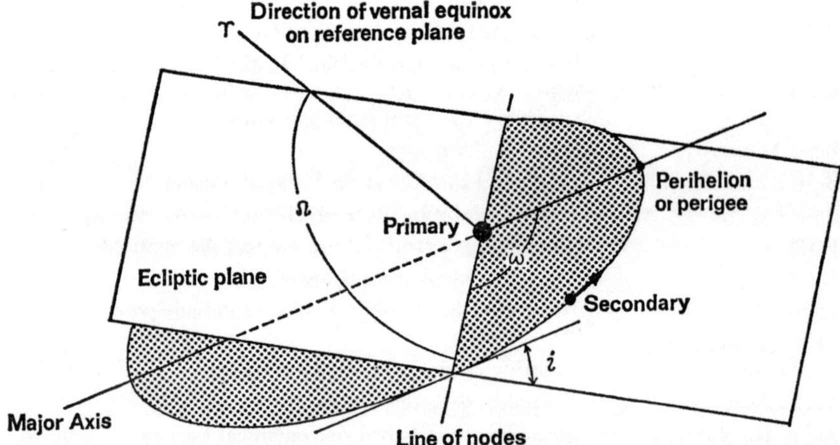

Fig. 51. The elements used to define the orientation in space of an elliptic orbit.

Appendix B

COPERNICUS, NICOLAUS (1473–1543), Polish astronomer, born February 19, at Thorn in Prussian Poland. The only work published by Copernicus on his own initiative was a Latin version of *Greek Epistles of Theophylact* (Cracow 1509). His treatise on coinage *De monetae cudendae ratione*, 1526, was written by order of King Sigismund I. His great work *De revolutionibus orbium* was extracted from him in 1540 and the first printed copy reached Frauenburg barely in time to be laid on its author's death-bed. Rheticus was his sole contemporary biographer and his work is lost to us. Rosen's *Three Copernican Treatises* (Dover 1959) contains an annotated Copernican bibliography of considerable wealth covering the years 1939–58. Dr Leopold Prowe's exhaustive *Nicolaus Coppernicus* (Berlin 1883–1918) embodies the outcome of 30 years research; see also Rybka, E., *Four hundred years of the Copernican heritage* (Cracow 1964); and Dingle, H., 'Copernicus', *The Observatory*, 65, 38 (1943); Dobson, J. F., and Brodetsky, S., 'Nicolaus Copernicus, Occasional Notes', R.A.S., No. 10, 1947 May).

AIRY, SIR GEORGE BIDDELL (1801–92) was born at Alnwick on 27 July 1801. He was educated at Colchester Grammar School and entered Trinity College, Cambridge, as a sizar in 1819. He graduated as senior wrangler in 1822. He was appointed Astronomer Royal in 1835 June, in succession to Pond. He resigned the office in 1881 and resided at White House, Greenwich, until his death. His most remarkable achievement amongst many in dynamical astronomy was his discovery of a new inequality in the motions of Venus and the Earth. He suspected this in correcting the solar tables of Delambre. The cause was found to lie in the near equality of the mean motions of Venus and the Earth. Eight times the mean motion of Venus is so close to thirteen times that of the Earth as to want but $1/240^{th}$ part of the Earth's mean motion, a difference of itself small yet one that receives in the differential equations a multiplier of 2×10^6. Airy postulated from these researches an inequality extending over 240 years. (*Phil. Trans.*, *122*, 67). For this work he received the medal of the Royal Astronomical Society in 1833. (See *The Times*, 5 January 1892).

KEPLER (1571–1630). Kepler's genius is not difficult to portray but his writings are voluminous[1] and he has not until recent years found a biographer. His manuscripts are now residing in the USSR Academy of Sciences in Leningrad.

Kepler's interest in nature was unusually wide. The abundant vintage of 1613 drew his attention to the defective methods for estimating the cubical contents of

---

[1] See *Johann Kepler—A tercentenary commemoration of his life and work*, The History of Science Society (Baltimore 1931). This work contains an excellent bibliography (BM 10709aaa 22). The Archives of the Academy of Science of the U.S.S.R. holds according to Kulikovsky (*Vistas in Astronomy*, vol. 9, p. 246) sixteen volumes of manuscripts of Kepler and four further volumes are in the Imperial Library of Vienna. Frisch produced the celebrated *Opera Omnia Joannis Kepler* in eight volumes between 1858 and 1871, but it was out of date by 1900 according to Hammer (*Vistas in Astronomy*, vol. 9, 261). A 'new' edition appeared under the editorship of W. van Dyck in 1937 and is to extend to twenty-four volumes; fifteen volumes were complete in 1967.

vessels and his essay *Nova Stereometria Doliorum*, Linz 1615, entitles him to rank among those who prepared the path to the discovery of the infinitesimal calculus.

Another example is his treatment of the motion of projectiles on the rotating earth equivalent to the formulation of the superposition principle of velocities. He deduced from the space filling symmetry of the honeycomb that its angles must be those of the rhombic dodecahedron.

He studied the snowflake[2] and wrote a beautiful treatise upon it (*Strena Sen de Nive Sexangula*).[3] In this work he deals also with the mysteries of the quincunx and the hexagon in seeking after a rational explanation of the five- and three-petalled flower. His pythagorean love for the five platonic bodies caused him to incorporate them into his system of the planetary world. Today this is dismissed as a nonsense, yet these bodies appear in the free developing minute organisms of the Radiolaria and one can but be surprised that we meet in the animal world the regular dodecahedron and icosahedron which surprisingly do not appear in crystalline forms.

He was aware of the Fibonacci series 1.1.2.3.5.8. used by Leonardo de Pisa nicknamed Fi Bonacci. The series was first so called by E. Lucas *c.* 1877. The general expression for the series is

$$U_n = \sqrt{\frac{1}{5}} \left\{ \left(\frac{1 + \sqrt{5}}{2}\right)^n - \left(\frac{1 - \sqrt{5}}{2}\right)^n \right\}$$

which converges to the Golden Mean of 1·618 and which to him was a symbol of creation. Today the Fibonacci series is seen to exist in the problems of phyllotaxis.

His *Somnium* dealing with a hypothetical voyage to the Moon gives a remarkable exposition of the solar system as seen from that vantage point. No English translation was available until 1965; see Lear, J., *Kepler's Dream*, trans. by Kirkwood, P. F. (University of California Press, 1965).

GALILEO (1564–1642) like Newton displayed a natural aptitude for the study of mechanical things and early in life constructed toy-machines. But unlike Newton, he was drawn also to poetry and the graphic arts. In 1581 he went to Pisa to study medicine but was more enthusastic for Euclid and was dubbed 'The Wrangler'. In 1609 he made three refracting telescopes able to magnify resp. 10, 20 and 30 diametres. With the third instrument he discovered the large moons of Jupiter. In 1610 he published his celebrated *Sidereus Nuncius*. His researches continued and in 1613 he published his *Letters on the Solar Spots*. In 1630 appeared his famous but ill-starred work, *Dialogo dei due massimi sistemi dei mondo*.

In 1636 he completed his *Dialoghi delle nuove scienze*. On the 8 January 1642 he closed his life having lived and worked in a period that almost exactly spanned the interval between the death of Michelangelo and the birth of Newton.

The first complete edition of Galileo's writings was published at Florence. Sixteen volumes. Florence 1842–56. Edited by Alberi. A 'national edition' in twenty volumes under the editorship of Favaro appeared at Florence in 1890. The Quadricentennial celebrations of Galileo's birth in February 1964 produced a number of reappraisals of his work; see Rose, J., 'Galileo, Galilei', *Nature, 201,* 653 (1964); Whitrow, G. J., 'Galileo's significance in the history of Astronomy', *Quat. J.R.A.S., 5,* 183 (1964); Hall, A. R., 'Galileo's System of the World', Ibid., *5,* 304 (1964); Kaplan, M. F., *Homage to Galileo* (M.I.T. Press, London 1966); Brecht's play and introductory notes to *The Life of Galileo* (Methuen paperback, 1963), are of interest, not least since he drew on the expertise of assistants of Niels Bohr to construct his vision of the Ptolemaic cosmology. See also Hartner, W., Galileo's contribution to Astronomy. *Vistas in Astronomy 11,* 31 (1969).

[2] See Schneer, C., 'Kepler's New Year Gift of a Snowflake', *Isis, 51,* 531 (1960).
[3] Reprinted 1966, Oxford University Press; C. Whyte Hardie, *The Six Cornered Snow Flake.*

NEWTON (1642–1727) was born at Woolsthorpe on Christmas Day 1642 (O.S.). As a boy he made windmills, waterclocks, kites and dials and invented a four-wheeled carriage moved by the rider. He was admitted to Trinity College on June 5 1761 as a subsizar and matriculated on 8 July. Little is known of his early years at Trinity, he was excused attendance at a course of lectures on logic having before entry read Sanderson's *Logic* to such purpose as to surprise his tutor. He is known to have bought a work on astrology at Stourbridge fair and his inability to understand part of it led him to study Euclid and later Descartes' *Geometry*. Writing later of his early studies in 1663–64 he wrote 'At such time I found the method of Infinite Series and computed the area of the hyperbola at Boothby, in Lincolnshire, to two and fifty figures by the same method.' The plague then caused him to return to Woolsthorpe where he invented *inter alia* the reflecting telescope described at the Royal Society meeting on the 8 July 1672. (See Philosophical Transactions No. 80.) At this time he wrote his paper *Analysis per Equationes Numero Terminorum Infinitas*.

In 1685–6 he composed almost the whole of his great work the *Principia*. In 1702 appeared his *Theory of the Moon's Motion* and in 1764 his *Optics* to which was added *Curvilinearum* and specifically *Tractatus de Quadratura* two treatises entitled generally *Tractatus duo de speciebus et magnitudine figurarum Curvarum* and *Enumeratio linearum tertii ordinis* based on a draft of 1671, the former containing an explanation of fluxions. He made extensive studies also into chemistry and theology see his *Tabula Quantitatum et Graduum Caloris, De Natura Acidorum, Observations on the Prophecies of Daniel and the Apocalypse of St. John, Lexicon Propheticum, Historical Account of two Notable Corruptions of the Scripture*.

There is no definitive edition of the work of Newton.[4] In 1779–85 there appeared a collected edition, *Isaaci Newtoni Opera Quae Exstant Omnia*, edited by Samuel Horsley, 5 vols. The standard life is Sir David Brewster's two volume work of 1855, *Memoirs of the Life Writings and Discoveries of Sir Isaac Newton* (New Ed. 1893). The Mathematical Society published a memorial volume in 1927 edited by W. J. Greenstreet. Professor Brodetsky's beautiful work *Sir Isaac Newton* (1927), and Andrade's work of the same title (1954) are worthy of close study. Newton's library is skilfully reconstituted in Villamil's book *Newton the Man* (1931), with an introduction by Einstein.

TYCHO BRAHE (1546–1601), born on 1546 December 14 at Knudstrup in Scania. His principal works are *Astronomiae Instauratae Progymnasmata*, 2 vols. (Prague 1602–3); *Astronomiae Instauratae Mechanica* (1598); *Epistolae Astronomicae* (1596); see Dreyer, J. L. E., *Tycho Brahe*, 1890 (Dover Reprint of 1963).

JOHN COUCH ADAMS (1819–92), English astronomer, was born at Lidcot farmhouse, Laneast, Cornwall, on 1819 June 5. In 1839 he entered as a sizar at St John's College Cambridge. He graduated B.A. in 1843 as Senior Wrangler. The discovery of Neptune by mathematical deduction was but one part of his researches into the more recondite parts of astronomical theory; see *The Scientific Papers of John Couch Adams* (1896, 1900); see also *Monthly Notices R.A.S.*, *53*, 184; *The Observatory*, *15*, 174, *Nature*, *34*, 565; *45* 301; *Edinburgh Review*, No. 381, 72.

[4] See Andrade, E.N., 'A Newton Collection', *Endeavour*, *12*, 68 (1953). The Babson Institute of America holds an impressive collection of Newtoniana. *A Descriptive Catalogue of the Grace K. Babson. Collection of the works of Sir Isaac Newton*. This supplements G. J. Gray's Bibliography of 1907. See also Whiteside, D. T., *The Mathematical Papers of Isaac Newton*, vol. I, 1664–6; vol. II, 1667–70 (1968). See the *Times Literary Supplement* 3461, June 27, 1968, and *The Listener*, October 19, 1961, p. 597. Einstein personally wrote a magnificent tribute to Newton at the 200th anniversary of his death, see the *Manchester Guardian*, 1927 March 19; A report appeared in *The Observatory* for 1927 May, pp. 146–53. *Nature* issued a special supplement at the time (*Nature* No. 2995, March 26th, 1927, pp. 21–60).

Appendix B

HALLEY (1656–1742), the Second Astronomer Royal, was born 1656 October 29 at Haggerston, Shoreditch, London. He was educated at St Paul's School and later studied at Queens' College, Oxford. In 1676 he communicated to the Royal Society a paper entitled *A Direct and Geometrical Method of Fixing the Aphelia and Eccentricity of the Planets.*

He observed the transit of Mercury across the Sun's disc of 1677 from St Helena, provided the money for the printing of Newton's *Principia;* produced the earliest star catalogue of the Southern Stars from telescopic observations, *Catalogus stellarum australiums* and predicted the return of the comet of 1682. The results of his cometary researches were gathered together in his *Synopsis Astronomiae Cometicae* 1705.

To these should be added some eighty-one miscellaneous papers in the *Philosophical Transactions* and his translations when Savilian Professor of Geometry at Oxford from the Arabic (which language he acquired for the purpose) of the treatise of Apollonius *De sectione rationis* with a restoration of his two lost books, *De Sectione spatii*, the *Conics* of Apollonius and the treatise by Serenus, *De sectione cylindri et coni* and an edition of the *Spherics* of Menelaus.

See *Edmond Halley, 1656–1742,* Memoirs of the British Astronomical Association, vol. 37. No. 3, Historical Section, 1956 Nov.; Armitage, A., *Edmond Halley* (Nelson, London 1966); Ronan, C. A., *Edmond Halley, Genius in Eclipse* (London 1970).

PIERRE SIMON LAPLACE (1749–1827), French mathematician and astronomer, was born at Beaumont-en-Auge in Normandy. His first published paper *Recherches sur le calcul integral*, Mélanges de la Soc. Roy de Turin (1766–69), at the age of twenty-four, earned him the title of 'The Newton of France'. In a paper read before the Academy of Sciences, on 1773 February 10 (*Mém présentés par divers savans*, tome VII (1776)) he announced his celebrated conclusion of the invariability of planetary mean motions, carrying the proof as far as the cubes of the eccentricities and inclinations. This was the first and most important step in the understanding of the stability of the solar system. His *Mécanique Céleste* ranks second only to the *Principia* of Newton. In 1842, by the intervention of Louis Philippe, 40,000 francs was granted for the publication of his works.

Ouevres de Laplace (1843–47), 7 vols; an English translation with copious elucidatory notes of the first 4 vols of the *Mecanique Celeste* by N. Bowditch was published at Boston, USA (1829–39) (BM 532 i 20–23).

JOSEPH LOUIS LAGRANGE (1736–1813), French mathematician, was born at Turin. He was weaned from a literary career by a reading of a tract by Halley and in his nineteenth year communicated to Euler his general method of dealing with 'isoperimetrical' problems, known later as the Calculus of Variations. At the age of twenty-six he was at the summit of European fame by his exposition of the new calculus. He won the 1764 prize of the Paris Academy of Science with his essay on the libration of the Moon and again in 1766, 1772, 1774 and 1778 with his essays on the Jovian System, the problem of three bodies, the secular equation of the Moon and the theory of cometary perturbations. His great work the *Mécanique analytique* was published in 1787. His works are collected in seven volumes under the title *Oeuvres de Lagrange, publiées sous les soins de M.J.A. Serret* (Paris 1867–77).

Appendix C

CATALOGUE OF COMETS

Some 101 short-period comets, having eccentricities less than unity are known (1971) and may properly be said to *belong* to the solar system. These are listed below.

62 periodic comets as of 1970. (N is the number of observed perihilion passages. P is the period in years)

39 periodic comets of *one* apparition only;

Name	N	P	Name	P
Encke	49	3·30	Wilson-Harrington	2·31
Grigg-Skjellerup	11	4·91	Helfenzrieder	4·51
Honda-Mrkos-Pajdušákova	4	5·21	Blanpain	5·10
Tempel (2)	14	5·26	du Toit (2)	5·28
Neujmin (2)	2	5·43	La Hire	5·38
Brorsen	5	5·46	Barnard (1)	5·40
Tuttle-Giacobini-Kresák	4	5·48	Schwassmann-Wachmann (3)	5·43
du Toit-Neujmin-Delporte	2	5·54	Grischow	5·44
Tempel-Swift	4	5·68	Brooks (1)	5·60
Tempel (1)	3	5·98	Lexell	5·60
Pons-Winnecke	17	6·30	Kulin	5·64
de Vico-Swift	3	6·31	Pigott	5·89
Kopff	10	6·31	Taylor	6·37
Giacobini-Zinner	9	6·41	Spitaler	6·37
Forbes	4	6·42	Harrington-Wilson	6·38
Schwassmann-Wachmann (2)	7	6·53	Curjumov Gerasimenko	6·55
Wolf-Harrington	4	6·54	Tsuchinshan (1)	6·62
Biela	6	6·62	Barnard (3)	6·63
Wirtanen	4	6·67	Giacobini	6·65
d'Arrest	11	6·67	Schorr	6·71
Perrine-Mrkos	5	6·71	Gunn	6·80
Reinmuth (2)	3	6·71	Swift (2)	7·22
Brooks (2)	10	6·72	Denning (2)	7·42
Tsuchinshan (2)	2	6·79	Metcalf	7·77
Harrington	2	6·80	Denning (1)	8·69
Arend-Rigaux	3	6·82	Swift (1)	8·92
Johnson	3	6·86	Klemola	10·90
Finlay	8	6·90	Wild	13·19
Borrelly	8	7·02	Peters	13·38
Daniel	5	7·09	du Toit (1)	14·79
Harrington-Abell	3	7·22	Perrine	16·4
Shajn-Schaldach	2	7·28	Pons-Gambart	63·83
Holmes	5	7·35	Ross	64·63
Faye	16	7·38	Dubiago	67·01
Whipple	6	7·46	de Vico	75·71
Ashbrook-Jackson	3	7·49	Väisälä (2)	85·52
Reinmuth (1)	5	7·60	Swift-Tuttle	119·6
Arend	3	7·79	Barnard (2)	128·3
Oterma	3	7·88	Mellish	145·3
Schaumasse	6	8·18		
Wolf	11	8·43		
Jackson-Neujmin	2	8·57		
Comas Solá	6	8·59		
Kearns-Kwee	2	8·95		
Väisälä (1)	3	10·46		
Neujmin (3)	2	10·95		
Gale	2	10·99		
Slaughter Burnham	2	11·64		
van Biesbroeck	3	12·41		
Tuttle	9	13·61		
Schwassmann-Wachmann (1)	3	16·10		
Neujmin (1)	4	17·97		
Crommelin	6	27·87		
Tempel-Tuttle	4	32·91		
Stephan-Oterma	2	38·96		
Westphal	2	61·73		
Brorsen-Metcalf	2	69·06		
Olbers	3	69·57		
Pons-Brooks	3	70·86		
Halley	29	76·04		
Herschel-Rigollet	2	156·0		
Grigg-Mellish	2	164·3		

130 long-period comets are also known (1966) and these moving in elliptical orbits are part of the solar system. Some 354 comets having eccentricities of unity and greater than unity move in parabolic and hyperbolic orbits. These are not a part of the solar system since from the nature of their orbits they do not return to the Sun. All of these comets are listed in *Catalogue of Cometary Orbits* and the Supplement, *Memoirs of the B.A.A.*, *39*, No. 3 (1961); *40* No. 2(1966); and Quat. J.R.A.S., *10*, 240 (1969); *11*, 221 (1970); *12*, 244, (1971).

Index

ACCRETION theory 171
Adam, M. G. 75
Adams, J. C. 85
Adams, W. S. *et al* 145
Airy, Sir George B. 85, 287
Alfvén, H. 166–9, 246, 248
Alexander, A. F. 99, 104
Allen, C. W. 135
Anaxagoras 21
Anaximander 20
Anaximenes 20
Anderson, J. D. 109
Andoyer 94
Angström (unit) 136
Angular momentum 166, 223, 237
Anorthosite event 169
Antichthon, counter Earth 21
Anti-matter 207
Antoniadi 83
Apollo missions 175–6, 180, 262
Apollonius 26
Archimedes 63–4
Ariel-Umbriel pair, satellites of Uranus 107
Aristarchus 24, 27, 214
   earthquake 160
Aristotle 19, 33, 52, 53, 184
Armillary spheres 66
Arrhenius, S. A. 126, 243
Ash, M. E. 109
Asteroid (minor planet) 183
   discovery of 193–7
   groups of 194–7
   nature of 197–8
   origin of 207–8
   position computed 222
   in solar scale 224, 225
   survey of 196–8
   see also astrobleme
Astropovich, I. S. 126–7
Astrobleme 203
Astronomical unit 73, 281–20

Atmosphere,
   Mercury, Venus 113–14
   Earth, Moon 159–63
   Mars 117
   Jupiter, Neptune, Saturn, Uranus 118
   Titan 97
   primeval 163, 256

BAILEY, J. M. 170–1
Baker, G. 210
Baldwin, R. B. 205
Barnard, E. E. 93
Barnard's star 267
Barringer crater 203
Barwell stones 201
Barycentre 150, 153
Behr, A. & Seidentopf, H. 125
Berlage, H. P. 243, 245
Bernal, J. D. 269
Bernoulli, J. 24
Bernstein, W. *et al* 159
Bethe, H. A. 137
Bickerton, A. W. 239
Biermann, L. 142
Binary sun hypothesis 243, 250
Birkeland, K. 243
Bjerknes, V. 135
Black body 115
Blackwell & Ingham 125, 126
Blagg relationship 107, 223, 225
Blue shift 125, 126
Blum, P. F. & Fahr, H. J. 142
Bode, J. E. 220
Bode-Titius relationship 220–2
   modifications 222–4, 226
   applications 107, 170, 171, 193
Boltzmann's constant 113
Brahe, Tycho 43–5, 61, 184, 214
   see also Tychonic system
Bremsstrahlung 148
Brookhaven experiment 139, 140
Brown's lunar theory 94

Index

Bruce, C. E. R. 134, 135
Bruno, G. 42
Buffon, G. L. 238
Bürgi, solar system models 65

CALCAGNINI, C. 34
Calendar recalculation 50
Calippus 24
Callisto (d.) 93–5
Cameron, A. G. 172
Capture hypothesis 166–72
Carbon-nitrogen fusion cycle (CNO) 138–42
Carrington, R. C. 133, 135, 144, 145
Cassini, J. D. 78, 79, 124–5, 215
Cassini division 82–3, 97–100, 102
Celestial mechanics 50, 53, 59–61, 82, 230
Ceres 194, 197
Challis, J. 85
Chamberlain, T. C. 239, 242
Chacaltaya station 125
Chapman, D. R. 210
Chapman layer 162
Chaucer, G. 32
Chladni, E. F. 127, 200
Chromosphere 131
Climatic zones 164–5
Clocks, astronomical 64–6
Coates, R. J. 180
Collision hypothesis 238, 239
Comets 183
    break-up of 189
    catalogue 291
    Copernicus observations 45
        Gibbon on 184
        Halley & other 186–8
    life history of 191
    nature of 186–8
    origin of 190–2, 193, 207–8
    tail momentum of 142–3
    Whipple model of 190
Commensurabilities of satellites 96–7, 104, 105, 107, 226–7, 233–4
Commoner, B. 257
Conel, J. E. & Holstrom, G. B. 179
Continental drift 163
Cook, (Capt.,) J. 215–16
Copernicus, N. 35, 42, 45, 287
Copernican hypothesis 52
    system 35–7, 39–41, 54–5
        model of 66–7
Coriolis force 147

Corona 131, 137, 138
Cowan, C. *et al* 207
Craters, lunar 177, 206
    terrestrial 203–7
Crepe ring 98
Crick, F. H. C. 257

DANTE, 16
Darwin, G. 172
Darwinism 274–82
D'Arrest, H. L. 99, 107
d'Azambuja, L. & M. D. 145
Deimos 79, 91, 92
de Lahire theorem 52–3
De Marcus, W. C. 121–3
Denisse, J. F. 148
Derham, W. 259–60
Dermott, S. F. 226
Descartes, R. 238
Deuterium 116
Diatoms 255
Dicke, R. H. 75, 146, 156, 180
Diurnal rotation of Earth 24, 34, 42, 45
    model showing 66
Dione d. 98, 99, 104, 233
D.N.A. 257, 273
Dollfus, A. 107, 159
Dondi clock 65
Drake, F. P. & Ewen, H. T. 83
Dreyer, J. L. E. 23
Dust, cosmic, fall of 127–8
    in zodiacal light 125

EARTH,
    atmosphere 159
        layers of 162–3
    craters 203–7
    distance from Sun 213–18
    gaseous tail 126–7
    interior 157, 160–1
    magnetic field 163
    obliquity of ecliptic 151
    origin, *see* Earth/Moon system
    planetary data tabulated 111
    pressure in 160
    radiation belt around 143–4
    rotation of 163
        variation 155–7
        see also diurnal rotation
    in solar scale 224–5
Earth/Moon system,
    origin theories, accretion 171
    capture 166–9

Ruskol's criticism 171–2
fission 172–5
nucleation 169
orbital velocity 170
rotation 150, 166
see also diurnal rotation
Earthquake 160
Easter, calculation 153
Eccentric system 26
Eclipse 153, 156
Elliptical systems 48–9, 53, 59–61
see also Keplerian system, Tychonic system
Empedocles 22
Encke (work on solar parallax) 217–18
Encke's division 100
Encounter hypotheses,
Sun–star 239, 243
Sun–interstellar body 246
Sun–smoke cloud 250
Epictetus 70
Epicyclic system 26–7, 34, 47, 53–4
see also Copernican system, Ptolemaic system
Equant 33
Eudoxus 22–5, 34
spheres, homocentric 25, 35, 45
Eugster, H. P. & Jones, B. F. 256
Europa (d.) 80, 93–5, 233
Evershed effect 134
Evolution, Darwinian 274–82
'Evolution' of primitive life 269–74
Explorer satellites 80, 143
Extra galactic nebulae 265

FAYE, H. A. E. A. 239
Fesenkov, V. G. 126–7
Fi Bonacci series 288
Fiction (*see* footnote) 265
Fission hypotheses 172–5
Flagstaff Observatory 70
Force, gravitational 59–61
of planetary motion 51–2
Fowler, W. A. 139
Fracastoro, G. 34
Fraunhofer spectrum 136

GALAXIES 265–8
Galileo, G. 52, 55, 58–9, 80, 82, 132, 144, 288
Galilean satellites 93, 94–5
Ganymede (d.) 93–5, 97, 233
Gas cloud hypothesis 237, 250, 251

Gas, loss from planets 112–13
interplanetary 126–7
see also atmosphere
Gegenschein 126–8
Geological magnetism 164–5
Geological periods 158–9
Geomagnetism 163–4
Gerstenkorn, H. 166–9
Gilvarry, J. J. 179
Gledhill, J. A. 80
Gleissberg, W. 133
Goethe 33
Goldreich, P. 233
Goldsborough, G. R. 98
Goodman, J. W. 101
Gravitation, Newton's calculation 59–61
modification 74–5
Gravitational theories 59–61, 75, 250
Great Red Spot 72, 80–1
Greek concept of solar system 19–29
Grene, M. 281–2

HAGIHARA, Y. 97, 229
Hale, G. E. 134
Hall, A. 75, 91
Halley, E. 289–90
comet 185–8 (period 186)
transit of Mercury 215
Hamlet 263
'Harmony of the spheres' 21
Hawley, J. 205
Heliocentric system 25, 42
Hell, J. 31
Hepburn diagrams 101–3
Heracleides 24
Heraclitus 21, 194
Herschel, Sir William 44, 83, 84, 183, 258
Sir John 85, 259
Hey, J. S. 148
Hide, R. 80–1, 157
Hills, J. G. 222
Hipparchus 26–7, 229
Homocentric spheres 25, 35, 45
Hooke, R. 78
Horrocks, J. 217
Hoyle, F. 236, 237, 250, 251, 266
& Wickramasinghe 122–3
Humboldt 124, 132
Hutchinson, D. R. *et al* 282
Huygens, C. 66, 71, 82, 214–15
Hyperion (d.) 104, 233

IAPETUS 106

Index

Imbrium collision 169, 206
Interplanetary space
   'dust' 127–8
   gas 126–7
   magnetic field 143
   probes 125, 142
Io (d.) 80, 93–5, 233
Ionosphere 162
Isotopes 252–3

JAMES, J. F. 125
Janus (d.) 98, 107
Jeans, J. H. 243–4
Jeffreys, H. 121, 153, 243–4
Jones, Sir Harold S. 218
Jovian planets, *see* Jupiter, Neptune, Saturn, Uranus
Jupiter 79–81
   atmosphere 118
   details tabulated 111
   great red spot 72, 80–1
   interior 122–3
   magnetosphere 80
   radio emission 80
   rotation 79, 81, 146
   satellites 79, 80, 93–7, 103
      orbital period relation 226–8
   Saturn commensurability 233
   in solar scale 222, 224
   temperature 118
   terrestrial view of 71, 79

KANT, I. 239, 240
Kaula, W. H. 164
Kelvin–Helmholtz contraction hypothesis 137
Kepler, J. 42, 43, 47–8, 287–8
   calendar recalculation 50
   Epicurean philosophy 266
   force maintaining planetary motion 51–2
   observation of Mars 193
   Mercury 71
Keplerian system, elliptical orbits 53
   mathematics of 48–9
   model of 67
   laws, first & second 52, 53
      third 50–1, 52, 214, 220
Kerkut, G. A. 276
Kirchoff, G. R. & Bunsen, R. W. 136
Kirkwood gaps 82, 99–100, 102, 195
Kitt Peak meteor shower 200
Kuiper, G. P. 92, 100–1, 108, 250

Kuroda, P. K. 135, 136, 253

LAGRANGE, J. L. 196, 231–3, 290
Laplace, P. S. 290
   relation 231–3
   solar system theory 239, 241
   theorems 231–3
Laplacian plane 91, 92
Lassel, W. 108
Leverrier, U. J. J. 74–5, 85–6
Lemniscate 24
Life, extra-terrestrial,
   evidence for 282–4
   light requirements for 283
   non-human 262
   statistical case for 263–4
   possible on Mars 255, 262–3
      on Venus 255
      in other galaxies 265–8
Life, terrestrial,
   primary organism 'evolution' 256–60, 268–74
   Darwinian—evolution 274–82
Local group of galaxies 265
Lodge, Sir Oliver 148
Lovejoy, A. D. 42, 259
Lowell 100
Luna 170
Lunar craters 177, 206–7
   rock 175–6
Lunik satellite 159
Lyttleton, R. A. 120–1, 161, 172–5, 191–3, 246, 247

McCALL, G. J. H. 177–8
McCrea, W. H. 172, 174, 236, 237, 238
MacDonald, G. F. J. 172
MacLaurin spheroids 173
Magnetic field,
   of Earth 163–5
   interplanetary 143
   sunspot 134–5
Magnitudes—definition 267
Mariner probes
   interplanetary dust 125, 142
   Mars 71, 79, 116, 117
   Venus 115, 116
   Phobos 92
Markowitz, W. 155
Mars 78–9
   atmosphere 116–17
   details tabulated 111
   interior 121

life possibility 255, 262–3
  mass & density 121
  polar caps 71, 117
  radio emission 79
  rotation 78
  satellites 79, 91–2
  in solar scale 224
  temperature 116
  used to establish solar system 43, 47–8
    Sun/Earth distance 215
Mascons 178–80, 181
Maxwell, J. C. 82, 97, 100
Mayer, T. 83
Mayr, E. 281
Mean motion 96–7, 104–5, 233
Medawar, Sir Peter 275–6
Medieval astronomy 30–2
  effect on thought 261
Mendelism 278–9
Mercury 73–6
  atmosphere 113–14
  details tabulated 111
  interior 121
  mass & density 73–4, 120–1
  perihelion shift 74–5, 76, 146
  radio emission 76
  rotation 56, 57
  in solar scale 222, 224
  terrestrial view 70–1
  transits 74
Meteor 183, 193
  shower 189, 199–201
Meteorite,
  causing lunar craters 177, 206
  terrestrial craters 203–7
  classification 201
  composition 201–3
  evidence of life in 282–3
  origin of 207–8
Meyer 99
Michelson–Morley experiment 55
Milky Way 130
Miller, S. L. 255
Mills, A. A. 177
Mimas (d.) 82, 98, 99, 104, 233
Minor planets see asteroids
Miranda 107, 108
Models of solar system 62–9
Moon,
  atmosphere 159
  craters 177, 206–7
  details tabulated 111
  interior 159, 161

  magnetic field 180
  mascons 178–80, 181
  origin see Earth/Moon system
  pressure in 161
  radio emission 180
  rocks 175–6
  rotation see Earth/Moon system
  surface 175–8
  water on 179
Mora, E. 94–5
Moulton, F. R. 242, 243
Müller, E. A. 137
Muller, P. M. & Sjorgren, W. L. 178, 179
Munk, W. H. 155
Muslim astronomers 28, 34
Mutation 279–81

NATURAL selection 275
Nature 15
Nebular hypothesis 238, 239
Neptune 84–7
  atmosphere 118
  density 122–3
  details tabulated 111
  discovery of 85–6
  interior 122–3
  rotation 84
  satellites 84, 108–9
  in solar scale 224–5
  temperature 118
Nereid 108
Neutrino 140–1
Newcomb, S. 104, 195
'New System' of Derham 259–60
Newton, H. W. 145
Newton, Sir Isaac 50, 53, 59–61, 74–5,
  186, 230, 289
Nilsson, H. 279
Nuclear reactions, solar 137–44, 252–3
Nucleation hypothesis 169

OBLATENESS, solar 75–6, 146
Obsidian 209
O'Leary, B. T. 267
Oort, J. H. 187, 190–1, 207
Oparin, A. I. 268
Orbit see elliptical system, epicyclic
  system
Orbital velocity hypothesis 170
Orgueil & Murray stones 282–3
Orowan, E. 169, 174, 223
Orrery 66–9

PAINE, T. 42
Palaeomagnetism 164–6
Pallas 194, 197
Palomar-Leiden survey 197, 198
Paradise Lost 185
Parsec-definition 267
Periodicity, of sunspots 132–3
    of rotation of sun 145–7
Perihelion shift of Mercury 74–5, 76, 146
    of Venus 77
Perseids 199
Peruvian Quipu 63
Peurbach 53, 56, 65
Phobos 79, 91, 92
Phoebe (d.) 106, 108
Photosphere 131, 136, 137, 138
Piazzi 194
Pioneer satellite 143
Pirie, M. W. 255
Planet 51–2, 223
    inferior *see* Mercury, Venus
    Earth *see* Earth
    superior *see* Mars, Jupiter, Saturn,
        Uranus, Neptune, Pluto
    see also epicyclic system, elliptical
        system
Planetarium 62, 65, 66, 67, 68
Plaskett, H. H. 146
Plato 22, 213
    model solar system 63
    polyhedra 46
Pluto 84, 119
    details tabulated 111
Poincaré 173
Polanyi, M. 55, 221
Polarity of sunspots 134
Poles, of Mars 71, 117
    of Sun 145, 174
Polyhedra 46
Popper, K. 276
Porter, J. G. 92, 94–5, 101, 189, 195
Poynting Robertson effect 124
'Primeval soup' 255, 271
Prior, G. T. 201
Probes *see* Mariner, Venera, Ranger,
    satellites space exploration
    planned 118–19
Proctor, R. A. 151
Proton-proton reaction 138–42
Proudhon 129
Ptolemy,
    'Almagest' translated 31

Earth-Sun distance 214
    Mercury orbit drawing 57
    epicyclic geostatic system *see* Ptolemaic
        system
Ptolemaic system 26–9, 33, 34
    Copernicus criticism 37–9
    Kepler criticism 45, 54
    Dondi clock model 65
Puccetti, R. 262
Pythagoras 20–1, 22

QUIPU 63

RABE, E. 71, 218–19
Radio carbon dating 165
Radio waves,
    to determine Earth-Sun distance 218
    from Moon 180
    from planets 76, 77, 79, 80, 83
Ranger probe 218
Reber, G. 148
Red shift 125, 126
Retrograde motion 24, 39, 77, 84, 106,
    108, 114
Rhea (d.) 98, 99, 104
Ring system, Saturn 82–3, 97–105
Roche, E. 100
Roche limit 92, 100, 168, 174
Röemer, O. 93, 283
Roosen 127
Rose–Tschermak–Brezima classification
    201
Ruskol, E. L. 171
Rossby waves 147
Rouse, C. A. 140, 141
Roxburgh, I. W. 75
Roy, A. E. & Ovenden, M. W. 96, 105,
    107, 225, 233
Rudolphine tables 74
Russel, H. N. 243

SAGAN, C. 115, 254–5
Salisbury F. B. 277
Sampson, R. A. 94
Satellites,
    of planets *see* individual planets
    meteorological 254
    space exploration 80, 143, 159, 180
Saturn,
    atmosphere 118
    details tabulated 111
    interior 122–3
    orbital period relation 226–8

radio emission 80
rings 82–3, 97–105
rotation 81, 146
satellites 82, 97–107
    commensurabilities 233
in solar scale 224–5
temperature 116
terrestrial observations 72
Scalar tensor gravitational theory 75
Schmidt, O. 225, 246, 249
Schopenhauer-stratagem 274
Schove, D. J. 131, 133
Schröter effect 71
Schwabe 132
Scripture & astronomy 42, 259–61
Season change 151–2
Severny, A. B. 135
Shapley, H. 265, 266
Sharanov, V. V. 70–1
Shelley 150
Sidereal period 24, 73, 111
Slichter, L. B. 166
Smoluchowski, R. 123
Solar
    flame 135
    magnetic hypothesis 243, 246
    nuclear reactions 137–44, 252–3
    oblateness 75–6, 146
    parallax definition 218
    spectrum 136
    wind 126, 142, 156, 181
Solar system,
    age 229
    origin theories 235–52
        nuclear clues 252–3
    scale 213–28
    stability 229–34
Solzhenitsyn 213
Soter & Ulrichs 114
Southworth, G. C. 148
Space exploration *see* probes
Spengler, O. 30, 282
Sporopollenin 283
Stability theorems 231–2
Star,
    Barnard's 267
    of Bethlehem 76
    fixed, in early astronomy 21, 25
    Sun as 236
    Uranus as 83
    Venus as 76
Stehli, F. G. 165
Stellar cluster hypothesis 237, 238, 250

Stipe, J. G. 178
Strasbourg clock 65
Sudbury basin 203–5
Sun,
    details tabulated 111
    distance from Earth 213–18
    layers 130–2
    limb darkening 146–7
    mass & volume 130
    nature of 129–37
    oblateness 75–6, 146
    poles 145, 174
    position 130
    radio emission 147–8, 149
    rotation 144–7
    source of energy of 137–44
    spots 129, 131–5
    temperature 135, 146–7, 152
    see also solar
Supernova 250, 266
Swedenborg 238, 258
Swift, J. (*voyage to Laputa*) 92
Synodic period 23

TAHITI 215
Taylor, Sir Geoffrey 81
Taylor, G. E. 84
Tektites 183, 209–12
Telescope,
    Galileo's 58
    modern 72, 73, 91, 93, 100, 106–8, 187,
        206
Tellurion 66
Ter Haar, D. 250
Terrestrial planets *see* Mercury, Venus,
        Earth, Mars
Tescereau, J. 107
Tethys (d.) 98, 99, 104, 233
Thales 19, 20
Thellier, E. & T. O. 164
Theophastrus 131
Thompson, W. R. 276
Three body problem 229
Time measurement 155, 157–9
Titan (d.) 97, 98, 99, 104, 233
Titania and Oberon pair 107
Transit of Mercury 74
    of Venus 77
        used to estimate Sun-Earth distance
        215–18
Triple Sun hypothesis 246
Triton 108
Turbulent eddies hypothesis 249

Turbulent contracting envelope theory
250
Tusi couple 36
Twain, M. 262
Tycho Brahe *see* Brahe
Tychonic system 44, 54–5
    model of 65

ULBRIGHT, T. L. 270
Ulrych T. J. 169
Urameton orrery, pl. 110
Uranus
    atmosphere 118
    density 122
    details tabulated 111
    discovery 83–4
    interior 122–3
    rotation 84, 146
    satellites 84, 107–8
    in solar scale 224
    temperature 118
    terrestrial observation 83
Urey, H. C. 137, 189, 206, 211

VAIHINGER, H. 265
Van Allen radiation belt 80, 142
Van de Kamp 267
Venera probes 115–16
Venus
    atmosphere 114–16
    details tabulated 111
    discovery 76
    Galileo's observation 55
    interior 125
    life possibility 255

mass and density 120–1
perihelion shift 77
radio emission 77
rotation 76–7
in solar scale 224, 225
temperature 115
terrestrial observation 71
transits 77
    used to estimate Sun-Earth distance
    215–18
Vesta, infra-red dia 197
Viking experiment 118
von Guericke, O. 55
von Weizsäcker, C. F. 222, 249
von Soemmerring 144
Vortex theory 238

WALKER, R. L. 107
Warner, B. 137
Watson-Crick theory 257
Wegener 163
Whipple, F. L. 92, 190, 193, 250
Widmanstatten structure 207
Wilkins, G. A. 109, 154, 157
Williams, I. P. & Cremin, A. W. 236
Wilson effect 134
Wise, D. V. 172
Wolfgang, R. 255
Wollaston 136
Woolfson, M. M. 222

ZECH 219
Zheleznyakov, V. V. 148
Zodiacal light 124–6
    false 126–7